AF679991

Plastic in the Environment: Measurement, Health Risks and Management

Plastic in the Environment: Measurement, Health Risks and Management

Ralph Matthews

SYRAWOOD
PUBLISHING HOUSE
New York

Published by Syrawood Publishing House,
750 Third Avenue, 9th Floor,
New York, NY 10017, USA
www.syrawoodpublishinghouse.com

Plastic in the Environment: Measurement, Health Risks and Management
Ralph Matthews

© 2023 Syrawood Publishing House

International Standard Book Number: 978-1-64740-381-2 (Hardback)

This book contains information obtained from authentic and highly regarded sources. All chapters are published with permission under the Creative Commons Attribution Share Alike License or equivalent. A wide variety of references are listed. Permissions and sources are indicated; for detailed attributions, please refer to the permissions page. Reasonable efforts have been made to publish reliable data and information, but the authors, editors and publisher cannot assume any responsibility for the validity of all materials or the consequences of their use.

Trademark Notice: Registered trademark of products or corporate names are used only for explanation and identification without intent to infringe.

Cataloging-in-Publication Data

Plastic in the environment : measurement, health risks and
management / Ralph Matthews.
p. cm.
Includes bibliographical references and index.
ISBN 978-1-64740-381-2
1. Plastics--Environmental aspects. 2. Plastic scrap--Environmental aspects.
3. Plastic scrap--Measurement. 4. Plastic scrap--Health aspects.
5. Plastic scrap--Management. I. Matthews, Ralph.
TD798 .P53 2023
363.728 8--dc23

TABLE OF CONTENTS

PREFACE

Plastic is a material that contains an organic substance of large molecular weight as an essential ingredient. Microplastics are the tiny plastic particles that are produced during commercial product development and from breakdown of larger plastics. The increasing concentration of plastics and microplastics in various ecosystems is a major environmental concern across the globe. Microplastics are classified into primary microplastics and secondary microplastics. The plastic fragments or particles that are less than or equal to 5.0 mm before they enter the environment are called primary microplastics, such as microfibers and microbeads. Environmental factors such as ocean waves and sun's radiation lead to the breakdown of larger plastic items, such as water bottles, into smaller parts which are called secondary microplastics. A major disadvantage of microplastics and plastics is their inability to break up into smaller particles. Several studies have suggested that the exposure to microplastics in humans can potentially cause toxicity through oxidative stress. This book covers in detail various topics related to plastics and microplastics in the environment as well as their measurement, health risks and management. It is appropriate for students seeking detailed information in this area of study as well as for experts.

This book has been the outcome of endless efforts put in by authors and researchers on various issues and topics within the field. The book is a comprehensive collection of significant researches that are addressed in a variety of chapters. It will surely enhance the knowledge of the field among readers across the globe.

It gives us an immense pleasure to thank our researchers and authors for their efforts to submit their piece of writing before the deadlines. Finally in the end, I would like to thank my family and colleagues who have been a great source of inspiration and support.

Ralph Matthews

1

The Microplastic Cycle and Plastic Pollution

Michael S. Bank and Sophia V. Hansson

Abstract The microplastic cycle was originally and formally introduced and defined as a novel concept and paradigm for understanding plastic pollution and its fluxes across ecosystem reservoirs. This concept has now been expanded to include macroplastic particles and links all aspects of the fate, transport, and effects of plastic pollution, including source-receptor models in the environment, and expanded on previously established perspectives that viewed the plastic pollution issue in a less integrated manner. The value of this paradigm is that this perspective integrates three basic scientific spheres: environmental chemistry, biology (i.e., trophic transfer), and human health. The goal of this chapter is to introduce readers to the microplastic pollution problem and to outline the microplastic cycle as a concept and holistic paradigm for addressing this ubiquitous environmental and potential public health problem. The specific objectives of this chapter were to (1) introduce this volume and its chapters by outlining the microplastic pollution issue in the context of the entire plastic cycle; (2) evaluate fluxes of microplastics across different ecosystem compartments, including the atmosphere, lithosphere, hydrosphere. and biosphere, including humans; and (3) provide insights on public policy and potential solutions to the microplastic pollution problem.

M. S. Bank (✉)
Institute of Marine Research, Bergen, Norway

University of Massachusetts Amherst, Amherst, MA, USA
e-mail: Michael.Bank@hi.no; mbank@eco.umass.edu

S. V. Hansson
Laboratoire Ecologie Fonctionnelle et Environnement – UMR-5245, CNRS, Université de Toulouse, Castanet Tolosan, France
e-mail: sophia.hansson@toulouse-inp.fr

1.1 Introduction

Microplastic pollution is a complex problem (Thompson et al. 2004; Windsor et al. 2019) that has considerable consequences for environmental and public health. This pollution issue is a classic transboundary example of how land-based pollution can become extremely widespread, even entering remote regions including pristine mountainous regions, wilderness areas, and the Arctic (Bergmann et al. 2019; Brahney et al. 2020) and the deepest trenches of the ocean (Jamieson et al. 2019). Because plastic pollution is physically visible, this issue has garnered significant interest from a wide array of stakeholders including scientists, policy makers, and especially the media and the public. The overall attention to this issue has been immense and possibly unlike any other pollution issue in the history of science (Sedlak 2017). As a result of this visibility and attention toward the plastic and microplastic pollution issue, new paradigms and holistic perspectives have emerged to evaluate, study, and manage (Borrelle et al. 2020; Lau et al. 2020; Bank et al. 2021) the plastic waste problem. Here we provide an outline for the chapters in this volume and shortly introduce the concept of the microplastic pollution cycle (Bank and Hansson 2019).

The microplastic cycle was originally and formally introduced and defined as a novel concept and paradigm for understanding plastic pollution and its fluxes across ecosystem reservoirs (Bank and Hansson 2019). This concept has now been expanded to include macroplastic particles (Lechthaler et al. 2020) and links all aspects of the fate, transport, and effects of plastic pollution, including source-receptor models (Waldschläger et al. 2020; Hoellein and Rochman 2021), in the environment and expanded on previously established perspectives that tended to view the plastic pollution issue in a less integrated manner. The value of this paradigm is that this perspective integrates three basic scientific spheres: environmental chemistry, biology (i.e., trophic transfer), and human health (Fig. 1.1).

The goal of this chapter is to introduce readers to the microplastic pollution problem and to outline the microplastic cycle as a concept and holistic paradigm for addressing this ubiquitous environmental and potential public health problem. The specific objectives of this chapter were to (1) introduce this volume and its chapters by outlining the microplastic pollution issue in the context of the entire plastic cycle; (2) evaluate fluxes of microplastics across different ecosystem compartments, including the atmosphere, lithosphere, hydrosphere, and biosphere, including humans; and (3) provide insights on public policy and potential solutions to the microplastic pollution problem.

Fig. 1.1 Conceptual figure of the microplastic cycle and the complex movements of plastic particles, and their associated chemicals, through different ecosystem compartments. (Adapted from Bank and Hansson 2019)

1.2 Fluxes of Microplastics Across Ecosystem Compartments

The field of plastic pollution is rapidly moving forward, and over the last few years, research efforts have advanced the understanding of microplastic pollution and the movement of microplastics from urban areas to rivers and lakes, river runoff and transport to the sea, as well as marine dispersion of microplastics across ocean basins and deep ocean layers (Horton et al. 2017; Peng et al. 2018; Hale et al. 2020). Several efforts have been made to summarize, critically review, and provide a larger perspective on the current status of microplastics in the environment (e.g., Sedlak 2017; Horton and Dixon 2018; Akdogan and Guven 2019; Wu et al. 2019; Zhang et al. 2020; Hale et al. 2020). Within this chapter, we therefore only present the issue of plastic pollution across ecosystem compartments in brief and refer to already published literature (e.g., Horton and Dixon 2018; Hale et al. 2020; Bank et al. 2021; Hoellein and Rochman 2021) and relevant chapters within this volume for more comprehensive and detailed descriptions.

1.3 Microplastic and Terrestrial Ecosystems

Despite the fact that the majority of the plastic consumption (the usage of plastic in maritime fishing being the exception) as well as all plastic production occurs on land (Horton and Dixon 2018), the terrestrial environment has received less attention (compared to, e.g., marine ecosystems) when it comes to research on plastic and microplastic pollution. This topic is covered in Chap. 4 (Kallenbach et al. this volume).

Sources and input of plastic to the terrestrial environment include traffic and vehicle tire abrasion (Kole et al. 2017; Evangeliou et al. 2020), domestic and household activities such as cosmetics and cleaning agents (Murphy et al. 2016), synthetic fibers from clothing and textile washing (Habib et al. 1998; Browne et al. 2011; Napper and Thompson 2016; Boucher and Friot 2017), coatings, paint, and preparatory painting activities such as abrasive blasting (Takahashi et al. 2012; Song et al. 2015; Chae et al. 2015) to name a few. Direct littering and inadequately managed waste, including industrial spillages and release from landfill sites (Sadri and Thompson 2014; Lechner et al. 2014; Mason et al. 2016; Murphy et al. 2016; Kay et al. 2018; Hale et al. 2020 and references therein) also contribute plastic to the terrestrial environment. This is also the case of intentionally or accidentally burning of plastics, i.e., plastics released via poorly controlled disposal through burning or via natural wildfires can release plastic particles to the atmosphere as well as the surrounding environment which will subsequently be transported into nearby waterways (Gullett et al. 2007; Asante et al. 2016; Ni et al. 2016; Hale et al. 2020).

Agricultural activities can also lead to a discharge of plastics to the terrestrial environment, either through improper disposal of wrapping and bale twine but also via sewage sludge applied to agricultural lands (Mahon et al. 2017; Corradini et al.

2019). For example, Nizzetto et al. (2016) estimated that in Europe alone, 125–850 tons of microplastic per million inhabitants per year are added to agricultural soils via sewage sludge applications. Combined with input from mismanaged waste and littering, the plastic stock stored within terrestrial ecosystems will either lead to a massive accumulation (Horton and Dixon 2018) or act as a source to other ecosystem compartments (Jambeck et al. 2015). It has indeed been shown that urban centers and resuspension of plastic particles in soil are the principal sources for plastics later deposited via wet deposition (Brahney et al. 2020).

1.4 Microplastic and Freshwater Ecosystems

Microplastic pollution in aquatic freshwater ecosystems is highly complex as its environmental compartments include ditches, streams, rivers, estuaries, temporary and permanent wetlands, ponds, dams, and lakes, all of which have different characteristics in terms of hydrology, chemistry, flora, and fauna as well as their surrounding watershed and land-use patterns. Furthermore, freshwater ecosystems can act as both a receiver, sink, and transporter of plastic pollution (Eerkes-Medrano et al. 2015; Horton and Dixon 2018; Li et al. 2018; van Emmerik and Schwarz 2020). For example, direct littering, as well as mismanaged waste or inadequate waste disposal, acts as sources of plastic to the aquatic environment through wind transport, atmospheric deposition, and/or surface runoff from adjacent lands (Horton et al. 2017; Xia et al. 2020). Hitchcock (2020) recently showed that storm events act as key drivers of microplastic contamination in aquatic systems. For example, microplastic abundance was >40-fold higher during, and directly after, a storm event compared to before. Similar results were also found by Xia et al. (2020) who showed that rainfall is a significant driver of environmental microplastic pollution to inland surface waters. It should be noted though that both studies concluded that it is not the rain directly that causes this increase in plastic input but rather the surface runoff caused by the rain events during which plastics (macro- and microscale) were transported from land to the associated aquatic ecosystems. This is further supported by the results of Boucher and Friot (2017), as well as Horton et al. (2017), who showed that storm drainage and urban runoff is often untreated and unfiltered, allowing macroplastic from littering as well as microplastics from, e.g., degraded wear of tires, vehicles, and road paint, to be washed directly into nearby aquatic systems.

Once deposited, the plastic may degrade from primary to secondary particles and be efficiently dispersed (Williams and Simmons 1996; Weinstein et al. 2016) or be retained in the sediment (Castañeda et al. 2014; Klein et al. 2015; Nizzetto et al. 2016). Furthermore, it has also been shown that rivers may act as major pathways in the transport of plastic from land to the ocean (Jambeck et al. 2015; Schmidt et al. 2017; Lebreton et al. 2017), even being referred to as "highways for microplastics" (Barbuzano 2019). For example, it has been estimated that rivers and estuaries release 0.47–2.75 million tons of plastic to the ocean on an annual basis (Schmidt et al. 2017; Lebreton et al. 2017).

Although much research has been focused on river ecosystems, it has also been shown that plastic pollution also occurs within ponds and lakes across the globe (Eriksen et al. 2013; Free et al. 2014; Baldwin et al. 2016; Vaughan et al. 2017; Alfonso et al. 2020). For example, in a recent study, Alfonso et al. (2020) concluded that microplastic pollution occurs even in lakes located in remote and relatively pristine areas such as the Patagonian Andes which is considered to be one of the most sparsely populated and remote regions of the world. However, in contrast to rivers, lakes and ponds are more likely to retain plastic that has been settled in the sediment, without further transport to the ocean, and would therefore likely accumulate plastic over time (Vaughan et al. 2017; Horton and Dixon 2018). Here the fate, distribution, and impacts of plastic pollution across a range of different particle size classes are discussed in Chaps. 4 and 7 (Kallenbach et al. this volume; Gomes et al. this volume).

1.5 Microplastic and Marine Ecosystems

Microplastic pollution in marine ecosystems is largely a result of terrestrial runoff and plastic industrial wastes, although abandoned fishing gear is also recognized as an important source (Xue et al. 2020). This topic is widely studied and is covered in more detail by seminal papers including Cole et al. (2011), Hidalgo-Ruz et al. (2012), Wright et al. (2013), Sharma and Chatterjee (2017), Choy et al. (2019), Isobe et al. (2019), Onink et al. (2019), Allen et al. (2020), Hale et al. (2020), Kane et al. (2020), van Sebille et al. (2020), as well as Chap. 5 (Lundebye et al. this volume). Human activities in the coastal zone including fishing, aquaculture (Lusher et al. 2017), tourism, and marine industry are also important sources of microplastic pollution in saltwater environments. Here the sources, fate, and transport dynamics and effects of plastic and microplastic pollution across a range of different size classes are discussed in Chaps. 4, 5, 7, 8, and 9 (Kallenbach et al. this volume; Lundebye et al. this volume; Gomes et al. this volume; Garrido Gamarro and Costanzo this volume; Marathe and Bank this volume).

The marine environment has a unique set of physicochemical conditions, ocean circulation patterns, pressure, and water column dynamics (Choy et al. 2019; Onink et al. 2019; Kane et al. 2020; van Sebille et al. 2020) that govern the sources, fate, and transport dynamics of microplastics in addition to other important aspects such as biofouling and biofilm production (Zettler et al. 2013), as well as release or adsorption of secondary contaminants (Sharma and Chatterjee 2017). Additionally, microplastics are made with a variety of polymers, have different molecular structures, and are extremely diverse regarding their size, shape, color, and density and are viewed as a complex suite of contaminants (Rochman et al. 2019). These different properties of microplastics influence their distribution, buoyancy and sinking properties, their fate, and transport dynamics within marine ecosystems and govern their bioavailability and trophic transfer to marine biota (Sharma and Chatterjee 2017). The concept of marine snow (e.g., the continuous settling of mostly organic particles from upper regions of the water column) is an important mechanism that can transport microplastics from the ocean's surface layer to deep pelagic and mesopelagic zones and may also enhance their bioavailability to biota inhabiting benthic habitats (Porter et al. 2018). Based on modeling simulations, Koelmans et al. (2017) estimated that 99.8% of aquatic plastic pollution since 1950 has settled beneath the ocean surface layer by 2016 with an additional ~9.4 million tons settling per year. Furthermore, while it is known that microplastics are transported to the seafloor by vertical settling from the surface, the spatial distribution, fate, and transport dynamics of microplastics are now understood to also be largely governed by sea bottom, thermohaline currents (Kane et al. 2020).

Microplastic in the ocean is a primary concern for ultimately two important and interrelated reasons. First, microplastics in the ocean can absorb and release toxic substances (Gouin et al. 2011) and are ingested by marine biota (Laist 1997; Cole et al. 2011; Wright et al. 2013), including seafood species (Smith et al. 2018). Microplastics are often found in high abundances in both the water column (Choy et al. 2019) and in deep-sea sediments (Kane et al. 2020) where they can then be taken up by biota. Second, the potential human health risks from the direct and indirect effects of microplastic pollution are also a primary concern (Bank et al. 2020; Barboza et al. 2018, 2020). However, while microplastic exposures have been reported to have negative effects on biota, ultimately many critical uncertainties regarding their complex toxicological profiles still remain, and overall much remains poorly understood (Hidalgo-Ruz et al. 2012; Wright et al. 2013; Kögel et al. 2020). Furthermore, the relationship between seafood safety is also not well understood although some recent investigations have identified important linkages between wild marine fish, microplastics, and toxic compounds such as bisphenol A (Barboza et al. 2020). These findings illustrate the importance and need for more comprehensive surveillance regarding the connection between seafood safety, human exposure, toxics, and overall food security (Barboza et al. 2018; Lundebye et al. this volume).

1.6 Microplastic and the Atmosphere

The fate and quantification of microplastics in the atmosphere are less explored compared to other ecosystem compartments, yet recent advancements have been made (Zhang et al. 2020). For example, recent studies have focused on microplastic occurrence in the atmosphere and have demonstrated significant microplastic atmospheric deposition in urban environments in, e.g., France (Dris et al. 2015, 2016; Gasperi et al. 2018), Germany (Klein and Fischer 2019), the UK (Stanton et al. 2019; Wright et al. 2020), Iran (Dehghani et al. 2017; Abbasi et al. 2019), and China (Cai et al. 2017; Liu et al. 2019).

However, like other environmental contaminants such as PAHs and metals, microplastic suspended in the atmosphere can also be subject to long-range transport and atmospheric deposition (Zhang et al. 2020 and reference therein). Studies based on microplastic in the atmosphere, in wet deposition and in soils, strongly indicate that the atmosphere may act as an important pathway in the dispersal of microplastic on a global scale by transporting microplastic from urban areas to remote locations (Dris et al. 2016; Peeken et al. 2018; Allen et al. 2019; Roblin et al. 2020). It has, for example, been recently shown that microplastic from atmospheric deposition can be found in remote areas such as the French Pyrenees (Allen et al. 2019) and the Alps (Ambrosini et al. 2019; Bergmann et al. 2019) but also in the Arctic (Bergmann et al. 2019; Zhang et al. 2019) and in ocean surface air (Liu et al. 2019; Wang et al. 2020). Further, Allen et al. (2019) showed that not only did atmospheric deposition of microplastic occur at their remote sampling site in the French Pyrenees (i.e., no urban populations or development within ≥ 95 km) but also that this deposition was comparable to the atmospheric deposition found in megacities such as Dongguan or Paris (Dris et al. 2016; Cai et al. 2017). Roblin et al. (2020) also showed that in four remote sites in Ireland, the majority (i.e., 70%) of the investigated anthropogenic and plastic microfibers were deposited via wet atmospheric deposition, whereas Brahney et al. (2020) showed that dry deposition of plastic also plays an important role in the global plastic cycle, especially when it comes to long-range and/or global transport.

Although microplastic particles exist in a diverse array of shapes, sizes, molecules, and molecular structures, their general low material density, small size, and high surface area enable them to easily enter and become suspended in the air (Dris et al. 2016; Abbasi et al. 2019; Wang et al. 2020). Anthropogenic activities in the terrestrial environment, such as direct littering, inadequately managed waste, industrial spillages, and release from landfill sites, are therefore all considered potential sources of microplastic to the atmosphere. However, although the ocean is generally perceived as a receiver and sink of macro- to nanoscale plastic (Eriksen et al. 2013, 2014; Isobe et al. 2019), deposited either directly (via, e.g., mismanagement of maritime fishing) or indirectly via river runoff or atmospheric deposition, it has recently been shown that the ocean may also act as a source of plastic back to the atmosphere via wind-driven sea spray formation and bubble burst ejection (Allen et al. 2020). As such, marine microplastic hotspots may therefore act not just as a sink but also as a source of microplastics to the atmospheric compartment contributing to long-range and terrestrial microplastic transport. This would mean a continuum of the transfer of plastic between ecosystem compartments and environmental reservoirs and that just as carbon, nitrogen, mercury, or lead, plastic too follows the pathway of full environmental and biogeochemical cycles (Bank and Hansson 2019).

1.7 Microplastic in Biota

Biological organisms, including humans, are important receptors of microplastics and are exposed via air, water, and ingestion of microplastics and through consuming the food items containing them (Cole et al. 2011; Wright et al. 2013; Gall and Thompson 2015; Anbumani and Kakkar 2018; Prinz and Korez 2020). The size class of plastic particles (Wright et al. 2013; Kögel et al. 2020) and association with other toxic compounds are recognized as an important concept regarding its overall toxicity, and microplastic particles are now viewed as a complex suite of contaminants (Rochman et al. 2019). Ecotoxicology and effects of microplastics on biota are synthesized in this volume in Chap. 7 (Gomes et al. this volume), and aspects of human health are covered in Chaps. 5 (Lundebye et al. this volume), 8 (Garrido Gamarro and Costanzo this volume), and 9 (Marathe and Bank this volume).

One of the primary issues confronting the assessment of microplastics in biota is the lack of standardized approaches, and in general this limits the progress regarding the potential abatement of microplastic pollution as well as the study of toxicological profiles which are inherently complex (Rochman et al. 2019; Koelmans et al. 2020). However, recently progress has been made regarding the development of probability-based models of species sensitivity distribution that correct for issues driven by the incompatibility of data and results from experiments caused by differences in the microplastic types used in effect studies compared to those that are truly environmentally relevant in natural settings (Koelmans et al. 2020). Moreover, of equal importance regarding the improvement of environmentally relevant exposure conditions (e.g., size and shape) for microplastic ecotoxicology studies is the need for verification of background contamination and addressing associated risks from inhibition of food or reduced nutrition, as well as internal and external physical damage from microplastics (de Ruijter et al. 2020).

The rise of microplastics as a ubiquitous pollutant has made human exposure, largely through ingestion and inhalation, inevitable, and little is known about the effects of microplastics on human health (Prata et al. 2020). Human exposure to microplastics is difficult to study especially considering that the critically needed, low-level exposure, clinical trials are complicated by the fact that no true controls groups exist due to everyone being exposed to plastic constituents over the course of their lifetime (Vandenberg et al. 2007; North and Halden 2013). Therefore, epigenetic and other comparable approaches will likely be required to further understand the potential health effects in humans. Increasingly there is a growing concern that the indirect effects of microplastic pollution may present considerable risks to human health. An important example of this is the role of microplastic pollution in antibiotic resistance (Parthasarathy et al. 2019; Laganà et al. 2019; Bank et al. 2020) which is synthesized in Chap. 9 (Marathe and Bank this volume). Lastly, there is also a great need for cellular and systemic toxicological investigations in humans as has been recently proposed by Yong et al. (2020).

1.8 Microplastics and Public Policy

Recently microplastic pollution has garnered significant interest from governments and policy makers, and policy aspects are covered in this volume in Chaps. 10 and 11 (Wagner this volume; Wingfield and Lim this volume). This issue is viewed as a planetary boundary threat (Galloway et al. 2017; Lam et al. 2018; Villarrubia-Gómez et al. 2018; Carney Almroth and Eggert 2019) and from a policy standpoint will ideally involve governance strategies (Vince and Hardesty 2017) that occur at local, regional, and global scales and that consider all ecosystem compartments (Bank et al. 2021). Additionally, a description and outline of the newly established Basel Convention Global Plastic Waste Partnership are presented in Chap. 10 (Wingfield and Lim this volume). The policy of microplastics has been reviewed by Sheavly and Register (2007), Pahl and Wyles (2017), Dauvergne (2018), Lam et al. (2018), Raubenheimer and McIlgorm (2018), Vince and Hardesty (2018), Black et al. (2019), as well as Carney Almroth and Eggert (2019) with recent IPCC style assessments also being undertaken by Borrelle et al. (2020) and Lau et al. (2020).

1.9 Conclusions

Microplastic and plastic pollution in general is an inherently complex issue. Moving ahead it is clear that small-scale and local efforts can have important global implications and can provide guidance regarding research and policy priorities. Of critical importance will be the estimation of fluxes and pools or the movement of microplastics and plastic particles across ecosystem compartments (Bank and Hansson 2019; Hoellein and Rochman 2021). These fluxes, and the microplastic concept in general, will serve as a critical foundation for global mass balance estimates and models (Bank et al. 2021). Such estimates and models can then be used to employ a structured approach, in the context of global environmental change processes, to support the identification of microplastic indicators, important pathways, mechanisms, and the general advancement of science and effective policymaking to holistically address this important environmental problem.

References

Abbasi S, Keshavarzi B, Moore F, Turner A, Kelly FJ, Dominguez AO, Jaafarzadeh N (2019) Distribution and potential health impacts of microplastics and microrubbers in air and street dusts from Asaluyeh County, Iran. Environ Pollut 244:153–164

Akdogan Z, Guven B (2019) Microplastics in the environment: a critical review of current understanding and identification of future research needs. Environ Pollut 254:113011

Alfonso MB, Scordo F, Seitz C, Mavo Manstretta GM, Ronda AC, Arias AH, Tomba JP, Silva LI, Perillo GME, Piccolo MC (2020) First evidence of microplastics in nine lakes across

Patagonia (South America). Sci Total Environ 733:139385. https://doi.org/10.1016/j.scitotenv.2020.139385.

Allen S, Allen D, Phoenix VR, Le Roux G, Durántez Jiménez P, Simonneau A, Binet S, Galop D (2019) Atmospheric transport and deposition of microplastics in a remote mountain catchment. Nat Geosci 12:339–344

Allen S, Allen D, Moss K, Le Roux G, Phoenix VR, Sonke JE (2020) Examination of the ocean as a source for atmospheric microplastics. PLoS One 15:e0232746

Ambrosini R, Azzoni RS, Pittino F, Diolaiuti G, Franzetti A, Parolini M (2019) First evidence of microplastic contamination in the supraglacial debris of an alpine glacier. Environ Pollut 253:297–301

Anbumani S, Kakkar P (2018) Ecotoxicological effects of microplastics on biota: a review. Environ Sci Pollut Res 25:14373–14396

Asante KA, Pwamang JA, Amoyaw-Osei Y, Ampofo JA (2016) E-waste interventions in Ghana. Rev Environ Health 31(1):145–148

Baldwin AK, Corsi SR, Mason SA (2016) Plastic debris in 29 Great Lakes tributaries: relations to watershed attributes and hydrology. Environ Sci Technol 50(19):10377–10385

Bank MS, Hansson SV (2019) The plastic cycle: a novel and holistic paradigm for the Anthropocene. Environ Sci Technol 53(13):7177–7179

Bank MS, Ok YS, Swarzenski PW (2020) Microplastic's role in antimicrobial resistance. Science 369(6509):1315

Bank MS, Swarzenski PW, Duarte CM, Rillig M, Koelmans B, Metian M, Kershaw P, Wright S, Provencher J, Sanden M, Jordaan A, Wagner M, Thiel M, Ok YS (2021) A global microplastic pollution observation system to aid policy. Environ Sci Technol 55(12):7770–7775

Barboza LGA, Vethaak AD, Lavorante BRBO, Lundeby A-K, Guilhermino L (2018) Marine microplastic debris: an emerging issue for food security, food safety and human health. Mar Pollut Bull 133:336–348

Barboza LGA, Cunha SC, Monteiro C, Fernandes JO, Guilhermino L (2020) Bisphenol A and its analogs in muscle and liver of fish from the North East Atlantic Ocean in relation to microplastic contamination. Exposure and risk to human consumers. J Hazard Mater 393:122419

Barbuzano J (2019) Rivers are a highway for microplastics into the ocean. Eos 100. https://doi.org/10.1029/2019EO130375. Published on 09 August 2019

Bergmann M, Mützel S, Primpke S, Tekman MB, Trachsel J, Gerdts G (2019) White and wonderful? Microplastics prevail in snow from the Alps to the Arctic. Sci Adv 5:eaax1157

Black JE, Kopke K, O'Mahony C (2019) A trip upstream to mitigate marine plastic pollution – a perspective focused on the MSFD and WFD. Front Mar Sci 6:689

Borrelle SB, Ringma J, Lavender Law K, Monnahan CC, Lebreton L, McGivern A, Murphy E, Jambeck J, Leonard GH, Hilleary MA, Eriksen M, Possingham HP, De Frond H, Gerber LR, Polidoro B, Tahir A, Bernard M, Mallos N, Barnes M, Rochman CM (2020) Predicted growth in plastic waste exceeds efforts to mitigate plastic pollution. Science 369(6510):1515–1518

Boucher J, Friot D (2017) Primary microplastics in the oceans: a global evaluation of sources. IUCN, Gland

Brahney J, Hallerud M, Heim E, Hahnenberger M, Sukumaran S (2020) Plastic rain in protected areas of the United States. Science 368:1257–1260

Browne MA, Crump P, Niven SJ, Teuten E, Tonkin A, Galloway T, Thompson R (2011) Accumulation of microplastic on shorelines worldwide: sources and sinks. Environ Sci Technol 45(21):9175–9179

Cai L, Wang J, Peng J, Tan Z, Zhan Z, Tan X, Chen Q (2017) Characteristic of microplastics in the atmospheric fallout from Dongguan city, China: preliminary research and first evidence. Environ Sci Pollut Res 24:24928–24935

Carney Almroth B, Eggert H (2019) Marine plastic pollution: sources, impacts, and policy issues. Rev Environ Econ Policy 13:317–326

Castañeda RA, Avlijas S, Simard MA, Ricciardi A, Smith R (2014) Microplastic pollution in St. Lawrence River sediments. Can J Fish Aquat Sci 71(12):1767–1771

Chae DH, Kim IS, Kim SK, Song YK, Shim WJ (2015) Abundance and distribution characteristics of microplastics in surface seawaters of the Incheon/Kyeonggi Coastal region. Arch Environ Contam Toxicol 69(3):269–278

Choy CA, Robison BH, Gagne TO, Erwin B, Firl E, Halden RU, Hamilton JA, Katija K, Lisin SE, Rolsky C, Van Houtan S, K. (2019) The vertical distribution and biological transport of marine microplastics across the epipelagic and mesopelagic water column. Sci Rep 9:7843

Cole M, Lindeque P, Halsband C, Galloway SC (2011) Microplastics as contaminants in the marine environment: a review. Mar Pollut Bull 62:2588–2597

Corradini F, Meza P, Eguiluz R, Casado F, Huerta-Lwanga E, Geissen V (2019) Evidence of microplastic accumulation in agricultural soils from sewage sludge disposal. Sci Total Environ 671:411–420

Dauvergne P (2018) Why is the global governance of plastic failing the oceans? Glob Environ Chang 51:22–31

de Ruijter VN, Redondo-Hasselerharm PE, Gouin T, Koelmans AA (2020) Quality criteria for microplastic effect studies in the context of risk assessment: a critical review. Environ Sci Technol 54:11692–11705

Dehghani S, Moore F, Akhbarizadeh R (2017) Microplastic pollution in deposited urban dust, Tehran metropolis, Iran. Environ Sci Pollut Res 24:20360–20371

Dris R, Gasperi J, Rocher V, Saad M, Renault N, Tassin B (2015) Microplastic contamination in an urban area: a case study in Greater Paris. Environ Chem 12:592–599

Dris R, Gasperi J, Saad M, Mirande C, Tassin B (2016) Synthetic fibers in atmospheric fallout: a source of microplastics in the environment? Mar Pollut Bull 104:290–293

Eerkes-Medrano D, Thompson RC, Aldridge DC (2015) Microplastics in freshwater systems: a review of the emerging threats, identification of knowledge gaps and prioritisation of research needs. Water Res 75:63–82

Eriksen M, Mason S, Wilson S, Box C, Zellers A, Edwards W, Farley H, Amato S (2013) Microplastic pollution in the surface waters of the Laurentian Great Lakes. Mar Pollut Bull 77:177–182

Eriksen M, Lebreton LCM, Carson HS, Thiel M, Moore CJ, Borerro JC, Galgani F, Ryan PG, Reisser J (2014) Plastic pollution in the world's oceans: more than 5 trillion plastic pieces weighing over 250,000 tons Afloat at Sea. PLoS One 9:e111913

Evangeliou N, Grythe H, Klimont Z, Heyes C, Eckhardt S, Lopez-Aparicio S, Stohl A (2020) Atmospheric transport is a major pathway of microplastics to remote regions. Nat Commun 11:3381

Free CM, Jensen OP, Mason SA, Eriksen M, Williamson NJ, Boldgiv B (2014) High-levels of microplastic pollution in a large, remote, mountain lake. Risk Anal 85:156–163

Gall SC, Thompson RC (2015) The impact of debris on marine life. Mar Pollut Bull 92:170–179

Galloway TS, Cole M, Lewis C (2017) Interactions of microplastic debris throughout the marine ecosystem. Nat Ecol Evol 1:1–8

Garrido Gamarro E, Costanzo V (this volume) Dietary exposure to plastic particles, plastic additives and contaminants through seafood consumption. In: Bank MS (ed) Microplastic in the environment: pattern and process. Springer, Cham

Gasperi J, Wright SL, Dris R, Collard F, Mandin C, Guerrouache M, Langlois V, Kelly FJ, Tassin B (2018) Microplastics in air: are we breathing it in? Curr Opin Environ Sci Health 1:1–5

Gomes T, Bour A, Coutris C, Almeida AC, Bråte IL, Wolf R, Bank MS, Lusher AL (this volume) Ecotoxicological impacts of micro- and nanoplastics in terrestrial and aquatic environments. In: Bank MS (ed) Microplastic in the environment: pattern and process. Springer, Cham

Gouin T, Roche N, Lohmann R, Hodges G (2011) A thermodynamic approach for assessing the environmental exposure of chemicals absorbed to microplastic. Environ Sci Technol 45:1466–1472

Gullett BK, Linak WP, Touati A, Wasson SJ, Gatica S, King CJ (2007) Characterization of air emissions and residual ash from open burning of electronic wastes during simulated rudimentary recycling operations. J Mater Cycles Waste Manag 9(1):69–79

Habib D, Locke DC, Cannone LJ (1998) Synthetic fibers as indicators of municipal sewage sludge, sludge products, and sewage treatment plant effluents. Water Air Soil Pollut 103:1–8

Hale RC, Seeley ME, La Guardia MJ, Mai L, Zeng EY (2020) A global perspective on microplastics. J Geophys Res Oceans 125:e2018JC014719

Hidalgo-Ruz V, Gutow L, Thompson RC, Thiel M (2012) Microplastics in the marine environment: a review of the methods used for identification and quantification. Environ Sci Technol 46:3060–3075

Hitchcock JN (2020) Storm events as key moments of microplastic contamination in aquatic ecosystems. Sci Total Environ 734:139436

Hoellein TJ, Rochman CM (2021) The "plastic cycle": a watershed-scale model of plastic pools and fluxes. Front Ecol Environ 2021. https://doi.org/10.1002/fee.2294

Horton AA, Dixon SJ (2018) Microplastics: an introduction to environmental transport processes. WIREs Water 5:e1268

Horton AA, Walton A, Spurgeon DJ, Lahive E, Svendsen C (2017) Microplastics in freshwater and terrestrial environments: evaluating the current understanding to identify the knowledge gaps and future research priorities. Sci Total Environ 586:127–141

Isobe A, Iwasaki S, Uchida K, Tokai T (2019) Abundance of non-conservative microplastics in the upper ocean from 1957 to 2066. Nat Commun 10:417

Jambeck JR, Geyer R, Wilcox C, Siegler TR, Perryman M, Andrady A, Narayan R, Law KL (2015) Plastic waste inputs from land into the ocean. Science 347(6223):768–771. https://doi.org/10.1126/science.1260352

Jamieson AJ, Brooks LSR, Reid WDK, Piertney SB, Narayanaswamy BE, Linley TD (2019) Microplastics and synthetic particles ingested by deep-sea amphipods in six of the deepest marine ecosystems on Earth. R Soc Open Sci 6:180667

Kallenbach EMF, Rødland ES, Buenaventura NT, Hurley R (this volume) Microplastics in terrestrial and freshwater environments. In: Bank MS (ed) Microplastic in the environment: pattern and process. Springer, Cham

Kane IA, Clare MA, Miramontes E, Wogelius R, Rothwell JJ, Garreau P, Pohl F (2020) Seafloor microplastic hotspots controlled by deep-sea circulation. Science 368:1140–1145

Kay P, Hiscoe R, Moberley I, Bajic L, McKenna N (2018) Wastewater treatment plants as a source of microplastics in river catchments. Environ Sci Pollut Res 25(20):20,264–20,267. https://doi.org/10.1007/s11356-018-2070-7

Klein M, Fischer EK (2019) Microplastic abundance in atmospheric deposition within the Metropolitan area of Hamburg. Germany Sci Total Environ 685:96–103

Klein S, Worch E, Knepper TP (2015) Occurrence and spatial distribution of microplastics in river shore sediments of the rhine-main area in Germany. Environ Sci Technol 49(10):6070–6. https://doi.org/10.1021/acs.est.5b00492

Koelmans AA, Kooi M, Law KL, Van Sebille E (2017) All is not lost: deriving a top-down mass budget of plastic at sea. Environ Res Lett 12:114028

Koelmans AA, Redondo-Hasselerharm PE, Mohamed Nor NH, Kooi M (2020) Solving the nonalignment of methods and approaches used in microplastic research to consistently characterize risk. Environ Sci Technol 54:12307–12315

Kögel T, Bjorøy Ø, Toto B, Bienfait AM, Sanden M (2020) Micro- and nanoplastic toxicity on aquatic life: determining factors. Sci Total Environ 709:136050

Kole PJ, Löhr AJ, Van Belleghem FGAJ, Ragas AMJ (2017) Wear and tear of tyres: a stealthy source of microplastics in the environment. Int J Environ Res Public Health 14:1265

Laganà P, Caruso G, Corsi I, Bergami E, Venuti V, Majolino D, La Ferla R, Azzaro M, Cappello S (2019) Do plastics serve as a possible vector for the spread of antibiotic resistance? First insights from bacteria associated to a polystyrene piece from King George Island (Antarctica). Int J Hyg Environ Health 222:89–100

Laist D (1997) Impacts of marine debris: entanglement of marine life in marine debris including a comprehensive list of species with entanglement and ingestion records. In: Coe J, Rogers D (eds) Marine debris: sources, impacts, and solutions. Springer, New York, pp 99–140

Lam C, Ramanathan S, Carbery M, Gray K, Vanka KS, Maurin C, Bush R, Palanisami T (2018) A comprehensive analysis of plastics and microplastic legislation worldwide. Water Air Soil Pollut 229:345

Lau WWY, Shiran Y, Bailey RM, Cook E, Stuchtey MR, Koskella J, Velis CA, Godfrey L, Boucher J, Murphy MB, Thompson RC, Jankowska E, Castillo A, Pilditch TD, Dixon B, Koerselman L, Kosior E, Favoino E, Gutberlet J, Baulch S, Atreya ME, Fischer D, He KK, Petit MM, Sumaila UR, Neil E, Bernhofen MV, Lawrence K, Palardy JE (2020) Evaluating scenarios toward zero plastic pollution. Science 369:1455–1461

Lebreton LCM, van der Zwet J, Damsteeg JW, Slat B, Andrady A, Reisser J (2017) River plastic emissions to the world's oceans. Nat Commun 8:15611

Lechner A, Keckeis H, Lumesberger-Loisl F, Zens B, Krusch R, Tritthart M, Glas M, Schludermann E (2014) The Danube so colourful: a potpourri of plastic litter outnumbers fish larvae in Europe's second largest river. Environ Pollut 188:177–181

Lechthaler S, Waldschläger K, Stauch G, Holger Schüttrumpf H (2020) The way of microplastic through the environment. Environment 7:73

Li J, Liu H, Chen JP (2018) Microplastics in freshwater systems: a review on occurrence, environmental effects, and methods for microplastics detection. Water Res 137:362–374

Liu K, Wu T, Wang X, Song Z, Zong C, Wei N, Li D (2019) Consistent transport of terrestrial microplastics to the ocean through atmosphere. Environ Sci Technol 53:10612–10619

Lundebye A-K, Lusher AL, Bank MS (this volume) Marine microplastics and seafood: implications for food security. In: Bank MS (ed) Microplastic in the environment: pattern and process. Springer, Cham

Lusher AL, Hollman PCH, Mendoza-Hill JJ (2017) Microplastics in fisheries and aquaculture: status of knowledge on their occurrence and implications for aquatic organisms and food safety. FAO Fisheries and Aquaculture Technical Paper. No. 615. Rome, Italy.

Mahon AM, O'Connell B, Healy MG, O'Connor I, Officer R, Nash R, Morrison L (2017) Microplastics in sewage sludge: effects of treatment. Environ Sci Technol 51:810–818

Marathe NP, Bank MS (this volume) The antibiotic resistance-microplastic connection. In: Bank MS (ed) Microplastic in the environment: pattern and process. Springer, Cham

Mason SA, Garneau D, Sutton R, Chu Y, Ehmann K, Barnes J, Fink P, Papazissimos D, Rogers DL (2016) Microplastic pollution is widely detected in US municipal wastewater treatment plant effluent. Environ Pollut 218:1045–1054

Murphy F, Ewins C, Carbonnier F, Quinn B (2016) Wastewater treatment works (WwTW) as a source of microplastics in the aquatic environment. Environ Sci Technol 50(11):5800–5808

Napper IE, Thompson RC (2016) Release of synthetic microplastic plastic fibres from domestic washing machines: effects of fabric type and washing conditions. Mar Pollut Bull 112(1):39–45

Ni HG, Lu SY, Mo T, Zeng H (2016) Brominated flame retardant emissions from the open burning of five plastic wastes and implications for environmental exposure in China. Environ Pollut 214:70–76

Nizzetto L, Langaas S, Futter M (2016) Correspondence—Pollution: do microplastics spill on to farm soils? Nature 537:488

North EJ, Halden RU (2013) Plastics and environmental health: the road ahead. Rev Environ Health 28:1–8

Onink V, Wichmann D, Delandmeter P, van Sebille E (2019) The role of Ekman currents, geostrophy, and stokes drift in the accumulation of floating microplastic. J Geophys Res Oceans 124:1474–1490

Pahl S, Wyles KJ (2017) The human dimension: how social and behavioural research methods can help address microplastics in the environment. Anal Methods 9:1404–1411

Parthasarathy A, Tyler AC, Hoffman MJ, Savka MA, Hudson AO (2019) Is plastic pollution in aquatic and terrestrial environments a driver for the transmission of pathogens and the evolution of antibiotic resistance? Environ Sci Technol 53:1744–1745

Peeken I, Primpke S, Beyer B, Gütermann J, Katlein C, Krumpen T, Bergmann M, Hehemann L, Gerdts G (2018) Arctic sea ice is an important temporal sink and means of transport for microplastic. Nat Commun 9:1505

Peng X, Chen M, Chen S, Dasgupta S, Xu H, Ta K, Du M, Li J, Guo Z, Bai S (2018) Microplastics contaminate the deepest part of the world's ocean. Geochem Perspect Lett 9:1–5
Porter A, Lyons BP, Galloway TS, Lewis C (2018) Role of marine snows in microplastic fate and bioavailability. Environ Sci Technol 52:7111–7119
Prata JC, da Costa JP, Lopes I, Duarte AC, Rocha-Santos T (2020) Environmental exposure to microplastics: an overview on possible human health effects. Sci Total Environ 702:134455
Prinz N, Korez Š (2020) Understanding how microplastics affect marine Biota on the cellular level is important for assessing ecosystem function: a review. In: Jungblut S, Liebich V, Bode-Dalby M (eds) YOUMARES 9 – the oceans: our research, our future. Springer, Cham, pp 101–120. ISBN: 978-3-030-20389-4
Raubenheimer K, McIlgorm A (2018) Can the basel and stockholm conventions provide a global framework to reduce the impact of marine plastic litter? Mar Policy 96:285–290
Roblin B, Ryan M, Vreugdenhil A, Aherne J (2020) Ambient atmospheric deposition of anthropogenic microfibers and microplastics on the Western Periphery of Europe (Ireland). Environ Sci Technol 54:11100–11108
Rochman CM, Brookson C, Bikker J, Djuric N, Earn A, Bucci K, Athey S, Huntington A, McIlwraith H, Munno K, De Frond H, Kolomijeca A, Erdle L, Grbic J, Bayoumi M, Borrelle SB, Wu T, Santoro S, Werbowski LM, Zhu X, Giles RK, Hamilton BM, Thaysen C, Kaura A, Klasios N, Ead L, Kim J, Sherlock C, Ho A, Hung C (2019) Rethinking microplastics as a diverse contaminant suite. Environ Toxicol Chem 38:703–711
Sadri SS, Thompson RC (2014) On the quantity and composition of floating plastic debris entering and leaving the Tamar Estuary, Southwest England. Mar Pollut Bull 81:55–60
Schmidt C, Krauth T, Wagner S (2017) Export of plastic debris by rivers into the Sea. Environ Sci Technol 51:12246–12253
Sedlak D (2017) Three lessons for the microplastics voyage. Environ Sci Technol 51:7747–7748
Sharma S, Chatterjee S (2017) Microplastic pollution, a threat to marine ecosystem and human health: a short review. Environ Sci Pollut Res Int 27:21530–21547
Sheavly SB, Register KM (2007) Marine debris & plastics: environmental concerns, sources, impacts and solutions. J Polym Environ 15:301–305
Smith M, Love DC, Rochman CM, Neff RA (2018) Microplastics in seafood and the implications for human health. Curr Environ Health Rpt 5:375–386
Song YK, Hong SH, Jang M, Han GM, Shim WJ (2015) Occurrence and distribution of microplastics in the sea surface microlayer in Jinhae Bay, South Korea. Arch Environ Contam Toxicol 69(3):279–287
Stanton T, Johnson M, Nathanail P, MacNaughtan W, Gomes RL (2019) Freshwater and airborne textile fibre populations are dominated by 'natural', not microplastic, fibres. Sci Total Environ 666:377–389
Takahashi CK, Turner A, Millward GE, Glegg GA (2012) Persistence and metallic composition of paint particles in sediments from a tidal inlet. Mar Pollut Bull 64:133–137
Thompson RC, Olsen Y, Mitchell RP, Davis A, Rowland SJ, John AW, McGonigle D, Russell AE (2004) Lost at sea: where is all the plastic? Science 304:838
van Emmerik T, Schwarz A (2020) Plastic debris in rivers. WIREs Water 7:e1398
van Sebille E, Aliani S, Law KL, Maximenko N, Alsina JM, Bagaev A, Bergmann M, Chapron B, Chubarenko I, Cózar A, Delandmeter P, Egger M, Fox-Kemper B, Garaba SP, Goddijn-Murphy L, Hardesty BD, Hoffman MJ, Isobe A, Jongedijk CE et al (2020) The physical oceanography of the transport of floating marine debris. Environ Res Lett 15:023003
Vandenberg LN, Hauser R, Marcus M, Olea N, Welshons WV (2007) Human exposure to bisphenol A (BPA). Reprod Toxicol 24:139–177
Vaughan R, Turner SD, Rose NL (2017) Microplastics in the sediments of a UK urban lake. Environ Pollut 229:10–18
Villarrubia-Gómez P, Cornell SE, Fabres J (2018) Marine plastic pollution as a planetary boundary threat—the drifting piece in the sustainability puzzle. Mar Policy 96:213–220
Vince J, Hardesty BD (2017) Plastic pollution challenges in marine and coastal environments: from local to global governance. Restor Ecol 25:123–128. https://doi.org/10.1111/rec.12388

Vince J, Hardesty BD (2018) Governance solutions to the tragedy of the commons that marine plastics have become. Front Mar Sci 5:214

Wagner M (this volume) Solutions to plastic pollution: a conceptual framework. In: Bank MS (ed) Microplastic in the environment: pattern and Process. Springer, Cham

Waldschläger K, Lechthaler S, Stauch G, Schüttrumpf H (2020) The way of microplastic through the environment – application of the source-pathway-receptor model (review). Sci Total Environ 713:136584

Wang X, Li C, Liu K, Zhu L, Song Z, Li D (2020) Atmospheric microplastic over the South China Sea and East Indian Ocean: abundance, distribution and source. J Hazard Mater 389:121846

Weinstein JE, Crocker BK, Gray AD (2016) From macroplastic to microplastic: degradation of high-density polyethylene, polypropylene, and polystyrene in a salt marsh habitat. Environ Toxicol Chem 35:1632–1640

Williams AT, Simmons SL (1996) The degradation of plastic litter in rivers: implications for beaches. J Coast Conserv 2:63–72

Windsor FM, Durance I, Horton AA, Thompson RC, Tyler CR, Ormerod SJ (2019) A catchment-scale perspective of plastic pollution. Glob Chang Biol 25:1207–1221

Wingfield S, Lim M (this volume) The United Nations Basel convention's global plastic waste partnership: history, evolution and progress. In: Bank MS (ed) Microplastic in the environment: pattern and process. Springer, Cham

Wright SL, Thompson RC, Galloway TS (2013) The physical impacts of microplastics on marine organisms: a review. Environ Pollut 178:483–492

Wright SL, Ulke J, Font A, Chan KLA, Kelly FJ (2020) Atmospheric microplastic deposition in an urban environment and an evaluation of transport. Environ Int 136:105411

Wu P, Huang J, Zheng Y, Yang Y, Zhang Y, He F, Chen H, Quan G, Yan J, Li T, Gao B (2019) Environmental occurrences, fate, and impacts of microplastics. Ecotoxicol Environ Saf 184:109612

Xia W, Rao Q, Deng X, Chen J, Xie P (2020) Rainfall is a significant environmental factor of microplastic pollution in inland waters. Sci Total Environ 732:139065

Xue B, Zhang L, Li R, Wang Y, Guo J, Yu K, Wang S (2020) Underestimated microplastic pollution derived from fishery activities and "Hidden" in deep sediment. Environ Sci Technol 54:2210–2217

Yong CQY, Valiyaveetill S, Tang BL (2020) Toxicity of microplastics and nanoplastics in mammalian systems. Int J Environ Res Public Health 17:1509

Zettler ER, Mincer TJ, Amaral-Zettler LA (2013) Life in the "plastisphere": microbial communities on plastic marine debris. Environ Sci Technol 47:7137–7146

Zhang Y, Gao T, Kang S, Sillanpää M (2019) Importance of atmospheric transport for microplastics deposited in remote areas. Environ Pollut 254:112953

Zhang Y, Kang S, Allen S, Allen D, Gao T, Sillanpää M (2020) Atmospheric microplastics: a review on current status and perspectives. Earth Sci Rev 203:103118

2

Collection and Analysis of Plastic Debris: An Analytical Chemistry Approach

Robert C. Hale, Meredith E. Seeley, Ashley E. King and Lehuan H. Yu

Abstract Approaches for the collection and analysis of plastic debris in environmental matrices are rapidly evolving. Such plastics span a continuum of sizes, encompassing large (macro-), medium (micro-, typically defined as particles between 1 μm and 5 mm), and smaller (nano-) plastics. All are of environmental relevance. Particle sizes are dynamic. Large plastics may fragment over time, while smaller particles may agglomerate in the field. The diverse morphologies (fragment, fiber, sphere) and chemical compositions of microplastics further complicate their characterization. Fibers are of growing interest and present particular analytical challenges due to their narrow profiles. Compositional classes of emerging concern include tire wear, paint chips, semisynthetics (e.g., rayon), and bioplastics. Plastics commonly contain chemical additives and fillers, which may alter their toxicological potency, behavior (e.g., buoyancy), or detector response (e.g., yield fluorescence) during analysis. Field sampling methods often focus on >20 μm and even >300 μm sized particles and will thus not capture smaller microplastics (which may be most abundant and bioavailable). Analysis of a limited subgroup (selected polymer types, particle sizes, or shapes) of microplastics, while often operationally necessary, can result in an underestimation of actual sample content. These shortcomings complicate calls for toxicological studies of microplastics to be based on "environmentally relevant concentrations." Sample matrices of interest include water (including wastewater, ice, snow), sediment (soil, dust, wastewater sludge), air, and biota. Properties of the environment, and of the particles themselves, may concentrate plastic debris in select zones (e.g., gyres, shorelines, polar ice, wastewater sludge). Sampling designs should consider such patchy distributions. Episodic releases due to weather and anthropogenic discharges should also be considered. While water

R. C. Hale (✉) · M. E. Seeley · A. E. King
Department of Aquatic Health Sciences, Virginia Institute of Marine Science, William & Mary, Gloucester Point, VA, USA
e-mail: Hale@vims.edu; meseeley@vims.edu; aeking@vims.edu

L. H. Yu
Department of Environmental Engineering & Ecology, School of Biology & Food Engineering, Guangdong University of Education, Guangzhou, China
e-mail: yulehuan@gdei.edu.cn

grab samples and sieving are commonplace, novel techniques for microplastic isolation, such as continuous flow centrifugation, show promise. The abundance of non-plastic particulates (e.g., clay, detritus, biological material) in samples interferes with microplastic detection and characterization. Their removal is typically accomplished using a combination of gravity separation and oxidative digestion (including strong bases, peroxide, enzymes); unfortunately, aggressive treatments may damage more labile plastics. Microscope-based infrared or Raman detection is often applied to provide polymer chemistry and morphological data for individual microplastic particles. However, the sheer number of particles in many samples presents logistical hurdles. In response, instruments have been developed that employ detector arrays and rapid scanning lasers. The addition of dyes to stain particulates may facilitate spectroscopic detection of some polymer types. Most researchers provide microplastic data in the form of the abundances of polymer types within particle size, polymer, and morphology classes. Polymer mass data in samples remain rare but are essential to elucidating fate. Rather than characterizing individual particles in samples, solvent extraction (following initial sample prep, such as sediment size class sorting), combined with techniques such as thermoanalysis (e.g., pyrolysis), has been used to generate microplastic mass data. However, this may obviate the acquisition of individual particle morphology and compositional information. Alternatively, some techniques (e.g., electron and atomic force microscopy and matrix-assisted laser desorption mass spectrometry) are adept at providing highly detailed data on the size, morphology, composition, and surface chemistry of select particles. Ultimately, the analyst must select the approach best suited for their study goals. Robust quality control elements are also critical to evaluate the accuracy and precision of the sampling and analysis techniques. Further, improved efforts are required to assess and control possible sample contamination due to the ubiquitous distribution of microplastics, especially in indoor environments where samples are processed.

Abbreviations

ABS Acrylonitrile butadiene styrene
AFM Atomic force microscopy
APPI Atmospheric pressure photoionization
ATR Attenuated total reflectance
BPA Bisphenol A
DART Direct analysis in real time
DCM Dichloromethane (methylene chloride)
DESI Desorption electrospray ionization
EA/IRMS Elemental analyzer/isotope ratio mass spectrometry
EDS Energy-dispersive X-ray spectroscopy
EM Electron microscopy

ESCA	Electron spectroscopy for chemical analysis
EVA	Ethylene vinyl acetate
FM	Fluorescence microscopy
FPA	Focal plane array
FR	Flame retardant
FTIR	Fourier transform infrared spectroscopy
GC	Gas chromatography
GPC	Gel permeation chromatography
HDPE	High-density polyethylene
HRMS	High-resolution mass spectrometry
IR	Infrared (spectroscopy)
LC	Liquid chromatography
LC-MS/MS	Liquid chromatography/tandem mass spectrometry
LDPE	Low-density polyethylene
LOD	Limit of detection
m/z	Mass-to-charge ratio
MALDI	Matrix-assisted laser desorption/ionization
MP	Microplastic
MPSS	Munich plastic sediment separator
MS	Mass spectrometry
MW	Molecular weight
NP	Nanoplastic
NR	Nile red
O-PTIR	Optical photothermal IR
PA	Polyamide
PC	Polycarbonate
PE	Polyethylene
PET	Poly(ethylene terephthalate)
PP	Polypropylene
PS	Polystyrene
PU	Polyurethane
PVC	Polyvinyl chloride
Py-GC/MS	Pyrolysis-gas chromatography/mass spectrometry
QA	Quality assurance
QC	Quality control
Q-TOF	Quadrupole time of flight
rpm	Revolutions per minute
RT	Room temperature
SEC	Size exclusion chromatography
SEM	Scanning electron microscopy
SFC	Supercritical fluid chromatography
TD-PTR-MS	Thermal desorption-proton transfer reaction-mass spectrometry
TED-GC/MS	Thermal extraction desorption-gas chromatography/mass spectrometry
TEM	Transmission electron microscopy
TGA	Thermogravimetric analysis

TMAH	Tetramethylammonium hydroxide
TOF	Time of flight
TOF-SIMS	Time-of-flight secondary ion mass spectrometry
UHMW	Ultrahigh molecular weight
UHPLC	Ultrahigh performance liquid chromatography
XPS	X-ray photoelectron spectroscopy
μFTIR	Micro-Fourier transform infrared spectroscopy
μRaman	Micro-Raman spectroscopy

2.1 Introduction

To date, the lack of sampling and analytical methods capable of adequately characterizing the diversity of plastic debris in the environment has handicapped studies of their distribution, fate, and consequences. Plastic debris in the environment exists in a continuum of sizes. Debris has been classified as macro- (>25 mm), meso- (5–25 mm), micro- (1 μm to 5 mm), and nanoplastic (<1 μm). Where not differentiated, the use of the term "microplastics" here will mean all particles <5 mm. In the environment, plastics fragment over time, rates varying depending on polymer composition and ambient conditions. As such, size distributions are not static. Most published methods have been designed to detect only a subset of microplastics (often those > 300 μm). Hence, resulting measurements are likely underestimates. Commonly, the number of particles detected in a sample (within a size range) or the identities of only select polymer types are reported, versus the complete plastic mass-based concentration. Readers should take these limitations into account when interpreting published studies. Plastics in the environment exhibit a range of properties and composition.

Representative sampling followed by comprehensive, accurate analysis of microplastics is a prerequisite for understanding their fate and biological consequences and for crafting effective solutions. When developing and applying methodologies, researchers must carefully consider study goals (Fig. 2.1). Both field and controlled (lab or mesocosm) approaches are needed to answer important questions. Controlled experiments typically utilize specific, preselected test plastics. A good understanding of plastic composition and properties is essential. Also, while the majority of lab studies employ un-weathered materials, plastics start to be altered once in use and following environmental release. The extent of weathering is a function of ambient conditions and duration, adding further complexity and variability to the microplastics to be analyzed (Luo et al. 2019, 2020; Zhang et al. 2021).

In studies of field-collected samples, the analytical methods typically applied do not encompass the complete range of plastic characteristics (e.g., polymer type, size, morphology). Polymers targeted are generally those manufactured in greatest abundance or commonly reported in surveys, e.g., those in single-use containers, such as polyethylene (PE), poly(ethylene terephthalate) (PET), and polystyrene (PS). This is comparable to prioritizing high production volume chemicals for monitoring, without factoring in their relative risks or potential to alter ecosystems. To elaborate,

Fig. 2.1 Researchers must first delineate their study goals and then select appropriate sampling and analysis approaches. For example, focusing on the detailed characteristics (e.g., size, shape, texture, composition, extent of weathering) of a few 10 μm microplastic particles (via, e.g., MALDI-MS) is informative (represented by the microscope icon). However, such a narrowly defined focus may not be compatible with a goal of assessing the range of diverse microplastics in, for example, an entire forested area. Further, ignoring large debris in favor of microplastics alone is problematic as the former will eventually degrade into many small fragments. Microplastics present at the time of sampling represent a snapshot of a dynamic situation. Further, remedies such as removal (and better prevention) of large plastic debris are critical. Documentation and removal of large debris also can be performed by those lacking sophisticated analytical tools, such as "citizen scientists." (Photo: Alaskan forest floor adjacent to a marine shoreline. Credit: Ted Raynor, GoAK.org)

single-use beverage containers are dominant plastic debris components. However, microplastics (mostly PE, polypropylene (PP), and PET) generated from these may exhibit modest chemical risk (Lithner et al. 2011), as these products were designed to present minimal threats to human health. In contrast, e-waste plastics typically contain percent concentrations by weight of persistent and toxic additives (Singh et al. 2020), such as flame retardants (Li et al. 2019) and metals (e.g., Cd, Cr, Hg, Pb, and Sb) at levels that may exceed hazardous waste guidelines (Turner et al. 2019). Hence, from an ecosystem health perspective, less abundant plastic products might be disproportionately impactful and worthy of prioritization for analysis.

Some have criticized laboratory-based studies for the use of "unrealistically" high microplastic concentrations. However, if existing measurements do not adequately represent the true levels present in the environment, this pronouncement may be hasty (Hale 2018; Covernton et al. 2019). In addition, environmental burdens are increasing at exponential rates, with an estimated doubling rate in, for example, coastal marine sediments of 15 years (Brandon et al. 2019). The quantification of microplastics in surface waters further illustrates this point. Most approaches to date have deployed sampling gear (e.g., plankton nets) with openings exceeding 300 μm. Thus, smaller particles may not be retained. Smaller microplastics are more difficult to detect but ironically may be more abundant in environmental samples (e.g., Enders et al. 2015). They may also present heightened toxicological impacts (von Moos et al. 2012; Kögel et al. 2020) due to their ability to infiltrate tissues (e.g., lung alveoli in mammals) and penetrate cell membranes (Prata et al. 2020). Small microplastics and nanoplastics also exhibit exaggerated surface areas and thus enhanced capacity for environmental interactions, including contaminant sorption (Wang et al. 2019).

Spatially, most published monitoring has focused on microplastics at the water's surface. Recently, interests in denser polymers and fibers and debris at depth have emerged. Studies on microplastics in air (Gasperi et al. 2018), soils (Ng et al. 2018), and sediments (Gomiero et al. 2019) are appearing in growing numbers. For example, Choy et al. (2019) observed in vertical transects off Monterey Bay, California, greater microplastic water concentrations between 200 and 600 m than at the surface. They reported that weathered PET and polyamide (PA) fibers (negatively buoyant polymers) dominated. Kane et al. (2020) reported up to 1.9 million microplastics (primarily fibers) m^{-2} in deepwater, sedimentary drift deposits in the Tyrrhenian Sea. Yu et al. (2018) observed that PET and cellulose-derived fibers were the major forms on southeastern US coastal beaches. Fibers present additional sampling and detection challenges due to their elongated shapes and small cross sections. Hence, analytical methods must be refined to accommodate these.

Additional subclasses of microplastics merit scrutiny. Paint chips and tire wear fragments have been less studied but are reported to be major components of microplastic debris in some environmental samples (Hale et al. 2020). These present novel analytical issues and will be discussed later in greater detail. Paint chips have been observed to be abundant in surface waters with substantial boat traffic (e.g., Imhof et al. 2016), as well as near shipyards (Turner 2010). Chips often exhibit distinct colors, facilitating visual identification. However, they may be quite small and contain high concentrations of additives that can confound spectra often used for polymer identification. Pigments can contain toxic organic or metallic compounds, so their identification and health consequences should be assessed (Turner 2010; Luo et al. 2020). Importantly, the ecological repercussions of natural particles (e.g., cellulose, chitin, and minerals) and processed bio-based (e.g., cellulose acetate, polylactic acid) versus fossil fuel-based plastics merit further evaluation. Their determination adds an additional layer of analytical considerations.

Recent reviews of microplastic analysis techniques have been published (e.g., Hidalgo-Ruz et al. 2012; Löder and Gerdts 2015; Van Cauwenberghe et al. 2015; Lusher et al. 2017; GESAMP 2019; Fu et al. 2020). Our goals in the subsequent sections of this chapter will be to present representative accepted, as well as some more novel approaches, and to describe challenges and conceptual elements.

2.2 About the Analytes

Being complex solids composed primarily of high molecular weight and low volatility polymers, sampling and analytical considerations for plastic debris diverge from those of more commonly monitored lower molecular weight contaminants, such as pesticides or metals. The latter are amenable to well-established and widely available techniques, such as gas chromatography/mass spectrometry (GC/MS) and atomic absorption/emission spectroscopy. For plastics, additional characteristics of interest exist, including polymer composition, particle shape, and size. The immense diversity of plastic products in commerce, and thus in the environment, makes their determination challenging (Hale 2017; Rochman et al. 2019). Weathering and abrasion during use and following discard may alter size distributions. Even in the lab, plastic fragmentation can occur. For example, Dawson et al. (2018) observed the generation of nanoplastics by the stomach and gastric mill of Antarctic krill that were fed a defined size class of microplastics in the lab. Aggressive treatment of embrittled microplastics during sample preparation may also affect size distributions.

Plastic composition is an essential factor when choosing sampling and analysis methods as it dictates their fate/behavior in the field and during preparation. Polymers are composed of repeating units or monomers. Plastics may also be composites (e.g., reinforced with fibers) or copolymers (mixture of different polymers). Polymer chains can differ in molecular weight within the same plastic. Chains may be compositionally homogeneous, i.e., consist of the same, or different monomers. The polymeric chains can be composed of, arranged, and chemically linked in various ways. The resulting materials may be amorphous or crystalline, which affects their properties. Residual monomers, as well as catalysts used in synthesis, may be retained in plastic products, adding heterogeneity. Plastics are often infused with additives to achieve the desired color, flame retardancy, flexibility, or other characteristics (Hahladakis et al. 2018). Additive levels, at times reaching percent by weight levels, can complicate the analysis of the plastics (Lenz et al. 2015) and alter their environmental fate and behavior. Fillers (e.g., calcium carbonate, clay, talc, carbon black) may also be incorporated to modify properties or reduce costs and may interfere with the spectroscopic analysis. Tires are an example of a complex product, consisting of natural or synthetic rubbers, polymeric and metallic fibers, carbon black, and a host of additives. Such materials may confound commonly applied identification techniques such as IR spectroscopy.

Once in the environment, plastic debris chemical composition may be modified by weathering, complicating analysis. For example, photooxidation can alter spectroscopic results by increasing the relative carbonyl to methylene absorbance of both polymers and additives (Su et al. 2019; Khaled et al. 2018). After release, debris from diverse sources will intermingle, creating complex heterogeneous mixtures.

Polymer type, form, and additive content affect physical behavior and toxicological outcomes. For example, Luan et al. (2019) reported that certain functional groups on PS resulted in differing effects during the key development stages of the clam *Meretrix meretrix*. Luo et al. (2019) attributed fluorescent additives, leached from polyurethane (PU) microplastics, to effects on microalgal photosynthesis. Hence, composition is important to determine analytically.

To date, the analysis of additives in plastic debris has been limited (e.g., Hermabessiere et al. 2017). However, extensive work has been done related to additive migration from packaging to food (Hahladakis et al. 2018). Considerable interest has also arisen regarding environmental contaminants that are polymer additives, e.g., flame retardants, in indoor dust and subsequent human exposure (Wu et al. 2007). Analysis of additives is important, but a detailed discussion is beyond the scope of this chapter.

2.3 Sampling

Depending upon study goals, a variety of environmental matrices have been chosen for sampling, for example, water, sediments/soils/dust, air, marine snow, plankton, and specific tissues of larger organisms (e.g., digestive tissues, gills, liver, muscle, etc.).

2.3.1 Aqueous Matrices

Such samples may include natural surface or drinking water, wastewater (influent, in process or effluent), or precipitation (rain, melted snow, or ice). For natural waters, the surface microlayer, water column, and sediment interstitial water may be of particular interest. Each type presents different challenges due to collection requirements and the level of matrix interferences, as well as the abundance and characteristics of the plastics therein. Historically, surface water has been most commonly evaluated due to ease of collection and the presumption that most plastics are buoyant. Recently, data showing substantial microplastics in other environmental compartments have been published (Kooi et al. 2017; Erni-Cassola et al. 2019). Knowledge of site characteristics, such as weather, season and flow patterns, and basin morphology, are critical to designing appropriate sampling and interpreting results.

Method selection criteria include their ability to retain and quantify the salient range of particle sizes (and shapes) and should be evaluated by the analyst (Koelmans et al. 2019). In their global review of small floating plastic debris, van Sebille et al. (2015) estimated that >90% of the surface water trawls contained meshes >330 µm. Hence, the sample particle distributions will differ from those in the field (Dai et al. 2018). Approaches that exclude small microplastics will underestimate the total abundances present (Pabortsava and Lampitt 2020; Covernton et al. 2019). For example, Dris et al. (2018) reported a 250-fold increase in fiber counts when sampling with an 80 µm versus a 330 µm mesh net. Collection methods may perform well for spherical microplastics but poorly for elongated fibers or fragments. Fibers can pass more readily through the mesh, depending on the angle of contact. Changes in particle collection efficiency over time due to blockage of openings can also occur (Prata et al. 2019).

Grab sampling of water may be employed to capture smaller particles, e.g., using buckets for surface and Niskin, Van Dorn, or other remote capture devices at depth.

In a novel study, Choy et al. (2019) used a remotely operated vehicle equipped with in situ samplers that pumped water (ranging from 1007 to 2378 m^3) through 100 μm mesh filters at selected depths up to 1000 m. Determination of coincident water characteristics (e.g., temperature, salinity, suspended solids, chlorophyll content) may also aid in interpreting microplastic results. Pumping water through a series of sieve(s) or meshes has also been explored (Prata et al. 2019), permitting larger, composite samples to be evaluated. Tamminga et al. (2019) observed orders of magnitude higher numbers of microplastics and more efficient collection of fibers by passing water through a cascade of filters compared to the collection with a manta net. However, comparatively rare debris may be missed due to the smaller volume of water sampled by pump or grab approaches versus towing nets across wide areas. For sampling surface microlayer microplastics, Ng and Obbard (2006) used a rotating drum while Song et al. (2014) sampled via a dipped mesh screen.

While continuous flow centrifugation (CFC) has been widely used for the sampling of suspended particulate matter, to date it has seen limited usage for the collection of microplastics. However, Leslie et al. (2017) collected and concentrated suspended riverine particulate matter by CFC and processed pooled concentrates using salt-based density separation techniques. They noted most microplastics in the suspended particulate matter were <300 μm. Hildebrandt et al. (2020) demonstrated CFC in the lab for the collection and pre-concentration of Pd-doped nanoplastics from ultrapure water and filtered and unfiltered Elbe River (Germany) water. One versus two centrifuges in series and various water flow rates were evaluated. They noted the possibility of removing high-density minerals from suspensions, as well as separating micro- from nanoplastics. Shipboard sampling and passage through the centrifuges would eliminate the need for storage containers, resultant nanoplastic sedimentation/surface adhesion losses, and reduced contamination potential. Compared to filter-based systems, CFC can be run continuously for days, allowing large volumes to be processed.

2.3.2 Air Samples

Microplastics in air are an emerging concern. Outdoors, airborne microplastics can be rapidly transported long distances. Indoors, human exposure via microplastic inhalation and ingestion may be particularly important due to the confined space, abundance of plastic products therein, and low air turnover. Citizens of developed countries often spend >90% of their time indoors. While data pertaining to microplastics remain limited, there is substantial literature on ambient particles in indoor and outdoor air (e.g., Whalley and Zandi 2016). These are typically collected on filters of varying porosity (Zhang et al. 2020a). Note that glass fiber filters are often used here and this matrix may enmesh and obscure microplastics, complicating later spectroscopic evaluation.

Dry and wet deposition of microplastics was recently evaluated by Brahney et al. (2020) in several remote US wilderness areas by initial collection in buckets. Wet samples were subsequently filtered through 0.45 µm polyethersulfone filters and dry material reacquired using a ceramic blade. Stationary high volume, portable personal, and passive samplers may also be used. Sommer et al. (2018) used a Sigma-2 passive sampler to collect airborne particulates near three German roadways. Particles were collected on a transparent adhesive acceptor surface, over 7 days. As in the case of water strata, the collection of air samples at different heights may yield particles of different characteristics (Quang et al. 2012). Akin to growing concerns over the toxicological consequences of small microplastics in water, inhalation of fine airborne particulate matter <10 µm (PM10) has long been recognized as a serious health concern due to its ability to infiltrate lung alveoli (e.g., Anderson et al. 2012). Approaches such as the breathing thermal manikin have been developed in an attempt to mimic human exposure (Vianello et al. 2019).

2.3.3 Sediments, Soils, and Dust

Microplastic contents of these matrices are of increasing concern. Bedded sediments are typically collected as a core or grab. Sediments integrate conditions over extended periods compared to surface water samples, but burdens can vary over short distances. Study goals drive the location and number of discrete samples. Sampling depth is a consideration for cores. As plastics have only become prevalent in the environment since the 1950s, investigations of distributions in sediment cores are rare. However, Brandon et al. (2019) reported plastic debris in a core from the Santa Barbara Basin spanning the period 1834–2009. These authors were limited to larger debris as visual sorting of candidate microplastics was utilized, followed by FTIR polymer identification of selected targets. The authors, after correcting for sample contamination, reported an exponential increase in plastic deposition from 1945 to 2009, with a doubling time of 15 years. Sediments were passed through a 104 µm mesh, so true microplastic concentrations were likely higher. In contrast to most surface water investigations, they also noted that fibers were the dominant form detected in their sediments.

Microplastic loads and particle characteristics vary widely in different sub-environments. For example, Haave et al. (2019) found that distributions of small (<100 µm) and large microplastics (>500 µm) in sediments of a Norwegian urban fjord differed spatially, with small microplastics preferentially observed in areas of higher organic matter deposition. In another example, Ceccarini et al. (2018) collected materials on a transect from subtidal sediments to supralittoral sand. They found large plastic fragments accumulated above the storm berm and higher-density polymer particles in the benthic sediments. Their work underscores the need for techniques that can generate results for the total amount of plastics present (inclusive of sizes normally below the limits of spectroscopic detection of discrete particles) in a sample. Other solids, such as municipal wastewater sludge, are increasingly being examined. Due to the surface skimming and sedimentation processes utilized, treatment sludges may contain >90% of microplastics that enter wastewater facilities (Mahon et al. 2017).

As plastics are primarily manufactured, used, and discarded on land (Hale et al. 2020), the soil is an important media to examine. However, compared to aquatic sediments, terrestrial soils have been less frequently considered. Möller et al. (2020) reviewed available soil sampling and microplastics analysis methods. The former included the use of stainless-steel scoops or shovels for surface samples and cores for samples at depth. They also emphasized choosing a sampling strategy (e.g., judgmental, random, grid, transect, or stratified) consistent with the study goals.

2.3.4 Biological Samples

The diversity of biological organisms is immense, and sampling will depend greatly on study objectives. More variables are in play for biological compared with abiotic media. Stationary organisms will better reflect local conditions than mobile/migratory species. Some organisms may preferentially ingest specific particle sizes (Ward et al. 2019). Small organisms may be composited (e.g., collected onto filters), while larger specimens may be collected (using nets, traps, or hook and line) and analyzed individually, in their entirety, or dissected. An important consideration is whether the microplastics reside within tissues proper or are associated with external or internal (e.g., digestive tract) surfaces. If not within tissues proper, toxicological risks may be less. Microplastics within digestive systems may pass through the body and be depurated. To remove digestive tract-entrained microplastics, the organism may be allowed to depurate gut contents, or the digestive tract manually flushed or removed in the lab. Disposition of microplastics within organisms and mode of meal preparation may also alter the likelihood of subsequent human exposure via ingestion, i.e., if the organism is first depurated, eaten in its entirety (e.g., many shellfish), or otherwise prepared (e.g., filleted finfish). Food preparation often differs regionally and between ethnic groups.

2.3.5 Sample Preservation

Preservation of microplastic samples is not commonly described, in part as most plastics are resistant to biodegradation. However, Courtene-Jones et al. (2017) examined freezing versus formaldehyde/ethanol preservation of microplastics in mussel tissue, reporting no differential effects of these treatments. While most plastics are recalcitrant, coincident sample constituents may be susceptible to decomposition, especially biological tissues. Hence, the lack of preservation may alter the concentration calculation, as the matrix weight is normally used in the denominator for sediments and tissues. Microplastic-containing samples are often held in oxidizing agents for extended periods of time as part of the purification process. For example, Song et al. (2014) digested solids, sieved from surface waters, with 34.5% H_2O_2 for 2 weeks. Semisynthetic polymers (manufactured from natural precursors) may be more labile and degraded. Common products generated from natural materials include cellophane packaging, cigarette filters, and rayon-based textiles. Cellulose acetate was reported to represent >50% of synthetic particles in landfill leachates (Su et al. 2019), deep-sea sediments (Lusher et al. 2013), ice cores (Obbard et al. 2014), and fish (Lusher et al. 2013).

2.4 Laboratory Processing

As a result of environmental weathering, field sampling, and lab preparation, physical changes in plastic debris may occur due to abrasion with instruments, sieves or sand grains, or sample freeze/thaw cycles (Klein et al. 2018). Biofilm formation and electrostatic interactions on surfaces facilitate agglomeration of microplastics, altering their apparent size and behavior in the environment and during collection and analysis (Rummel et al. 2017; Michels et al. 2018; Lapointe et al. 2020). Depending on matrix complexity, a sequence of preparative steps is typically employed, commonly organic matter digestion and density-based separation. Steps can be divided into sample preparation, microplastic concentration, matrix purification, microplastic size separation, and particle detection (Fig. 2.2). Some methods focus on evaluating the characteristics of individual particles (e.g., those applying vibrational spectroscopic techniques such as Raman or IR spectroscopy), while others focus on the bulk, weight-based concentration of polymers present in the sample (e.g., pyrolysis-gas chromatography/mass spectrometry (Py-GC/MS)) or solvent extraction. As such, sample preparation may be dictated by the characteristics of the detection technique (discussed later).

2.4.1 Sample Preparation

Researchers may process constituents into particle size classes (so-called binning) by passage through a series of increasingly fine sieves or filters. This is especially useful if the ultimate detection approach does not yield individual particle characteristics (e.g., Py-GC/MS). For example, Gomiero et al. (2019) separated microplastics by sequential passage through 250, 100, 40, and 10 μm stainless-steel sieves. The samples had previously been subjected to oxidative and enzymatic cleanup, as well as density-based separation steps (discussed below). Bulk separation/characterization techniques (e.g., solvent extraction, followed by spectroscopy or thermogravimetry) may provide an estimate of total microplastics that encompass contributions from particles smaller than what even sophisticated analytical instruments can detect (typically 10–20 μm, Raman down to 1 μm). Notably, Gomiero et al. (2019) reported that the 40–100 μm fraction, a size range below which is often reported in the literature, contributed most to the total polymer

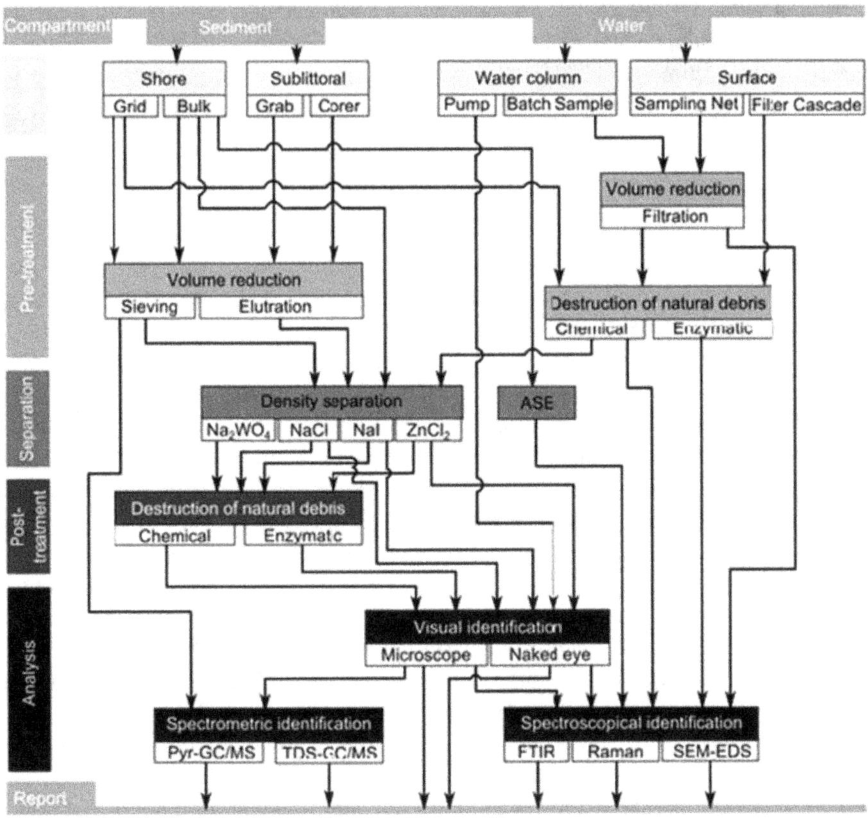

Fig. 2.2 Possible strategies described in the literature for the analysis of plastic debris in sediment and water samples, from sampling to reporting of the results. The sample preparation here is split into the pretreatment, density separation, and the posttreatment of microplastics. Fourier transform infrared spectroscopy (FTIR), scanning electron microscopy energy-dispersive X-ray spectroscopy (SEM-EDS), pyrolysis- or thermal desorption-gas chromatography/mass spectrometry (Py-GC/MS, TDS-GC/MS), and others may be deployed for the plastic analysis. When ASE separation is utilized, chromatographic analysis techniques not listed here, as well as spectrometric identification, may also be employed.

quantity of their field samples. It should be noted that in that study, the field-collected sediments were initially homogenized before further treatment in "a standard stainless-steel orbital mixer (approx. 20 rpm, 10 min at RT) using the K-beater knife." The sediments were predominantly fine sand. Plastics in the environment may be embrittled by weathering, such as ultraviolet (UV) oxidation (Song et al. 2017; Khaled et al. 2018). Hence, it is possible that abrasive lab homogenization techniques may further fragment brittle microplastics. To their credit, Gomiero et al. (2019) performed a series of procedural validations using spiked unweathered microspheres of three sizes and polymer types (PE, PP, and polyvinyl chloride, PVC). But such polymer types are less vulnerable to alteration by caustic treatments than PA and polycarbonate (PC). In addition, microspheres may be size-fractionated more consistently and be less vulnerable to fragmentation than plastic films or fibers. Further supporting the above concern about abrasive handling fragmenting plastic debris, Efimova et al. (2018) used a laboratory rotating mixer and coarse beach sediment to intentionally generate secondary microplastics in the laboratory.

2.4.2 Chemical and Enzymatic Digestion

Field samples often contain inorganic (e.g., clay minerals) and organic particles (e.g., detritus) that can interfere with the plastic analysis. The removal requirements for such particles vary, depending on the matrix (e.g., water, tissue, or sediments). Plastic debris from the field quickly accumulates an organic coating that may alter its chemical composition, properties (McGivney et al. 2020), and behavior (e.g., promoting aggregation or increasing their apparent density) or confound later spectroscopic analysis. A host of chemical agents have been employed to eliminate such films. Hydrogen peroxide (at different concentrations and temperatures) has been commonly used to oxidize labile organics. Duration of contact varies from hours to days. Some researchers have utilized Fenton's reagent (a solution of H_2O_2 and Fe^{2+}; Tagg et al. 2017). Alternatively, acids (e.g., HCl, HNO_3, formic) and bases (e.g., NaOH, KOH) have been employed, especially for digesting biological tissues. Repeated treatments and concurrent heating may be required to fully oxidize labile organic matter. However, PA and ester-based polymers (e.g., PET) appear more vulnerable to degradation by these treatments (Karami et al. 2017; Hurley et al. 2018). Wolff et al. (2019) reported substantial alterations of ethylene vinyl acetate (EVA), PU, and PA after exposure to treatment with H_2O_2, NaClO, $ZnCl_2$, and hexane. Hence, purification methods should initially be validated for the targeted polymer types.

Enzymatic digestion has also been applied to samples. Enzymes typically cause less polymer degradation than caustic agents. Unfortunately, such procedures can be complex (e.g., including detergents, proteases, lipases, chitinases, and cellulases), time-consuming (>10 days), and expensive. Enzyme treatments may also be augmented with caustic treatments to enhance the removal of interferences (Löder et al. 2017). Increased steps and handling of samples also enhance the potential for introduction of contaminants, loss of targeted plastics, or alteration of their physical characteristics.

2.4.3 *Physical Separation of Plastics from the Matrix: Filtration and Sieving*

To permit more facile detection, plastics are typically retrieved from air, aqueous, or solid phases and concentrated on surfaces, e.g., filters or sieves. In some cases, particulates may then be transferred from the initial filter to a second for optimal spectroscopic analysis. Analyst goals must guide desired filter characteristics: plastic size retention, filter composition and thickness/structure, and compatibility with the chemical agents used and with the instrumental detection approach to be employed. Filter configuration (e.g., punched holes, fiber weaves, or sintered metal disks) may control particle size/shape retention and the ease of retrieval of retained plastics. As this step can be labor-intensive and result in microplastic loss, refinements are beneficial. In this context, Nakajima et al. (2019a) developed a stainless-steel sieve (32 μm mesh) apparatus resistant to common oxidizing agents but compact enough to submerge in a glass beaker. They reported its use reduced the number of collections, rinsing, and transfer steps and generated better microplastic recoveries than widely used filter-based methods.

Researchers should ensure filter constituent materials do not contribute plastics or interfere with spectroscopic detection of targeted polymers. Filters/sieves themselves can be constructed of a range of materials, including quartz, stainless steel, nylon, cellulose, silicon, silver membrane, gold-coated PC, alumina-based membrane, and Teflon™ (Löder et al. 2015; Oßmann et al. 2017; Wolff et al. 2019; Wright et al. 2019; Käppler et al. 2015). Wright et al. (2019) collected inhalable microplastics of several polymer types on a variety of filter types (quartz, polytetrafluoroethylene, alumina, cellulose, and silver membrane) and then evaluated their compatibility with Raman imaging. Best results were obtained with silver membrane filters. Käppler et al. (2015) investigated filter materials for FTIR and recommended a silicon-based membrane. Particulates can become entrained in fibrous surfaces such as quartz fibers (Wright et al. 2019), and some materials may release fragments or interfere with the spectra of the targeted polymers. Consequences will vary depending on the detection scheme applied (e.g., visible light, FTIR, or Raman) and the polymer types targeted.

2.4.4 *Density and Other Physical Separation*

Flotation of plastics and sedimentation of dense inorganic particulates (ranging from 1.6 to >2.4 g cm^{-3}) is a common purification step, especially for sediments, sludges, and water samples. While some common polymers (e.g., PE and PP) exhibit densities lower than water, others are near neutral buoyancy or denser (PVC, PC, PS, PET, PA, ABS, tire rubber: 1.0–1.4 g cm^{-3}; Teflon: 2.2 g cm^{-3}). Note: the presence of intact air pockets in foamed polymers (e.g., PS or PU) will increase buoyancy. Water surface tension can keep even dense plastics at the surface for extended periods. The presence of polymer fillers and additives can also alter plastic behavior. For example, the densities of carbon black, calcium carbonate ($CaCO_3$), and glass fiber fillers typically exceed 2 g cm^{-3}. To achieve flotation of a range of polymer types, analysts have typically used concentrated saline solutions consisting of NaCl, NaI, sodium polytungstate, $ZnCl_2$, and $ZnBr_2$ (GESAMP 2019). Choices are based on effectiveness, cost, and safety. The behavior of the microplastics in said solutions should be carefully monitored. For example, Rodrigues et al. (2019) observed that salt can deposit on the surface of microplastics, increasing their overall density, leading to their sinking and loss.

A variety of devices have been used for the gravity separation of plastics from denser particulates. These include simple glass funnels (Rodrigues et al. 2019) to the elaborate stainless-steel Munich Plastic Sediment Separator (MPSS; Imhof et al. 2012). Retention of plastics in the settled solids, as well on the container sides, may decrease their recovery. Again, the presence of polymeric material in the construction of settling apparatuses could lead to possible sample contamination. Nakajima et al. (2019b) engineered a simple all-glass separator, the JAMSTEC microplastic-sediment separator (JAMSS unit). It is comprised of two glass plates (Fig. 2.3), the upper consisting of an open cylinder and the lower, a chamber with capacities of 30, 60, or 100 ml. The design is based on the Combined Plate or Utermöhl Chamber, long-used for examining settled phytoplankton. The approach is to settle the solids into the lower container and then to isolate these from the overlying water column (containing the more buoyant microplastics) by sliding the upper plate. The plastics can then be poured onto a filter or sieve, followed by water washes of the chamber to dislodge any adhering microplastics.

Wang et al. (2018) evaluated the recovery of polystyrene nano- and microplastics from sewage sludge and soil. They noted that 100 μm microbeads were effectively recovered by $ZnCl_2$ solution-based flotation, but smaller beads were not. They also evaluated flotation efficiency as a function of time, indicating substantial periods were needed to reach 90% for particles <5 μm. Möller et al. (2020) reviewed several extraction methods for removing plastics from soils, including electrostatic separation for dry solids, oil extraction (utilizing the lipophilicity of the plastics), froth flotation (using a stream of air), various density-based approaches, and magnetic separation using lipophilic nanoparticles functionalized with iron.

2.4.5 *Solvent Extraction*

If research goals do not require the visualizing and counting/characterization of individual particles, but rather quantifying the total mass of plastic in a sample, solvent dissolution/extraction of plastics from a sample matrix may be appropriate. Polymer solubility must be initially established. Solvents that are most effective for a given polymer typically have similar solubility parameters, and solvation typically increases with temperature (Miller-Chou and Koenig 2003). Separation of undesired co-extractives is typically necessary. As an example, Ceccarini et al. (2018)

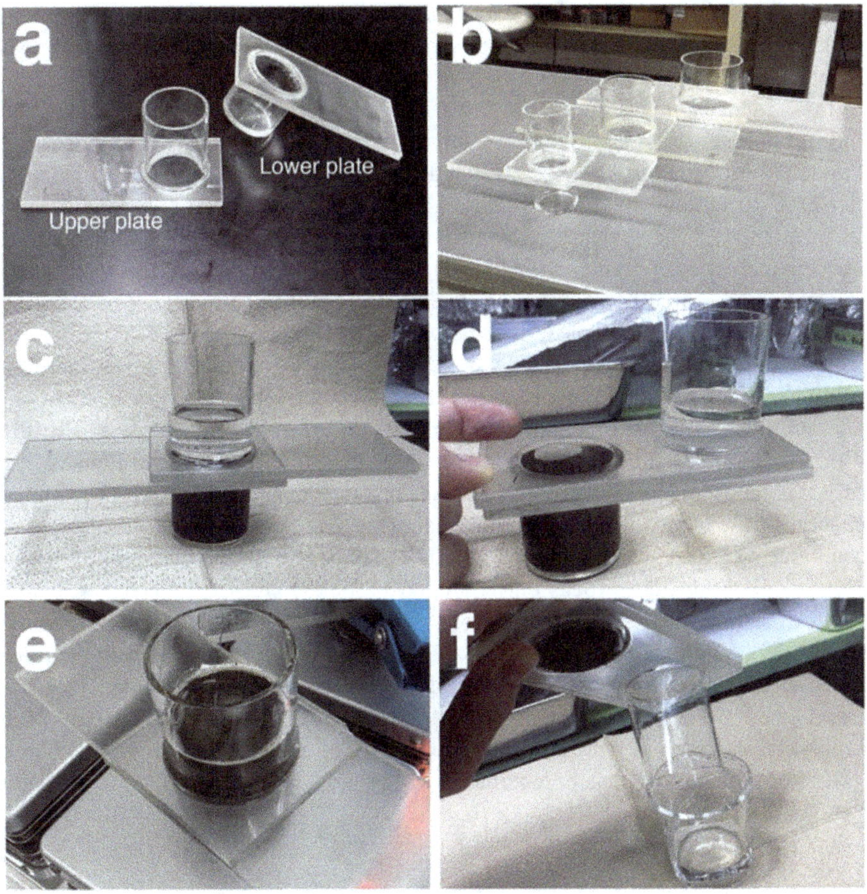

Fig. 2.3 JAMSTEC microplastic-sediment separator (JAMSS) unit. (**a**) The upper plate (left) incorporates an open glass tube, while the lower plate (right) includes a cylindrical glass container. (**b**) Small, middle, and large models of assembled JAMSS, consisting of a cylindrical container of 30, 60, and 100 ml volume, respectively. (**c**) JAMSS during density flotation with sediment in the lower container. (**d**) Separation of sediment and supernatant by sliding the two plates against each other. (**e**) JAMSS can be placed on a magnetic stirrer to ensure the sediments are well mixed during microplastic flotation. (**f**) Microplastics in the supernatant in the upper tube are poured out and rinsed from the internal walls of the tube.

extracted dried sand collected along a beach transect by first refluxing at ~37 °C with dichloromethane (DCM). This was followed by a second extraction of the sand in the same device with xylenes at 135–140 °C to obtain remaining, less-degraded polyolefins. Molecular size distributions of the extracted polymers were determined by gel permeation (also known as size exclusion (SEC)) liquid chromatography (GPC). 1H NMR, FTIR, and Py-GC/MS analyses of extracts were also performed. It is noteworthy that the authors observed greater amounts of DCM extractable residues (presumably degraded polymerics) in the more UV-exposed dune and backshore sands than in the foreshore sediments, highlighting the role of polymer weathering.

In another example, Fuller and Gautam (2016) pre-extracted municipal waste and soil samples in a pressurized fluid solvent extractor (Dionex ASE-350). Between 2 and 10 g of dried sample were extracted. They first were extracted with methanol at 100 °C to remove soluble fats and oils. The solids were then re-extracted with DCM at 180 °C. The solvent was removed by evaporation, yielding a solid residue, wherein the various polymers from the original sample were intermixed. FTIR analysis indicated spatial homogeneity in the resulting solidified plastic. The resulting FTIR spectra may be complex, as they will represent a composite of the different polymers present. If the extract is sufficiently free of nonplastic co-extractives, gravimetric determination of polymer mass allows a concentration determination. The authors (Fuller and Gautam 2016) performed recovery studies with PE, PP, PVC, PS, and PET. They suggested the method could be applicable to PU and PC, based on their detection in field-collected samples. Advantages include the simplification of cleanup/isolation procedures and the ability to automate analysis and quantitate total plastics, regardless of particle size. Again, most spectroscopy-based microscopic approaches focus on individual particle counts and are either instrument, time-intensive, or ineffective for microplastics below about 20 μm (Wolff et al. 2019). While data on microplastic shape and size were not obtained by Fuller and Gautam (2016), pre-separation of solids by passage through a sequence of different sized sieves could yield insightful data on particle size characteristics. The simplicity and ease in automation of the extraction method would provide some relief from the increased sample numbers.

2.5 Microplastic Detection and Instrumentation

Synthetic polymers are complex, typically high molecular weight organic molecules, and thus share attributes with natural dissolved (DOM) and particulate organic matter (POM). Hence, consideration of analytical techniques useful in studies of those materials may be fruitful (e.g., Materić et al. 2020). Being commercial products, extensive analytical work has also been done on polymers for developmental and quality control purposes and may also serve as a rich source of techniques. Indeed, considerable literature from the industrial plastic perspective has been available for years (e.g., Hakkarainen 2012). When selecting a mode of detection, the range of analytes (e.g., polymer types and sizes) to be included and particle characteristics (weight-based concentrations, particle abundances, or shapes) to be measured must be considered. In theory, it is desirable to obtain as complete as possible a suite of plastic particle characteristics. However, this is rarely achieved, even when using highly sophisticated instrumentation, due to the diversity of microplastic morphologies and polymeric compositions. Tradeoffs between detailed characterization of a limited number of particles and large sample throughput must be made. Complex

Table 2.1 The capability of different instrumental approaches for analyzing microplastics. Researchers can follow the color code to identify what methods fulfill their desired data requirements. The cells denoted by "possible" indicate that it can be accomplished upstream of the analysis. For example, it is feasible to quantify particle count in ATR-FTIR by physically interacting with and counting individual particles, while FPA-μFTIR can count small microplastics contained in the field of view. Notably, no single technique is capable of addressing all questions of interest. † Surface weathering and biofilm may be characterized. Solvent extraction of matrices (e.g., sediments) can be used to coalescence microplastics of all sizes into a single mass, including particles below the size detection capabilities (generally <10 μm) of individual target-based techniques

Goals and Capabilities of Various Instrumental Microplastic Detection Schemes

Number of particles	Polymer contribution by weight	Identify polymers ≤333 μm
Particle size	Additives/associated compounds	Automation
Polymer type	Identify polymers ≤10-20 μm	Laboratory processing required

Technique	Particle count	Particle size	Polymer Type	Polymer weight	Additives/associated compounds	Limit of Detection	Automation	Chemical Digestion Required
ATR-FTIR	Possible	Possible	Yes	Possible	Possible	-	No	No, if minimal biological growth. Yes, if biological growth: rinsing or chemical digestion.
μFTIR	Possible	Yes	Yes	No	Possible	≥10-20 μm	No	Yes: chemical digestion
FPA-μFTIR	Yes	Yes	Yes	No	Possible	≥10-20 μm	Yes	Yes: chemical digestion
Raman	Yes	Yes	Yes	No	No	1 μm	Possible (scanning mode)	Yes: chemical digestion
O-PTIR	Yes	Yes	Yes	No	No	≤10-20 μm	No	Yes: chemical digestion
Py-GC/MS	No	Possible	Yes	Yes	Yes	0 μm	Possible	No, unless high biological growth (chemical digestion)
TED-GC/MS	No	Possible	Yes	Yes	Yes	0 μm	Possible	No
MS	No	Possible	Yes	Yes	Yes	≥10-20 μm†	No	Yes: polymer pre-extraction (e.g. ASE)
XPS	Possible	Possible	Yes	Possible	No†	-	No	No, if biofilm desired. Yes, if biofilm not desired: digestion.

analytical schemes (due to financial, temporal, and manpower resource limitations) restrict the number of samples that can be examined. Several detection techniques described here are themselves hybrid approaches, i.e., consisting of an initial analyte introduction or separation process (e.g., chromatography, thermal desorption, pyrolysis), followed by compositional measurement proper (e.g., MS, FTIR, or Raman). Each technique should be selected based on its ability to answer the desired research questions. A table detailing discussed techniques, as well as their limitations in addressing different research questions (Table 2.1), is presented.

2.5.1 *Visual Identification*

Many investigations of plastics in the environment rely on the initial visual identification of particles using light microscopy, e.g., dissecting scopes. Such equipment is widely available. In the course of such studies, relatively large plastic debris may be removed with forceps for additional evaluation. Plastic debris identification criteria used by human observers include shape, color, texture, and absence of internal structures. However, such decisions are vulnerable to error depending on observer experience, matrix, and particle characteristics. Visual assessment of melting characteristics may be useful by contacting the particle with a hot needle (discussed later). Reliability of identification drops with decreasing target size but smaller particles will typically outnumber larger ones. Microplastics often are homogeneous in hue (although differential weathering may alter this) and lack internal structures or organelles. Filamentary structures are common in nature, so fiber identification imparts additional concerns. Fibers present narrow cross-sectional areas for examination and thus may be misidentified. Transparent particles may be overlooked using light microscopy. An interesting advanced technique is the use of "optical tweezers" for shepherding nanoplastics in liquids. Gillibert et al. (2019) demonstrated this, in combination with Raman microscopy, on a range of micro- and nanoplastics in fresh and saltwater, as well as particles that exhibited a thin biofilm.

2.5.2 *Dyes and Fluorescence Microscopy*

A number of stains have been evaluated for visualizing microplastics. Lipophilic Nile red has become a popular choice (Maes et al. 2017). This dye fluoresces, facilitating the detection and counting of small particles (Erni-Cassola et al. 2017). Prior digestion of coincident natural organic matter (e.g., cellulose and chitin) in the sample is recommended, as these polymers may also absorb dye to varying degrees. PE, PP, PS, and PA absorb the Nile red and fluoresce intensely; less hydrophobic polymers (e.g., PC, PET, PVC, and PU) absorb less and exhibit less intensity (Erni-Cassola et al. 2017). As brightness and particle size affect detectability, smaller particles and less lipophilic polymer types are more difficult to quantify by this approach due to their fainter signals. The fluorescent dye technique facilitates the identification of individual particles and shapes but is less diagnostic for polymer type. Image analysis software allows calculation of approximate plastic mass, based on the particle area and assumed density. As polymer thickness and composition (and hence density) are unknown, this approach can encompass considerable error. Flow cytometry, widely used for cell counting in phytoplankton and hematology research, has occasionally been used in microplastic-related lab experiments (e.g., Summers et al. 2018; Woods et al. 2018; Fu et al. 2020) but less so in field monitoring efforts. In this technique, particles in suspension are focused "single file" with a sheath fluid and passed by a laser. Impinging light may then be scattered forward or sideways, as a function of the size and granularity of the particle, respectively. Additionally, absorption may occur and particle fluorescence measured. Treatment of microplastics with Nile red might be advantageous here. Prior digestion of samples with chemical agents to remove biofilms or oxidize biogenic particulates might also facilitate detection. A major limitation of using dyes and fluorescence microscopy, however, is that the dye may interfere with subsequent polymer identification. Further, it has been shown that fluorescent compounds may leach from the polymer in tissues, so fluorescence itself may not be a reliable indicator of particle location (Schür et al. 2019).

It should be noted that fluorescent pigments, dyes, and optical brightening or whitening agents are widely used in the plastic and textile industries (Christie 1994).

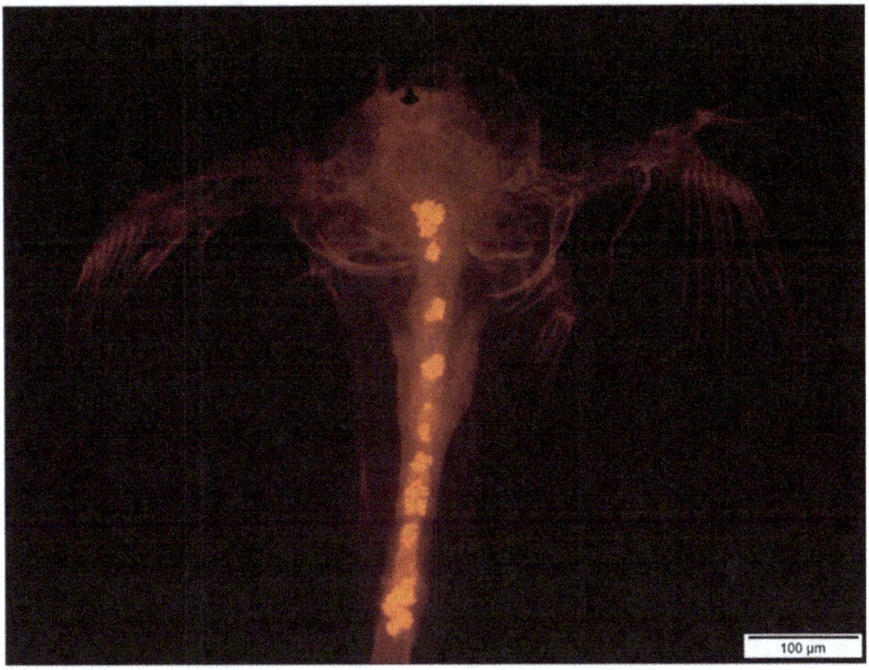

Fig. 2.4 Image of polyurethane microplastics (<53 µm) ingested by brine shrimp nauplii (*Artemia* sp.). Additives within the polyurethane elicited a fluorescent response. Imaged on an Olympus FV1200 laser scanning confocal microscope. Credit: Hamish Small (VIMS) and Virginia Worrell (Virginia Governor's School)

Thus, certain types of plastics may be detected without the addition of dyes. Dehghani et al. (2017) employed fluorescence microscopy to assess microplastics in urban dust samples. Bulk street sweeping samples were collected from the central district of Tehran, Iran. Fluorescent particles and fibers were visible in all samples. Hale et al. (2020) observed that colored PU foam commonly used in gymnastic pits fluoresced strongly. Additionally, <53 µm microplastics (produced by cryogenic fragmentation of bulk foam) were readily ingested by brine shrimp larvae in the lab. These microplastics were easily observed within the digestive tract by fluorescence microscopy (Fig. 2.4).

2.5.3 *Electron Microscopy (EM)*

While not typically used for direct polymer identification, scanning electron microscopy (SEM) is a powerful technique for delineating minute structural features, including deformities in, as well as colonizing organisms on, plastic debris surfaces (Zettler et al. 2013; Gniadek and Dąbrowska 2019). EM exhibits orders of magnitude greater spatial resolution (~0.0004 μm) than light microscopy (~0.3 μm). This derives from the differences in wavelengths between visible light (400–700 nm) and the high-energy electrons (0.001–0.01 nm) used for illumination (Girão et al. 2017). This concept also contributes to the resolution limitations of IR spectroscopy (2.5–20 μm). High-energy electron beams in EM are, however, capable of altering/damaging specimens, so care must be taken when imaging. Instrumentation and sample preparation costs for EM are also much greater than for light microscopy. The technique is not suitable for inspecting large microscopic fields or for rapid sample throughput. EM can be combined with energy-dispersive X-ray spectroscopy (EDS) to provide additional elemental composition information of targets (Girão et al. 2017). For example, Wang et al. (2017) applied SEM/EDS to evaluate elemental signals from selected particles. They succeeded in identifying the presence of chlorine in PVC microplastics, as well as ruling out nonplastic minerals in samples. Fluorinated polymers may also be amenable to this technique. Fries et al. (2013) identified the presence of Ba, S, O, and Zn associated with Wadden Sea microplastics with SEM/EDS. They also detected TiO_2 nanoparticles, which they theorized were used as white pigments or UV blockers in plastics. Ghosal and Wagner (2013) used SEM/EDS to identify Br and Sb, components of polymer organic and inorganic flame retardant additives, as well as TiO_2 in residential dusts containing microplastics. Further, they applied micro-Raman spectroscopy (μRaman) to associated particles and identified PE. They suggested ingestion of such microplastics could be an important human exposure pathway for these additives. Care must be exercised to remove possible matrix interferences from samples. For example, Br and Cl ions are common in seawater and residues may complicate conclusions.

2.5.4 Chromatography

Chromatography is widely used to separate components of complex mixtures by their properties, including polarity, solubility, volatility, and molecular size. Systems typically consist of a sample inlet, chromatographic column, and a detector. Analytes chemically or physically interact with a stationary phase contained within the column and are transported through the column by a mobile phase (e.g., a gas, liquid, or supercritical fluid). Analyte retention time (or volume) and detector response (e.g., a characteristic spectrum) are used for identification. Gas chromatography (GC) requires volatilization of analytes and employs a gaseous mobile phase. As such, GC has limited applicability to the direct analysis of low volatility polymers. However, it is invaluable for separating volatile constituents (e.g., additives), as well as polymer thermal degradates (see section on pyrolysis). Its facile coupling with low-cost mass spectrometers greatly expands its value.

Liquid chromatography (LC) in the form of size exclusion liquid chromatography (SEC), also known as gel permeation chromatography (GPC), has considerable utility in polymer analysis. Analyte elution time is a function of the pore/exclusion size of the media. Traditionally, larger molecules emerge earlier from the column than smaller molecules (which take a more circuitous path through the media). Hence, it has efficacy for both purification and characterization of polymeric materials. SEC has been widely used in polymer science, engineering, and product quality control, less so, to date, in the delineation of plastic contamination of the environment. Advancement in MS interfaces capable of handling liquid eluents has increased the power of this technique. LC has also been coupled to a variety of spectroscopic detectors, and their use is now appearing in the plastic debris-related literature. For example, Hintersteiner et al. (2015) evaluated molecular weight distributions of olefinic microplastics, following isolation from the sample matrix (in this case, personal care products). They dissolved the microplastics in 1,2,4-trichlorobenzene at 160 °C and separated the polymers by molecular size using high-temperature SEC with IR spectroscopy detection. Calibration was accomplished using PS standards ranging from 700 to 2 million g mol^{-1}. LC typically exhibits lower resolution capabilities than GC. However, the development of ultrahigh performance (UHP) LC instruments, equipped with columns containing particles of extremely small size, has reduced this disadvantage.

Supercritical fluid chromatography (SFC) often exhibits higher-resolution capabilities than LC and has been used to characterize polymers (e.g., Takahashi 2013). In SFC the pressure and temperature of a gas (e.g., CO_2) are manipulated in such a way that it exhibits properties intermediate to a gas and liquid. Commercially available SFC equipment has become widely available, as are improved interfaces compatible with modern detectors. However, no published references were found to indicate it has yet been used for the analysis of microplastics obtained from environmental media.

2.5.5 *Infrared (IR) Spectroscopy*

IR spectroscopy, commonly performed with a Fourier transform infrared (FTIR) spectrometer, is an established technique for identification of polymeric materials. IR spectroscopy is a nondestructive technique, allowing reanalysis of the same material. The two primary modes most commonly applied for microplastic analysis are transmittance and reflectance (Chen et al. 2020). The long wavelengths of IR radiation limit the spatial resolution of FTIR to 2.7 μm, but instrument limitations reduce the spatial resolution to about 10–20 μm, depending on the design. Compositional identification is achieved by comparing the sample spectrum with a known reference polymer. FTIR may be suitable for the identification of colored microplastics whose pigments may fluorescence and interfere with Raman spectra. FTIR may also be valuable for monitoring the degree of weathering of polymers, e.g., the development of hydroxyl and carbonyl groups (Cai et al. 2018).

For larger plastic debris (>500 μm), FTIR can be paired with an attenuated total reflectance (ATR) accessory, which measures the surface composition of materials. The IR from an ATR typically penetrates the polymer to a depth of 0.5–2 μm (Li et al. 2018b), problematic for heterogeneous, layered materials. The measurement requires physical contact of a crystal (composed of germanium, zinc selenide, silicon, or diamond) with the targeted material. This requirement can lead to logistical issues with some materials, e.g., small micro- or nanoplastics, and is not amenable to samples with numerous targets. However, ATR-FTIR has been used to identify larger microplastics in Arctic deep-sea sediments (Bergmann et al. 2017) and in fish from the African Great Lakes (Biginagwa et al. 2016).

Micro-FTIR (μFTIR) combines vibrational spectroscopy and microscopy, allowing the analysis of smaller particles. Use in transmission or reflectance mode permits application using membrane filters with minimal sample preparation. Transmission mode requires an IR-translucent substrate, while the reflectance mode can be applied to thick and more opaque samples (Li et al. 2018b). Application to irregularly shaped particles can result in non-interpretable spectra, due to refractive errors (Harrison et al. 2012). Li et al. (2018a) used μFTIR to demonstrate that supermarket-purchased mussels in the UK contained microplastics and concluded that their quantification should be included in food safety management measures.

Manual repositioning of a μFTIR stage across a viewing field containing numerous particles is subject to human bias, tedious, and time-consuming. Field-derived samples may contain hundreds of candidate particles. As a consequence, commercial instruments have been developed to automate this process and generate spectral image maps. However, the process can still be slow (hours to days), as a function of required resolution and sample complexity. That being said, μFTIR analysis is typically faster than μRaman (described below). Instrument cost increases with increased capabilities and in the present market may be beyond many lab budgets.

The use of focal plane array (FPA) μFTIR detectors allows for analysis of multiple particles at one time with high resolution. Such FTIR imaging can produce a detailed, high-throughput analysis of total microplastics on a filter. However, particles with irregular shapes may still not be suitable for FPA-μFTIR imaging. Additionally, only a few μFTIR instruments can analyze particles <20 μm. Fibers, due to their narrow cross-sectional areas, also present challenges. However, Tagg et al. (2015) successfully demonstrated that FPA-based μFTIR imaging could identify a range of microplastics in wastewater. Primpke et al. (2017) developed an automated image analysis method using FPA-μFTIR to provide particle identity, count, and size of microplastics in complex matrices, increasing data quality and ease of data interpretation. Later, Primpke et al. (2018) generated a FTIR reference library for automated analysis of microplastics. Other reference libraries and automated sampling software are available, such as siMPle (simple-plastics.eu; Liu et al. 2019; Primpke et al. 2019).

2.5.6 Raman Spectroscopy

Akin to FTIR, Raman is a nondestructive, vibrational spectroscopic technique increasingly used in the analysis of plastic debris. Unlike IR spectroscopy, where the absorbance of radiation by molecules is measured, in Raman a narrow wavelength laser is used to excite the surface of a particle, and the photons resulting from inelastic scattering (Raman scattering) are detected and recorded as a spectrum. The spectrometer may be interfaced with a microscope (designated μRaman). Compared to FTIR, Raman is adept at identifying nonpolar functional groups (e.g., aromatic, C—H, etc.; Käppler et al. 2016). Hence, IR spectroscopy and Raman are complementary techniques. μRaman typically has greater spatial resolution potential than μFTIR, down to about 1 μm. However, fluorescence from irradiated materials can overwhelm the relatively weak Raman scattering. But the analyst may be able to minimize fluorescence by choosing a different excitation wavelength. Both FTIR and Raman are typically used to analyze particles on a filter or other surface, following pretreatment of sample to eliminate interfering materials. The filter or window on which the samples are placed may limit the wavenumber range possible in transmission mode.

Raman can identify plastics from a variety of environmental matrices. However, to date available studies are heavily skewed towards water and sediments (most recently reviewed in Erni-Cassola et al. 2019). Recently, Wright et al. (2019) applied μRaman spectral imaging to the identification of microplastics in inhalable air, while simultaneously evaluating routine air quality monitoring metrics. Fortin et al. (2019) characterized microplastics <10 μm by μRaman in water from an advanced wastewater treatment facility. Cabernard et al. (2018) reported that μRaman identified more particles in the 10 to 500 μm range. Particle agglomeration and losses were observed, suggesting the need for surrogate spiking and percent recovery studies in the microplastic analysis.

Researchers are developing open-access reference libraries for Raman spectra. For example, Munno et al. (2020) established a spectral library of plastic particles (SLoPP) encompassing 148 diverse reference spectra. Their SLoPP-Environmental or SLoPP-E libraries included spectra of 113 particles collected globally. This addition is important as weathering can modify spectral characteristics. When compared to manufacturer reference libraries, 63% of particles tested registered the strongest matches using SLoPP or SLoPP-E, illustrating the utility of reference libraries created specifically with microplastic identification in mind.

Two notable studies have compared μRaman to FPA-μFTIR for analysis of microplastics in environmental samples. Käppler et al. (2016) found that Raman was superior at identifying PVC particles, while polyesters and particles with high dye content or fluorescence were more accurately identified with FPA-μFTIR. Due to this and its greater resolution, μRaman detected significantly more particles than FPA-μFTIR. Yet, when methods were automated to reduce the analysis time from 38 h to 90 min (closer to the 20-min sampling time of FPA-μFTIR), the same number of particles was identified. This underscored the opinion that longer processing times are necessary to reap the advantages of Raman. Cabernard et al. (2018), in a study of North Sea surface waters, arrived at similar conclusions. These authors reported that the Raman signal was obstructed by highly pigmented particles, limiting identification, particularly for rubber and ethylene vinyl acetate (EVA) microplastics. Käppler et al. (2016) identified TiO_2 in inorganic particles by Raman. The TiO_2 peak was also identified as a white pigment, along with an acrylic resin using

FTIR. Imhoff et al. (2016) also identified paint particles using Raman by their characteristic high pigment content.

2.5.7 *Scanning Probe Microscopy (SPM)*

SPM is a family of technologies that has been widely used in materials science. Atomic force microscopy (AFM) uses a mechanical cantilever to physically measure the surface topography of a sample at low micron and even sub-nanometer scales (Dazzi et al. 2012). The interaction volume between the physical probe and sample may be on the picometer scale.

The high resolution of AFM can be invaluable for characterizing minute physical features, as small as an atom. However, instruments can typically only interrogate micron-sized spatial areas and must be used in workspaces carefully engineered to

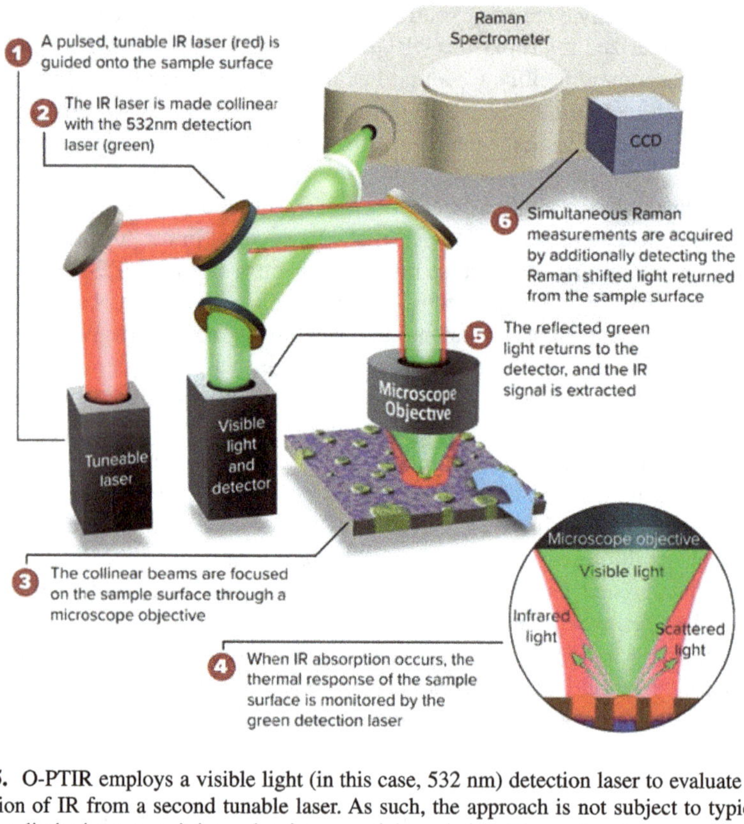

Fig. 2.5. O-PTIR employs a visible light (in this case, 532 nm) detection laser to evaluate target absorption of IR from a second tunable laser. As such, the approach is not subject to typical IR refraction limitations, permitting sub-micron spatial resolution. The system here was combined with a Raman spectrometer, allowing the collection of complementary spectral information. IR and Raman spectra were obtained from several locations (shown as green, blue, and purple colors) across a 4 × 14 μm microplastic particle present in NIST SRM#2585 (indoor dust). Spectra were consistent with a polymethacrylate polymer. (Images courtesy of Photothermal Spectroscopy Corp)

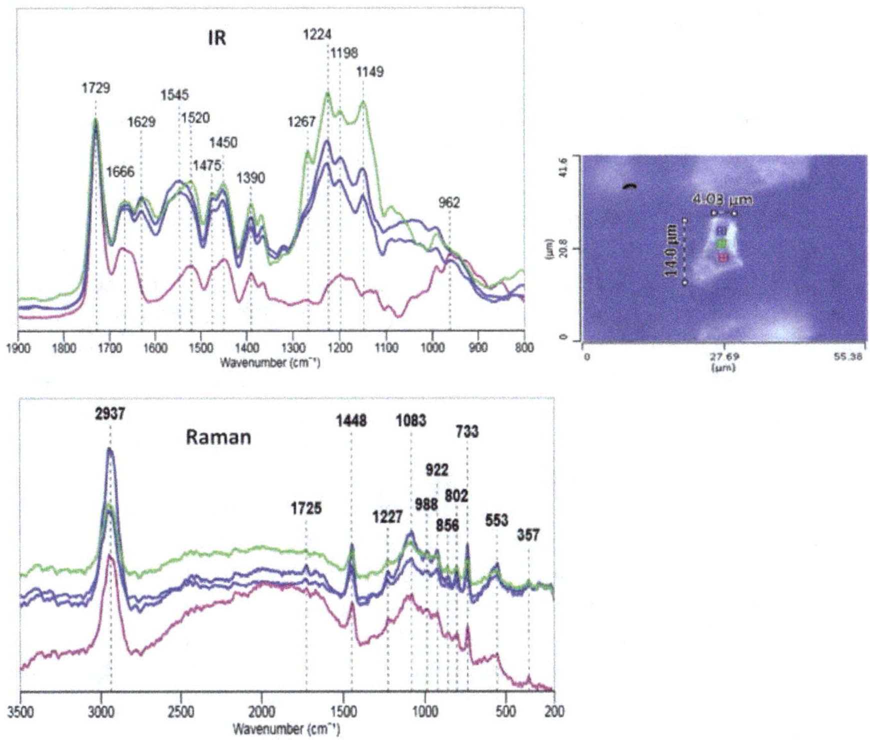

Fig. 2.5 (continued)

eliminate ambient vibrations. In AFM-IR, the thermal expansion of materials following the incidence of IR radiation is measured (Fu et al. 2020).

Optical photothermal IR (O-PTIR) is a novel technique that uses a visible laser light source, rather than a mechanical cantilever, to evaluate thermal expansion of the targeted material following illumination with a collinear, tunable mid-IR laser (Fig. 2.5). This approach allows the extraction of a signal that closely approximates widely available FTIR spectra. Spatial resolution is <1 μm, far lower than conventional IR. Since the system is not based on IR transmittance, thicker samples can be evaluated than with conventional IR absorption. It is also compatible with irregular surfaces. The system can be coupled with Raman spectroscopy, allowing the collection of Raman and IR spectroscopy data for the same particle. As these techniques are complementary, identification capabilities are increased.

Merzel et al. (2019) recently applied AFM-IR, O-PTIR, and fluorescence microscopy to image and chemically interrogate nanometer-sized PS beads taken up and retained by freshwater mussels during an in-laboratory exposure. They noted, while sensitive, AFM-IR analysis was time-consuming and applicable only to the immediate surface of the material examined. The presence of a surface biofilm also interfered with spectra acquisition for the underlying polymer. In contrast, O-PTIR was faster and less vulnerable to biofilm interference.

2.5.8 *Mass Spectrometry (MS)*

MS is a powerful tool for the identification and quantification of organic materials. The technique measures the mass-to-charge ratio (m/z) of ions generated from the fragmentation of an analyte. A variety of ionization approaches, differing in energy, have been developed, depending on the material under study. MS with specialized interfaces can be utilized for the identification of polymers, degradation products, and additives. Interfaces include chromatographic (i.e., GC and LC) and thermal techniques (i.e., pyrolysis (Py), thermogravimetric analysis (TGA), thermal extraction desorption (TED)) for better characterization of different type of plastics. Spectra acquisition is very rapid, on the order of milliseconds, depending on the instrument design and desired mass range. MS is popular to identify and quantify both targeted and untargeted additives (e.g., plasticizers (Peters et al. 2018), flame retardants (Hale et al. 2002; Khaled et al. 2018), and pigments (Imhof et al. 2016)) in plastic debris, as well as surface-sorbed contaminants.

High-resolution and tandem (where two or more mass analyzers are coupled) MS instruments are becoming more widely available. However, MS (with the exception of pyrolysis applications) for the analysis of plastics present in environmental matrices has to date been limited. This is due to the relatively low mass range of most MS compared to that of many polymers. MS units capable of high molecular weight analyses tend to be expensive and complex and may require considerable operator skill.

Limited or no sample preparation is a desirable feature. In some cases, ambient (direct) analysis of plastics is feasible by probing surfaces with an energetic beam. However, these techniques may not be applicable to molecules with molecular weights >3000 Da (Schirinzi et al. 2019). In other approaches polymers are dissolved, separated from the matrix (via vaporization, pyrolysis, or liquid chromatography), and then detected by MS.

Schirinzi et al. (2019) evaluated several MS-based techniques for the analysis of PS microplastics obtained from natural waters, including SEC/atmospheric pressure photoionization (APPI) MS, direct analysis in real-time (DART), matrix-assisted laser desorption ionization (MALDI), and desorption electrospray ionization (DESI). They reported that SEC/APPI-MS exhibited the greatest sensitivity, with a detection limit of about 20 pg. MALDI-MS has been explored for the detection of various synthetic polymers (Weidner and Trimpin 2011). A challenging aspect of MALDI-MS analysis is the identification and discrimination of plastic fragments in the presence of coincident interferences, e.g., biofilms. Lin et al. (2020) recently used MALDI-MS to identify and quantify PS particles ranging from 100 nm to 4 mm. Signals were enhanced by thermal pretreatment, enabling higher quantification accuracy. Lin et al. (2020) examined the feasibility of such methods to analyze commercial plastic products, as well as microplastics from river water and fish. MALDI-MS thus is a promising tool for the evaluation of limited numbers of microplastics in samples (Huppertsberg and Knepper 2018).

In a novel study, Wang et al. (2017) depolymerized ester-containing polymers (i.e., PC, PET) using alkali and heat. They then subjected the resulting products (bisphenol A and *p*-phthalic acid) to liquid chromatography/tandem mass spectrometry (LC-MS/MS). Sample matrices included wastewater sludge, marine sediments, indoor dust, and shellfish. As contributions of bisphenol A and *p*-phthalic acid from non-polymer sources were possible, samples were also assessed for their pre-depolymerization levels to differentiate pre-existing contamination by these constituents. The authors reported particularly high concentrations of PC (246 mg/kg) and PET (430 mg/kg) in the dust sample.

Materić et al. (2020) applied thermal desorption-proton transfer reaction-mass spectrometry (TD-PTR-MS) to detect nanoplastics of several polymer types in 0.2 µm-filtered water samples derived from snow cores from the Austrian Alps. TD-PTR-MS has mainly been employed for volatile compounds but has recently been extended to semivolatiles. Materić et al. (2020) used chemical ionization via hydronium ions to produce low fragmentation ions. Evaporation/sublimation of constituents was achieved by ramping from 35 to 350 °C at 40°/min.

Time-of-flight secondary ion mass spectrometry (TOF-SIMS) is capable of providing compositional and spatial distribution information for plastics. Excitation of a surface by a focused ion beam causes an emission of secondary ions and clusters from the sample. The TOF analyzer then measures the exact mass of these, allowing for compositional determination. The resulting data can be used to create images of very thin polymer surfaces (on the order of nm). For example, Jungnickel et al. (2016) applied this technique to image 10 µm PE particles.

Direct analysis in real time (DART) is an ion source that uses a heated helium, argon, or nitrogen plasma stream to desorb and excite molecules from surfaces. These may then be drawn into the inlet of a high-resolution MS, permitting exact mass measurements and subsequent constituent identification. Zhang et al. (2020b) introduced selected microplastics into a thermal desorption/pyrolysis inlet connected to a DART ionization device interfaced with a Q Exactive™ hybrid quadrupole-orbitrap MS. They reported the detection of both additives (plasticizers, antioxidants, and cross-linking agents) and polymers (PE, PP, PET, PS, polyester, PA). Multivariate statistical evaluations of the ions produced from the thermal desorption and pyrolysis processes were used to establish identifications.

Elemental analyzer/isotope ratio mass spectrometry (EA/IRMS) is widely used in geochemistry to establish the origin of organic matter using carbon isotopes and to evaluate food authenticity (e.g., via nitrogen isotope patterns). Berto et al. (2017) demonstrated that carbon isotopic composition is sufficient to discriminate fossil fuel-derived polymers (e.g., high- and low-density PE) from plant-derived bioplastics in commercial products. This method also was advantageous for testing darkly colored samples, which are problematic in some spectroscopic techniques. However, EA/IRMS alone provides limited information about the specific type, shape, size, and composition of MPs.

Fourier transform ion cyclotron resonance mass spectrometry (FTICR-MS) has long been used for the characterization of synthetic polymers (e.g., Brenna and Creasy 1991), petroleum residues (Chen et al. 2016), and natural organic matter (Riedel and Dittmar 2014). FTICR-MS is a promising technique, allowing the accurate mass analysis of high molecular weight species. However, it has not been widely applied to the issue of plastic debris to date. Instruments are costly.

Inductively coupled plasma MS (ICP-MS) has occasionally been employed to characterize a range of elements, particularly metals (common in pigments) associated with plastics. Samples are typically digested using strong acids prior to analysis. See the pigment discussion below for more details.

2.5.9 *Thermal Analysis Techniques*

A variety of thermal analysis techniques are applicable to microplastics. A simple approach is the application of a hot needle to selected particles to evaluate melting potential (Silva et al. 2018). Natural materials do not typically melt or curl. Peñalver et al. (2020) recently reviewed a number of thermal analysis techniques. These range from gravimetric measurements across time and temperature gradients, to more sophisticated hyphenated techniques. Gravimetric measurements, being rather nonspecific, are vulnerable to the presence of interferences from nonplastic materials. Detection limits may be problematic as many approaches (e.g., pyrolysis) are sample size limited. Pre-concentration of microplastics may reduce this shortcoming.

Pyrolysis-GC/MS (Py-GC/MS) has proven to be a valuable technique for microplastic characterization. This destructive method thermally deconstructs samples over one or multiple temperature ranges. Polymers that produce characteristic degradates may thus be identified and quantitated. Further, GC peaks for additives or sorbed hydrophobic organic pollutants may be identified using an additional thermal desorption step. Py-GC/MS has typically been employed for discrete plastic particles. More recently, it has been adapted to analyze mixtures, allowing higher-throughput analysis. As it is not dependent on particle size, Py-GC/MS is a promising technique for nanoplastic analysis (Mintenig et al. 2018; Ter Halle et al. 2017), provided sufficient material is input. Unlike FTIR or Raman methods, however, pyrolysis by itself does not permit the acquisition of particle counts, shape, or size data, unless other techniques (or pre-sieving into size classes) are employed prior to pyrolysis.

A sample is typically introduced to the pyrolyzer in a glass thermal desorption tube or stainless-steel cup. These can accommodate small pieces of plastic, generally less than 1.5 mm diameter. The development of larger pyrolysis chambers would advance this technique, allowing characterization of microplastics that have been concentrated on a filter. For polymer analysis, high temperatures (~700 °C) are required. At lower temperatures (590 °C), pyrolysis is possible with thermochemolysis, established by spiking the sample with, for example, ~10 µl of tetramethylammonium hydroxide (Gomiero et al. 2019; Fischer and Scholz-Böttcher 2017). Pyrolysis thermally degrades the polymer, yielding organic pyrolysates, which are characterized via subsequent GC/MS. Comparison of the chromatographic fingerprint to that of known polymers can reveal polymer type and peak integration of one or more marker compounds used to obtain analyte mass (Fischer and Scholz-Böttcher 2017). For example, in a study of plastic pollution analysis using Py-GC/MS, Fries et al. (2013) presented a PE chromatogram containing a characteristic series of n-alkanes, n-alkenes, and n-alkadienes (also shown in Ter Halle et al. 2017). A comprehensive list of characteristic decomposition products for different polymers can be found in Fischer and Scholz-Böttcher (2017). Fries et al. (2013) were able to identify eight polymer types collected from sediments using this technique, including PE, PP, PS, and PA. Likewise, Doyen et al. (2019) and Gomiero et al. (2019) identified multiple polymers in plastics extracted from beach sediments, while Hendrickson et al. (2018) used Py-GC/MS to validate polymer identification in surface waters of Lake Superior. Py-GC/MS has the ability to characterize the relative contribution of different polymers in complex, layered polymers, or copolymers.

Complex samples with multiple unknown polymers may be analyzed using Py-GC/MS. For example, Ter Halle et al. (2017) used ultrafiltration to concentrate nanoplastics in seawater and then validated their presence using dynamic light scattering analysis. The thermal degradation products were compared to known polymer standards, and advanced statistics were applied to estimate the relative percent contribution of up to three polymers per sample. Weight-based estimates were not possible using this approach. An alternative statistical approach can be found in Zhang et al. (2020b). The complexity of these analyses was underscored by the shared pyrolysis products from multiple compounds (Fischer and Scholz-Böttcher 2017). Further, coincident matrices in the sample can contribute thermal decomposition products, such as styrene from chitin (Fischer and Scholz-Böttcher 2017). Sample digestive pretreatment may be effective in eliminating much of this interference.

An additional capability of Py-GC/MS is investigating polymer additives or particle-sorbed contaminants. To detect volatile or semi-volatile organic additives, thermal desorption is typically employed prior to pyrolysis. Here, a sample is first heated to 350 °C, for example, in order to release the more volatile constituents. Cryogenic cooling is employed to trap thermal decomposition products prior to GC and subsequent pyrolysis of the sample. Fries et al. (2013) used this "double-shot" technique to detect phthalates, benzaldehyde (flavoring substance), and 2,4-di-*tert*-butylphenol (antioxidant) in plastics. Alternatively, in place of separate thermal desorption and pyrolysis steps, compositional data may be collected over a temperature ramp whereby thermally desorbed compounds will elute early, while pyrolysis decomposition products elute later. Zhang et al. (2020a, b) demonstrated this approach, in conjunction with high-resolution MS. As such, Py-GC/MS can simultaneously investigate complex organic compounds within/on plastics not feasible using IR or Raman spectroscopic techniques.

A related technique to Py-GC/MS is thermal extraction desorption GC/MS (TED-GC/MS). This approach employs thermal decomposition over a longer period of time, utilizing a thermal desorption unit (Duemichen et al. 2019). This process is amenable to larger samples and reduces the need for pretreatment. Eisentraut et al. (2018) used TED-GC/MS for the analysis of street runoff and sludge samples. Street runoff was sieved and reconcentrated on glass fiber filters, while sludge was homogenized with a ball mill. In addition to traditionally identified thermoplastics (i.e., PE, PP, etc.), these authors reported the presence of tire wear particles using marker compounds (elastomers, antioxidants, and vulcanization agents). Overall, TED-GC/MS and Py-GC/MS hold promise for the characterization of plastic debris, independent of size, and for increasing sample throughput.

Thermogravimetric analysis (TGA) is a method where the loss in sample mass is evaluated as a function of temperature, under specific atmospheric conditions. Without additional detectors such as MS or IR, qualitative information provided is limited. Sample pretreatment is typically modest. The relatively high sample amounts employed in TGA units (about 200 times higher during TED-GC/MS than Py-GC/MS) enable the measurement of heterogeneous matrices, reducing sample representativeness concerns (Dümichen et al. 2015; Elert et al. 2017). However, further investigations of the applicability of this method for matrices with high concentrations of impurities, such as natural organic matter, are needed. Mengistu et al. (2019) utilized a simultaneous thermal analyzer, a TGA interfaced with a FTIR, to systematically evaluate responses from tire granules and sediments amended with tire granules. Here, both sample mass losses over a temperature and time program and IR spectral information were obtained. The formulated sediments (wetted to 50% by volume) consisted of 5% organic matter (conifer bark), 75% quartz sand, and 20% kaolinite clay.

2.5.10 X-ray Photoelectron Spectroscopy (XPS)

XPS, also known as electron spectroscopy for chemical analysis (ESCA), is a technique for qualitative and quantitative measurement of elemental composition and chemical/electronic state of elements in materials. X-ray irradiation of a surface produces spectra denoting spatial and depth distribution. The kinetic energy and number of electrons that escape from the surface (1–12 nm) of the material are simultaneously measured. Hernandez et al. (2017) used XPS to evaluate the chemical composition of PE nanoplastics in personal care products, while Lu et al. (2018) investigated the aqueous aggregation of PS microspheres under varying pH, humic acid, and ionic conditions.

XPS can also provide valuable information on chemical changes occurring in the first atomic layer (<3 nm) of polymeric surfaces. Tian et al. (2019) examined the formation of C–O groups on PS nanoplastics after 48 h of UV irradiation. In contrast, no significant changes were observed by FTIR analysis. Future application of XPS to the study of micro- or nanoplastic surface weathering under environmental conditions may be insightful. Furthermore, XPS may facilitate a deeper understanding of biofilm formation on plastics. In this context, Feng et al. (2018) explored changes via XPS in extracellular polymeric substances released by microorganisms in activated sludge.

2.6 Microparticle Classes of Emerging Concern

Many researchers exclude biogenically-derived or partially synthetic plastic debris (e.g., cellulose-derived products such as rayon) from their consideration of "microplastics" in the environment. However, organismal effects may occur via physical mechanisms, irrespective of the particle's precursors (fossil fuel, bio-based, or wholly natural). For example, negative impacts may arise from physical (e.g., blocking of digestive tracts or overall reduction of caloric value of ingested materials) rather than chemical interactions (Wright et al. 2013). Suaria et al. (2020) reported that naturally occurring and man-made cellulosic fibers outweighed by >10-fold synthetic microfibers in the waters from six ocean basins. Reed et al. (2018) reported that rayon fibers were the most abundant microplastic class detected in marine sediments near the Rothera Research Facility (Antarctica). These likely arose from textiles released via wastewater. Further, biopolymers have been touted as a replacement for fossil fuel-based polymers. However, the mechanisms of toxicity of microparticles remain uncertain.

Analysis of natural precursor-derived particles (e.g., rayon) can be problematic due to their vulnerability to caustic sample preparation techniques commonly used for wholly synthetic microplastics. In addition, their chemical similarity to natural detritus may confound subsequent instrumental analysis. Surface coatings and tire wear are two additional classes of particles that are environmentally prevalent. However, to date, they have not garnered the level of attention paid to other microplastics. In part, this arises from their complex nature and difficulties in their analysis.

2.6.1 Surface Coatings/Paints

Buildings, roadway markings, and vessels are frequently treated or painted. Modern surface coatings often have polymeric components. Fragments are released as a result of abrasion and weathering. Paints may be formulated with pigments containing metals (e.g., Ba, Cd, Cr, Cu, Fe, Hg, Pb, Sn, Ti, and Zn) and organic constituents. Specialty coatings may contain intentionally toxic chemicals such as antifoulants and microbicides. Takahashi et al. (2012) observed that paint particles constituted up to 0.2% of cores of UK estuarine sediments. Here, particles were manually separated and weighed. The authors noted that metallic entities were rather homogeneously distributed vertically in the cores. They hypothesized the metals, being more water-soluble, may have been released from the pigment particles.

Song et al. (2014) investigated microplastics in the waters of Jinhae Bay, Korea. They observed that paint particle abundance exceeded that of other microplastic classes and that most paint particles were small, less than 100 μm. Alkyd resins and poly(acrylate/styrene) from fiberglass were common. Particles concentrated in the water's surface microlayer, a zone of considerable physical and biological activity. Microplastics were collected on GF/F glass fiber filters and identified by ATR-FTIR. Imhof et al. (2016) investigated microplastics and paint particles in freshwater lake sediments. They reported that paint particles were small: 1–50 μm. Hence, they exhibited high surface area to volume ratios. Samples were treated with hydrogen peroxide and sulfuric acid to remove organic interferences. They were stirred for 1 week and collected on 2.2 μm quartz fiber filters (Whatman QM-A). Microplastics were identified by μRaman. Particle abundances were determined by counting selected regions of the filter. However, this approach may result in errors if the particles are heterogeneously distributed on the filter. The authors also noted that the pigment particles were brittle. Hence, rigorous physical treatment of samples could distort the original particle size distributions. In some cases, the authors were unable to identify associated polymers in the particles. Imhof et al. (2016) determined metal content by inductively coupled plasma MS. Cabernard et al. (2018) evaluated particles from North Sea surface waters by μRaman and ATR-FTIR and reported many large (>500 μm) "varnish" fragments (suggesting precursors derived from nonfossil fuel sources). Smaller particles were collected on a gold-coated mirror for μRaman or gold-coated PC filters for FTIR in reflectance mode. They also encountered colored particles that exhibited substantial fluorescence, interfering with Raman analysis. Aggregation of particles on the filter, reducing apparent particle numbers, was observed. The presence of dark pigments and fibers was an additional analytical challenge. Cabernard et al. (2018) also noted long analysis time, residual organics interfering with spectra, and the presence of salt precipitates from lab procedures compromised results.

2.6.2 *Tire Particles*

Tire wear fragments have been reported by some to be among the most abundant synthetic microparticles, especially near roadways (Kole et al. 2017; Sommer et al. 2018; Hale et al. 2020). In addition, substantial amounts of scrap tires are recycled, including use on playgrounds and athletic fields. In addition to the rubber infill, such fields may incorporate polymer fibers to simulate grass blades, as well as polymeric carpet backing (Cheng et al. 2014). Tire scrap and other waste plastics are also being incorporated into asphalt pavements. Materials are subsequently subjected to additional weathering and fragmentation to varying degrees. Fragments may be transported by runoff to surface waters. From there they may be transported by runoff to surface waters. Particle composition varies as a function of the manufacturer and may include natural rubber, carbon black, ABS plastic, metal and fiberglass belts, and other materials. A variety of techniques (Leads and Weinstein 2019) have been used to quantify tire wear, including visual examination, Py-GC/MS, SEM-EDX (scanning electron microscopy-energy-dispersive X-ray analysis), FTIR (for ABS), and investigation of specific chemical markers, such as benzothiazole.

Analysis of tire particles using spectroscopic methods (Sommer et al. 2018) can be impeded by near-complete absorption of IR light due to filler components in tire fragments (e.g., carbon black) and strong fluorescence (Eisentraut et al. 2018). Thermoanalytical methods, like Py-GC/MS and TED-GC/MS, are an alternative approach. These techniques utilize markers such as decomposition products and vulcanizing agents for identification. Eisentraut et al. (2018) used TED-GC/MS to simultaneously measure microplastics originating from thermoplastics and tire wear abrasion products in environmental matrices. Unice et al. (2012) developed a protocol to analyze and quantify tire particles in environmental matrices using Py-GC/MS.

Optimal markers should be selective, stable, and easily detectable. Benzothiazole-based vulcanization agents have previously been used as chemical markers to estimate presence (Spies et al. 1987). However, their suitability in quantification (and identification) of tire particles has been questioned due to their water solubility/leachability and reactivity (Eisentraut et al. 2018; Wagner et al. 2018; Klöckner et al. 2019). Zn as an elemental marker has potential due to its high concentration in tires. Unfortunately, there are numerous sources of Zn (Klöckner et al. 2019). Thus, using elemental markers like Zn for the detection of tire particles in environmental matrices may be suitable only if coincident interferences are removed prior to analysis of the fragments of interest.

Natural rubber (NR) is the main elastomeric constituent of truck tires but has been found in other tire materials (Wagner et al. 2018). NR decomposition products include dimers, trimers, and tetramers of isoprene. However, coincident natural sample constituents may exhibit similar decomposition products (Eisentraut et al. 2018). Thus, analysis of NR in environmental samples may be problematic if such extraneous organics cannot be excluded. Styrene-butadiene rubber (SBR), a synthetic petroleum-based product, has also received attention as a chemical marker compound. SBR is abundant in tires, exhibits limited leaching, and has few non-tire sources (Eisentraut et al. 2018).

Sommer et al. (2018) used multiple approaches to identify tire-derived aerosol particulates collected near German roadways. These included transmitted light microscopy and SEM-EDX, as well as diagnostic particle axial ratios and volumes. The authors noted that most tire wear particles exhibited rounded, elongated shapes, with adhering road and brake wear particles. Mengistu et al. (2019) developed a method for detection and quantification of tire wear particles in sediments that entails the use of FTIR, simultaneous thermal analysis (STA), and parallel factor analysis (PARAFAC). STA and FTIR were first used to generate data matrices, which then provided data for the PARAFAC. PARAFAC was used to decompose the overlying components in the spectral data into groups of substances for easier analysis. Mengistu et al. (2019) speculated that with further development and incorporation of PARAFAC with FTIR analysis, the method proposed could be automated for faster analysis of tire wear particles in sediment samples.

2.7 Quality Assurance and Quality Control

The development of appropriate goals and hypotheses is the first and most critical step in plastic debris studies. However, to date many published works have relied on opportunistic field studies or biased sampling designs. This is due in part to the contemporaneous inception of plastics manufacturing (circa 1950) and the recent discovery of the widespread environmental distribution of microplastics. The significance of results must be evaluated in the context of the quality and extent of the sampling, accuracy, and inclusiveness of the analysis methods applied.

de Ruijter et al. (2020) discussed important quality criteria for evaluating microplastic risk assessments. Among those most pertinent to analytical studies were delineating particle size, shape, polymer type, source of microplastic, appropriate reporting units, chemical purity, effectiveness of lab preparation steps, and verification of background contamination and replication. The selection of appropriate study materials to meet study goals is essential. On account of the wide range of plastics made, care must be taken when extrapolating findings from a small subset of plastics to the wide range in commerce. In some lab studies, the basis for polymer types chosen includes concordance with those used in previous studies (a standardization mindset) or their facile acquisition (i.e., commercially available or already in-hand materials). For example, low- and high-density polyethylenes (LDPE and HDPE, respectively) are common in surface waters. However, ultrahigh molecular weight (UHMW) PE has been used as a representative polymer in some influential studies (e.g., Teuten et al. 2007; Bakir et al. 2012) to evaluate microplastic sorption of water-borne organic pollutants. However, UHMWPE is a high crystallinity polymer. It is used in niche applications such as surgical implants in humans and high-precision mechanical gears due to its extreme hardness, inertness, and durability. UHMWPE's prevalence in the environment is likely very low. Its environmental behavior, versus widely used LDPE and HDPE, merits examination. Additive and filler packages may also differ. Unfortunately, plastic manufacturers often provide an incomplete list of ingredients of plastic products or declare composition "confidential business information." Accordingly, researchers on occasion have published lab exposure or behavior studies without describing the chemical composition (polymer type or additives present) of the microplastics used (e.g., Ogonowski et al. 2016; Barboza et al. 2018a, b; Pacheco et al. 2018; Martins and Guilhermino 2018). This makes the applicability of results uncertain. Hildebrandt et al. (2019) cautioned that the actual size ranges of commercial microplastic products may differ from supplier specifications and should be verified before use in recovery exercises. They also noted the use of nonspherical microplastics would be more germane in validation exercises, as secondary microplastics are most abundant in the environment.

Contamination of samples during collection and analysis is problematic due to the ubiquitous presence of plastics, especially indoors (e.g., Wesch et al. 2017; Catarino et al. 2018). Where feasible, sampling equipment, materials/reagents used in preparation, and storage containers should be free of plastic components to reduce potential sample contamination. Nets and lines commonly consist of polymeric (e.g., PE, PP, and PA) materials, which may shed fragments (Welden and Cowie 2017). The use of natural fiber sampling nets is an option. Glass containers and metal sieves may be used. However, container closures (e.g., caps) may be a source of microplastic contamination, as shown for bottled water (Winkler et al. 2019). Cotton or non-shedding clothes are recommended for persons collecting and analyzing samples. Nylon is a common component of clothes, protective apparel (e.g., gloves), and water purification apparatuses. Airborne microplastics are very abundant indoors (Catarino et al. 2018) where lab preparation occurs. Hence, samples should be covered where possible. Wesch et al. (2017) noted that the use of a clean bench with particle filtration reduced sample contamination by >96%. Harsh sample treatment (e.g., exposure to abrasion and caustics) may reduce analytical interferences, allowing better spectra to be obtained, but can fragment particles, altering the sample's original size distribution.

Inclusion of lab and field blanks, as well as positive controls (amending matrices with microplastic standards representing the targeted polymer types, sizes, and shape characteristics) is necessary for quantitative approaches. These should be passed through all procedural steps to evaluate possible contamination and analyte losses (Koelmans et al. 2019). Blank results should be reported. The use of multiple devices or containers during sample processing, exposure to particle-laden ambient air, and addition of preservatives increase the potential for contaminant introduction. Reagents (e.g., oxidizers and preservatives) may be passed through filters to reduce particulates. But the filters themselves must not introduce interferences, and laboratory blanks should be inspected (Koelmans et al. 2019). For example, Fortin et al. (2019) suggested that some sub-10 μm microplastics detected in highly treated wastewater effluent samples evaluated may have arisen from polymeric materials used in the purification equipment itself.

Calls have been made to standardize microplastic sampling and analysis methods to facilitate comparisons between studies. However, sampling and analytical approaches for microplastics remain a "work in progress." Thus, coalescing on immature methods that fail to adequately identify the range of microplastics is ill-advised. Provencher et al. (2020) discussed the concept of method harmonization versus strict standardization. These authors also point out the responsibility of manuscript reviewers and journal editors in ensuring the appropriateness of the methods, data, and language used by authors.

The question of units for expressing microplastic measurements is critical. Commonly, studies simply report the number of a limited range of particle sizes (e.g., those between 300 and 5000 μm) present. However, this may be misleading (Simon et al. 2018; Rivers et al. 2019). As discussed previously, particle size characteristics are subject to change (i.e., plastic debris fragments over time into smaller particles). Importantly, all enumerations to date are likely underestimates due to a failure to report small micro- and nanoplastics. Studies should strive to determine and report mass-based concentrations of the plastics present. Masses may be estimated by determining particle volumes (e.g., by flow cytometry), although a lack of density measurements will be a source of error in such calculations. However, most polymer densities reside within a range of only about 10%. Alternatively, areas of particles derived from microspectroscopic assessments could be determined and summed by polymer type. Such area totals, as well as estimated volumes and associated masses, could then be calculated (Simon et al. 2018). This would be a significant step forward towards determining mass balances of MPs in the environment.

2.8 Conclusion

Over the last 70 years, plastic debris has entered all global environmental compartments, resulting in the formation of complex mixtures of chemically and physically distinct particles. This reality poses extraordinary sampling and analysis challenges. The researcher's first and most critical task is to formulate appropriate study goals. This is essential, as no single currently available sampling or analysis protocol is capable of capturing, identifying, and quantitating the full range of plastic debris present (Fig. 2.1; Table 2.1). To date, polymer analysis priorities have largely been guided by plastic production statistics, preexisting data on debris distribution (often skewed towards selected locales, such as beaches and the water's surface), and ease of sampling and detection. This has begun to change.

Risk at the individual, population, and ecosystem levels should be a driving force for studies and thus for prioritizing analysis method development. But such data remain incomplete. The study of impacts may be well served by controlled exposures. However, even in lab studies, greater diligence regarding the identities and properties of the study materials is warranted (Kögel et al. 2020). The first step to investigate environmental plastic pollution, field sampling, is critical. Careful preparation and highly sophisticated detection instrumentation cannot rescue what inappropriate sampling has missed or compromised. Researchers must recognize that plastic debris in the environment is composed of an immense diversity of polymers of varying properties. Polymers are also commonly augmented with additives or fillers, contributing up to percent by weight of the final plastic product. Their presence may affect plastic debris toxic potential and environmental behavior, as well as their recovery during sampling and subsequent sample preparation (and responses during instrumental detection). Size, shape, and surface texture of debris also affect the behavior in the field and during analysis. Small micro- (<20 μm) and nanoplastics are particularly difficult to characterize and enumerate and accordingly have been rarely reported (Kögel et al. 2020). Ironically, these are likely the most abundant, in part due to the continuous fragmentation of plastic debris in the environment.

The recognition that small microplastics are important is driving sampling towards more inclusive approaches, e.g., the use of pumps and filter arrays to capture and separate them into size classes. However, this may come at the cost of extended analysis time and representative sampling (due to a small sample size).

Techniques such as continuous flow centrifugation are promising in this arena. Automation is desirable from the perspective of analysis cost, accuracy, and precision. Complex, labor-intensive treatment schemes also increase the opportunity for analyte loss and sample contamination.

Powerful analytical techniques (e.g., MALDI-MS and AFM) to probe the surface properties of plastics exist. However, these are expensive and inappropriate to quantify large numbers of microplastic targets. μFTIR is becoming more available and is amenable to the automation of particle searching. FPA detectors further reduce analysis time. However, conventional FTIR does not resolve particles below about 10 μm. μRaman has capabilities to detect smaller particles (<1 μm) but is more expensive, time-consuming, and vulnerable to interference from fluorescence. In terms of evaluating total plastic in samples, approaches such as solvent extraction and thermal analysis hold promise. They can encompass submicron plastic debris, but their qualitative power is limited unless enhanced with supplemental techniques. By themselves, they do not provide plastic debris size, shape, or texture data, although sample pre-sieving permits separation into particle size bins. But this, in turn, increases the number of samples to be analyzed. However, these techniques may be readily automated and, compared to more sophisticated approaches, are inexpensive and rugged.

As daunting as the above seems, researchers concerned about plastics in the environment may be able to take advantage of some, to date, largely untapped resources, i.e., expertise and research in other disciplines. These include polymer chemists in the academic and commercial fields, the professionals who actually design and formulate plastics. As such, they possess substantial experience characterizing polymers, as well as knowledge of their behavior. Unfortunately manufacturers often deem compositional details of plastic products confidential. This is a complex issue, involving legal and business concerns (i.e., trade secrets). In some cases these may be fundamentally less important than associated toxicological issues. However, suggestions to remedy this are beyond the scope of this analytical chemistry-tasked chapter. Plastics are engineered to fulfill the performance requirements of specific applications at as low a cost as feasible, maximizing financial returns. Regrettably, postconsumer environmental safety and fate have not always been adequately evaluated and incorporated into the calculation. These considerations must evolve if we are to successfully tackle growing global plastic contamination. Bioplastics are also entering the market in increased volumes, due to concern over finite fossil fuels and the expectation that these materials are more "eco-friendly." This expands the diversity of analytes and their similarities to naturally occurring polymeric materials present further challenges for analytical methods.

Acknowledgement This is Contribution 3906 of the Virginia Institute of Marine Science, William & Mary.

References

Anderson JO, Thundiyil JG, Stolbach A (2012) Clearing the air: a review of the effects of particulate matter air pollution on human health. J Med Toxicol 8(2):166–175. https://doi.org/10.1007/s13181-011-0203-1

Bakir A, Rowland SJ, Thompson RC (2012) Competitive sorption of persistent organic pollutants onto microplastics in the marine environment. Mar Pollut Bull 64(12):2782–2789. https://doi.org/10.1016/j.marpolbul.2012.09.010

Barboza LGA, Vieira LR, Branco V, Carvalho C, Guilhermino L (2018a) Microplastics increase mercury bioconcentration in gills and bioaccumulation in the liver, and cause oxidative stress and damage in *Dicentrarchus labrax* juveniles. Sci Rep 8(1):15655. https://doi.org/10.1038/s41598-018-34125-z

Barboza LGA, Vieira LR, Branco V, Figueiredo N, Carvalho F, Carvalho C, Guilhermino L (2018b) Microplastics cause neurotoxicity, oxidative damage and energy-related changes and interact with the bioaccumulation of mercury in the European seabass, *Dicentrarchus labrax* (Linnaeus, 1758). Aquat Toxicol 195:49–57. https://doi.org/10.1016/j.aquatox.2017.12.008

Bergmann M, Wirzberger V, Krumpen T, Lorenz C, Primpke S, Tekman MB, Gerdts G (2017) High quantities of microplastic in Arctic deep-sea sediments from the HAUSGARTEN observatory. Environ Sci Technol 51(19):11000–11010. https://doi.org/10.1021/acs.est.7b03331

Berto D, Rampazzo F, Gion C, Noventa S, Ronchi F, Traldi U, Giorgi G, Cicero AM, Giovanardi O (2017) Preliminary study to characterize plastic polymers using elemental analyser/isotope ratio mass spectrometry (EA/IRMS). Chemosphere 176:47–56. https://doi.org/10.1016/j.chemosphere.2017.02.090

Biginagwa FJ, Mayoma BS, Shashoua Y, Syberg K, Khan FR (2016) First evidence of microplastics in the African Great Lakes: recovery from Lake Victoria Nile perch and Nile tilapia. J Great Lakes Res 42(1):146–149. https://doi.org/10.1016/j.jglr.2015.10.012

Brahney J, Hallerud M, Heim E, Hahnenberger M, Sukumaran S (2020) Plastic rain in protected areas of the United States. Science 368(6496):1257–1260. https://doi.org/10.1126/science.aaz5819

Brandon JA, Jones W, Ohman MD (2019) Multidecadal increase in plastic particles in coastal ocean sediments. *Science Advances,* 5(9), eaax0587. https://doi.org/10.1126/sciadv.aax0587

Brenna JT, Creasy WR (1991) High-molecular-weight polymer analysis by laser microprobe Fourier transform ion cyclotron resonance mass spectrometry. Appl Spectrosc 45(1):80–91. https://doi.org/10.1366/0003702914337786

Cabernard L, Roscher L, Lorenz C, Gerdts G, Primpke S (2018) Comparison of Raman and Fourier transform infrared spectroscopy for the quantification of microplastics in the aquatic environment. Environ Sci Technol 52(22):13279–13288. https://doi.org/10.1021/acs.est.8b03438

Cai L, Wang J, Peng J, Wu Z, Tan X (2018) Observation of the degradation of three types of plastic pellets exposed to UV irradiation in three different environments. Sci Total Environ 628–629:740–747. https://doi.org/10.1016/j.scitotenv.2018.02.079

Catarino AI, Macchia V, Sanderson WG, Thompson RC, Henry TB (2018) Low levels of microplastics (MP) in wild mussels indicate that MP ingestion by humans is minimal compared to exposure via household fibres fallout during a meal. Environ Pollut 237:675–684. https://doi.org/10.1016/j.envpol.2018.02.069

Ceccarini A, Corti A, Erba F, Modugno F, La Nasa J, Bianchi S, Castelvetro V (2018) The hidden microplastics: new insights and figures from the thorough separation and characterization of microplastics and of their degradation byproducts in coastal sediments. Environ Sci Technol 52(10):5634–5643. https://doi.org/10.1021/acs.est.8b01487

Chen H, Hou A, Corilo YE, Lin Q, Lu J, Mendelssohn IA, Zhang R, Rodgers RP, McKenna AM (2016) 4 years after the *Deepwater Horizon* spill: molecular transformation of Macondo well oil in Louisiana salt marsh sediments revealed by FT-ICR mass spectrometry. Environ Sci Technol 50(17):9061–9069. https://doi.org/10.1021/acs.est.6b01156

Chen Y, Wen D, Pei J, Fei Y, Ouyang D, Zhang H, Luo Y (2020) Identification and quantification of microplastics using Fourier-transform infrared spectroscopy: current status and future prospects. Curr Opin Environ Sci Health 18:14–19. https://doi.org/10.1016/j.coesh.2020.05.004

Cheng H, Hu Y, Reinhard M (2014) Environmental and health impacts of artificial turf: a review. Environ Sci Technol 48:2114–2129. https://doi.org/10.1021/es4044193

Choy CA, Robison BH, Gagne TO, Erwin B, Firl E, Halden RU, Hamilton JA, Katija K, Lisin SE, Rolsky C, Van Houtan KS (2019) The vertical distribution and biological transport of marine microplastics across the epipelagic and mesopelagic water column. Sci Rep 9(1):7843. https://doi.org/10.1038/s41598-019-44117-2

Christie RM (1994) Pigments, dyes and fluorescent brightening agents for plastics: an overview. Polym Int 34(4):351–361. https://doi.org/10.1002/pi.1994.210340401

Courtene-Jones W, Quinn B, Murphy F, Gary SF, Narayanaswamy BE (2017) Optimisation of enzymatic digestion and validation of specimen preservation methods for the analysis of ingested microplastics. Anal Methods 9(9):1437–1445. https://doi.org/10.1039/c6ay02343f

Covernton GA, Pearce CM, Gurney-Smith HJ, Chastain SG, Ross PS, Dower JF, Dudas SE (2019) Size and shape matter: a preliminary analysis of microplastic sampling technique in seawater studies with implications for ecological risk assessment. Sci Total Environ 667:124–132. https://doi.org/10.1016/j.scitotenv.2019.02.346

Dai ZF, Zhang HB, Zhou Q, Tian Y, Chen T, Tu C, Fu CC, Luo YM (2018) Occurrence of microplastics in the water column and sediment in an inland sea affected by intensive anthropogenic activities. Environ Pollut 242:1557–1565. https://doi.org/10.1016/j.envpol.2018.07.131

Dawson AL, Kawaguchi S, King CK, Townsend KA, King R, Huston WM, Bengtson Nash SM (2018) Turning microplastics into nanoplastics through digestive fragmentation by Antarctic krill. Nat Commun 9(1):1001. https://doi.org/10.1038/s41467-018-03465-9

Dazzi A, Prater C, Hu Q, Chase DB (2012) AFM-IR: combining atomic force microscopy and infrared spectroscopy for nanoscale chemical characterization. Appl Spectrosc 66(12):1365–1384. https://doi.org/10.1366/12-06804

de Ruijter VN, Redondo-Hasselerharm PE, Gouin T, Koelmans AA (2020) Quality criteria for microplastic effect studies in the context of risk assessment: A critical review. Environ Sci Technol 54(19). https://doi.org/10.1021/acs.est.0c03057

Dehghani S, Moore F, Akhbarizadeh R (2017) Microplastic pollution in deposited urban dust, Tehran metropolis, Iran. Environ Sci Pollut Res 24(25):20360–20371. https://doi.org/10.1007/s11356-017-9674-1

Doyen P, Hermabessiere L, Dehaut A, Himber C, Decodts M, Degraeve T, Delord L, Gaboriaud M, Moné P, Sacco J, Tavernier E, Grard T, Duflos G (2019) Occurrence and identification of microplastics in beach sediments from the Hauts-de-France region. Environ Sci Pollut Res 26(27):28010–28021. https://doi.org/10.1007/s11356-019-06027-8

Dris R, Gasperi J, Rocher V, Tassin B (2018) Synthetic and non-synthetic anthropogenic fibers in a river under the impact of Paris megacity: sampling methodological aspects and flux estimations. Sci Total Environ 618:157–164. https://doi.org/10.1016/j.scitotenv.2017.11.009

Duemichen E, Eisentraut P, Celina M, Braun U (2019) Automated thermal extraction-desorption gas chromatography mass spectrometry: A multifunctional tool for comprehensive characterization of polymers and their degradation products. J Chromatogr A 1592:133–142. https://doi.org/10.1016/j.chroma.2019.01.033

Dümichen E, Barthel AK, Braun U, Bannick CG, Brand K, Jekel M, Senz R (2015) Analysis of polyethylene microplastics in environmental samples, using a thermal decomposition method. Water Res 85:451–457. https://doi.org/10.1016/j.watres.2015.09.002

Efimova I, Bagaeva M, Bagaev A, Kileso A, Chubarenko IP (2018) Secondary microplastics generation in the sea swash zone with coarse bottom sediments: laboratory experiments. Front Mar Sci 5. https://doi.org/10.3389/fmars.2018.00313

Eisentraut P, Dümichen E, Ruhl AS, Jekel M, Albrecht M, Gehde M, Braun U (2018) Two birds with one stone – fast and simultaneous analysis of microplastics: microparticles derived from thermoplastics and tire wear. Environ Sci Technol Lett 5(10):608–613. https://doi.org/10.1021/acs.estlett.8b00446

Elert AM, Becker R, Duemichen E, Eisentraut P, Falkenhagen J, Sturm H, Braun U (2017) Comparison of different methods for MP detection: what can we learn from them, and why asking the right question before measurements matters? Environ Pollut 231(Pt 2):1256–1264. https://doi.org/10.1016/j.envpol.2017.08.074

Enders K, Lenz R, Stedmon CA, Nielsen TG (2015) Abundance, size and polymer composition of marine microplastics ≥ 10 μm in the Atlantic Ocean and their modelled vertical distribution. Mar Pollut Bull 100(1):70–81. https://doi.org/10.1016/j.marpolbul.2015.09.027

Erni-Cassola G, Gibson MI, Thompson RC, Christie-Oleza JA (2017) Lost, but found with Nile red: A novel method for detecting and quantifying small microplastics (1 mm to 20 μm) in environmental samples. Environ Sci Technol 51(23):13641–13648. https://doi.org/10.1021/acs.est.7b04512

Erni-Cassola G, Zadjelovic V, Gibson MI, Christie-Oleza JA (2019) Distribution of plastic polymer types in the marine environment; A meta-analysis. J Hazard Mater 369:691–698. https://doi.org/10.1016/j.jhazmat.2019.02.067

Feng LJ, Wang JJ, Liu SC, Sun XD, Yuan XZ, Wang SG (2018) Role of extracellular polymeric substances in the acute inhibition of activated sludge by polystyrene nanoparticles. Environ Pollut 238:859–865. https://doi.org/10.1016/j.envpol.2018.03.101

Fischer M, Scholz-Böttcher BM (2017) Simultaneous trace identification and quantification of common types of microplastics in environmental samples by pyrolysis-gas chromatography-mass spectrometry. Environ Sci Technol 51(9):5052–5060. https://doi.org/10.1021/acs.est.6b06362

Fortin S, Song B, Burbage C (2019) Quantifying and identifying microplastics in the effluent of advanced wastewater treatment systems using Raman microspectroscopy. Mar Pollut Bull 149:110579. https://doi.org/10.1016/j.marpolbul.2019.110579

Fries E, Dekiff JH, Willmeyer J, Nuelle MT, Ebert M, Remy D (2013) Identification of polymer types and additives in marine microplastic particles using pyrolysis-GC/MS and scanning electron microscopy. Environ Sci Process Impacts 15(10):1949–1956. https://doi.org/10.1039/c3em00214d

Fu W, Min J, Jiang W, Li Y, Zhang W (2020) Separation, characterization and identification of microplastics and nanoplastics in the environment. Sci Total Environ 721:137561. https://doi.org/10.1016/j.scitotenv.2020.137561

Fuller S, Gautam A (2016) A procedure for measuring microplastics using pressurized fluid extraction. Environ Sci Technol 50(11):5774–5780. https://doi.org/10.1021/acs.est.6b00816

Gasperi J, Wright SL, Dris R, Collard F, Mandin C, Guerrouache M, Langlois V, Kelly FJ, Tassin B (2018) Microplastics in air: are we breathing it in? Curr Opin Environ Sci Health 1:1–5. https://doi.org/10.1016/j.coesh.2017.10.002

GESAMP (2019) Guidelines for the monitoring and assessment of plastic litter and microplastics in the ocean. GESAMP Joint Group of Experts on the Scientific Aspects of Marine Environmental Protection, London. https://environmentlive.unep.org/media/docs/marine_plastics/une_science_dvision_gesamp_reports.pdf

Ghosal S, Wagner J (2013) Correlated Raman micro-spectroscopy and scanning electron microscopy analyses of flame retardants in environmental samples: A micro-analytical tool for probing chemical composition, origin and spatial distribution. Analyst 138(13):3836–3844. https://doi.org/10.1039/c3an00501a

Gillibert R, Balakrishnan G, Deshoules Q, Tardivel M, Magazzù A, Donato MG, Maragò OM, Lamy de La Chapelle M, Colas F, Lagarde F, Gucciardi PG (2019) Raman tweezers for small microplastics and nanoplastics identification in seawater. Environ Sci Technol 53(15):9003–9013. https://doi.org/10.1021/acs.est.9b03105

Girão AV, Caputo G, Ferro MC (2017) Application of scanning electron microscopy-energy dispersive X-ray spectroscopy (SEM-EDS). Compr Anal Chem 75:153–168. https://doi.org/10.1016/bs.coac.2016.10.002

Gniadek M, Dąbrowska A (2019) The marine nano- and microplastics characterisation by SEM-EDX: the potential of the method in comparison with various physical and chemical approaches. Mar Pollut Bull 148:210–216. https://doi.org/10.1016/j.marpolbul.2019.07.067

Gomiero A, Øysæd KB, Agustsson T, van Hoytema N, van Thiel T, Grati F (2019) First record of characterization, concentration and distribution of microplastics in coastal sediments of an urban fjord in Southwest Norway using a thermal degradation method. Chemosphere 227:705–714. https://doi.org/10.1016/j.chemosphere.2019.04.096

Haave M, Lorenz C, Primpke S, Gerdts G (2019) Different stories told by small and large microplastics in sediment – first report of microplastic concentrations in an urban recipient in Norway. Mar Pollut Bull 14:501–513. https://doi.org/10.1016/j.marpolbul.2019.02.015

Hahladakis JN, Velis CA, Weber R, Iacovidou E, Purnell P (2018) An overview of chemical additives present in plastics: migration, release, fate and environmental impact during their use, disposal and recycling. J Hazard Mater 344:179–199. https://doi.org/10.1016/j.jhazmat.2017.10.014

Hakkarainen M (2012) Mass spectrometry of polymers – new techniques. Springer, Berlin/Heidelberg, 212p. https://doi.org/10.1007/978-3-642-28041-2

Hale RC (2017) Analytical challenges associated with the determination of microplastics in the environment. Anal Methods 9:1326–1327. https://doi.org/10.1039/c7ay90015e

Hale RC (2018) Are the risks from microplastics truly trivial? Environ Sci Technol 52(3):931. https://doi.org/10.1021/acs.est.7b06615

Hale RC, La Guardia MJ, Harvey E, Matt Mainor T (2002) Potential role of fire retardant-treated polyurethane foam as a source of brominated diphenyl ethers to the US environment. Chemosphere 46(5):729–735. https://doi.org/10.1016/S0045-6535(01)00237-5

Hale RC, Seeley ME, La Guardia MJ, Mai L, Zeng EY (2020) A global perspective on microplastics. J Geophys Res Oceans 125(1):1–40. https://doi.org/10.1029/2018jc014719

Harrison JP, Ojeda JJ, Romero-González ME (2012) The applicability of reflectance micro-Fourier-transform infrared spectroscopy for the detection of synthetic microplastics in marine sediments. Sci Total Environ 416:455–463. https://doi.org/10.1016/j.scitotenv.2011.11.078

Hendrickson E, Minor EC, Schreiner K (2018) Microplastic abundance and composition in western Lake Superior as determined via microscopy, Pyr-GC/MS, and FTIR. Environ Sci Technol 52(4):1787–1796. https://doi.org/10.1021/acs.est.7b05829

Hermabessiere L, Dehaut A, Paul-Pont I, Lacroix C, Jezequel R, Soudant P, Duflos G (2017) Occurrence and effects of plastic additives on marine environments and organisms: A review. Chemosphere 182:781–793. https://doi.org/10.1016/j.chemosphere.2017.05.096

Hernandez LM, Yousefi N, Tufenkji N (2017) Are there nanoplastics in your personal care products? Environ Sci Technol Lett 4(7):280–285. https://doi.org/10.1021/acs.estlett.7b00187

Hidalgo-Ruz V, Gutow L, Thompson RC, Thiel M (2012) Microplastics in the marine environment: A review of the methods used for identification and quantification. Environ Sci Technol 46(6):3060–3075. https://doi.org/10.1021/es2031505

Hildebrandt NL, Voigta N, Zimmermann T, Reese A, Proefrock D (2019) Evaluation of continuous flow centrifugation as an alternative technique to sample microplastic from water bodies. Mar Environ Res 151:104768. https://doi.org/10.1016/j.marenvres.2019.104768

Hildebrandt L, Mitrano DM, Zimmermann T, Pröfrock D (2020) A nanoplastic sampling and enrichment approach by flow centrifugation. Front Environ Sci 8:89. https://doi.org/10.3389/fenvs.2020.00089

Hintersteiner I, Himmelsbach M, Buchberger WW (2015) Characterization and quantitation of polyolefin microplastics in personal-care products using high-temperature gel-permeation chromatography. Anal Bioanal Chem 407(4):1253–1259. https://doi.org/10.1007/s00216-014-8318-2

Huppertsberg S, Knepper TP (2018) Instrumental analysis of microplastics—benefits and challenges. Anal Bioanal Chem 410(25):6343–6352. https://doi.org/10.1007/s00216-018-1210-8

Hurley RR, Lusher AL, Olsen M, Nizzetto L (2018) Validation of a method for extracting microplastics from complex, organic-rich, environmental matrices. Environ Sci Technol 52:7409–7417. https://doi.org/10.1021/acs.est.8b01517

Imhof HK, Schmid J, Niessner R, Ivleva NP, Laforsch C (2012) A novel, highly efficient method for the separation and quantification of plastic particles in sediments of aquatic environments. Limnol Oceanogr Methods 10(7):524–537. https://doi.org/10.4319/lom.2012.10.524

Imhof HK, Laforsch C, Wiesheu AC, Schmid J, Anger PM, Niessner R, Ivleva NP (2016) Pigments and plastic in limnetic ecosystems: A qualitative and quantitative study on microparticles of different size classes. Water Res 98:64–74. https://doi.org/10.1016/j.watres.2016.03.015

Jungnickel H, Pund R, Tentschert J, Reichardt P, Laux P, Harbach H, Luch A (2016) Time-of-flight secondary ion mass spectrometry (ToF-SIMS)-based analysis and imaging of polyethylene microplastics formation during sea surf simulation. Sci Total Environ 563–564:261–266. https://doi.org/10.1016/j.scitotenv.2016.04.025

Kane IA, Clare MA, Miramontes E, Wogelius R, Rothwell JJ, Garreau P, Pohl F (2020) Seafloor microplastic hotspots controlled by deep-sea circulation. Science 368(6495):1140–1145. https://doi.org/10.1126/science.aba5899

Käppler A, Windrich F, Löder MGJ, Malanin M, Fischer D, Labrenz M, Eichhorn KJ, Voit B (2015) Identification of microplastics by FTIR and Raman microscopy: a novel silicon filter substrate opens the important spectral range below 1300 cm^{-1} for FTIR transmission measurements. Anal Bioanal Chem 407(22):6791–6801

Käppler A, Fischer D, Oberbeckmann S, Schernewski G, Labrenz M, Eichhorn K-J, Voit B (2016) Analysis of environmental microplastics by vibrational microspectroscopy: FTIR, Raman or both? Anal Bioanal Chem 408(29):8377–8391. https://doi.org/10.1007/s00216-016-9956-3

Karami A, Golieskardi A, Choo CK, Romano N, Ho YB, Salamatinia B (2017) A high-performance protocol for extraction of microplastics in fish. Sci Total Environ 578:485–494. https://doi.org/10.1016/j.scitotenv.2016.10.213

Khaled A, Rivaton A, Richard C, Jaber F, Sleiman M (2018) Phototransformation of plastic containing brominated flame retardants: enhanced fragmentation and release of photoproducts to water and air. Environ Sci Technol 52(19):11123–11131. https://doi.org/10.1021/acs.est.8b03172

Klein S, Dimzon IK, Eubeler J, Knepper TP (2018) Analysis, occurrence, and degradation of microplastics in the aqueous environment. In: Wagner M, Lambert S (eds) Freshwater microplastics: emerging environmental contaminants? Springer, Cham, pp 51–67

Klöckner P, Reemtsma T, Eisentraut P, Braun U, Ruhl AS, Wagner S (2019) Tire and road wear particles in road environment – quantification and assessment of particle dynamics by Zn determination after density separation. Chemosphere 222:714–721. https://doi.org/10.1016/j.chemosphere.2019.01.176

Koelmans AA, Mohamed Nor NH, Hermsen E, Kooi M, Mintenig SM, De France J (2019) Microplastics in freshwaters and drinking water: critical review and assessment of data quality. Water Res 155:410–422. https://doi.org/10.1016/j.watres.2019.02.054

Kögel T, Bjorøy Ø, Toto B, Bienfait AM, Sanden M (2020) Micro- and nanoplastic toxicity on aquatic life: determining factors. Sci Total Environ 709(5817):136050. https://doi.org/10.1016/j.scitotenv.2019.136050

Kole PJ, Löhr AJ, Van Belleghem FGAJ, Raga AMJ (2017) Wear and tear of tyres: A stealthy source of microplastics in the environment. Int J Environ Res Public Health 14(10):1265. https://doi.org/10.3390/ijerph14101265

Kooi M, Nes E Hv, Scheffer M, Koelmans AA (2017) Ups and downs in the ocean: effects of biofouling on vertical transport of microplastics. Environ Sci Technol 51(14):7963–7971. https://doi.org/10.1021/acs.est.6b04702

Lapointe M, Farner JM, Hernandez LM, Tufenkji N (2020) Understanding and improving microplastic removal during water treatment: impact of coagulation and flocculation. Environ Sci Technol 54(14):8719–8727. https://doi.org/10.1021/acs.est.0c00712

Leads RR, Weinstein JE (2019) Occurrence of tire wear particles and other microplastics within the tributaries of the Charleston Harbor estuary, South Carolina, USA. Mar Pollut Bull 145:569–582. https://doi.org/10.1016/j.marpolbul.2019.06.061

Lenz R, Enders K, Stedmon CA, MacKenzie DMA, Nielsen TG (2015) A critical assessment of visual identification of marine microplastic using Raman spectroscopy for analysis improvement. Mar Pollut Bull 100(1):82–91. https://doi.org/10.1016/j.marpolbul.2015.09.026

Leslie HA, Brandsma SH, van Velzen MJM, Vethaak AD (2017) Microplastics en route: field measurements in the Dutch river delta and Amsterdam canals, wastewater treatment plants, North Sea sediments and biota. Environ Int 101:133–142. https://doi.org/10.1016/j.envint.2017.01.018

Li JN, Green C, Reynolds A, Shi HH, Rotchell JM (2018a) Microplastics in mussels sampled from coastal waters and supermarkets in the United Kingdom. Environ Pollut 241:35–44. https://doi.org/10.1016/j.envpol.2018.05.038

Li JY, Liu HH, Chen JP (2018b) Microplastics in freshwater systems: A review on occurrence, environmental effects, and methods for microplastics detection. Water Res 137:362–374. https://doi.org/10.1016/j.watres.2017.12.056

Li HR, La Guardia MJ, Liu HH, Hale RC, Mainor TM, Harvey E, Sheng GY, Fu JM, Peng PA (2019) Brominated and organophosphate flame retardants along a sediment transect encompassing the Guiyu, China e-waste recycling zone. Sci Total Environ 646:58–67. https://doi.org/10.1016/j.scitotenv.2018.07.276

Lin Y, Huang X, Liu Q, Lin ZY, Jiang GB (2020) Thermal fragmentation enhanced identification and quantification of polystyrene micro/nanoplastics in complex media. Talanta 208:120478. https://doi.org/10.1016/j.talanta.2019.120478

Lithner D, Larsson A, Dave G (2011) Environmental and health hazard ranking and assessment of plastic polymers based on chemical composition. Sci Total Environ 409(18):3309–3324. https://doi.org/10.1016/j.scitotenv.2011.04.038

Liu F, Olesen KB, Borregaard AR, Vollertsen J (2019) Microplastics in urban and highway stormwater retention ponds. Sci Total Environ 671. https://doi.org/10.1016/j.scitotenv.2019.03.416

Löder MGJ, Gerdts G (2015) Methodology used for the detection and identification of microplastics—a critical appraisal. In: Bergmann M, Gutow L, Klages M (eds) Marine anthropogenic litter. Springer, Cham, pp 201–227

Löder MGJ, Kuczera M, Mintenig S, Lorenz C, Gerdts G (2015) Focal plane array detector-based micro-Fourier-transform infrared imaging for the analysis of microplastics in environmental samples. Environ Chem 12(5):563–581. https://doi.org/10.1071/EN14205

Löder MGJ, Imhof HK, Ladehoff M, Löschel LA, Lorenz C, Mintenig S, Piehl S, Primpke S, Schrank I, Laforsch C, Gerdts G (2017) Enzymatic purification of microplastics in environmental samples. Environ Sci Technol 51(24):14283–14292. https://doi.org/10.1021/acs.est.7b03055

Lu SH, Zhu KR, Song WC, Song G, Chen DY, Hayat T, Alharbi NS, Chen CL, Sun Y (2018) Impact of water chemistry on surface charge and aggregation of polystyrene microspheres suspensions. Sci Total Environ 630:951–959. https://doi.org/10.1016/j.scitotenv.2018.02.296

Luan L, Wang X, Zheng H, Liu L, Luo X, Li F (2019) Differential toxicity of functionalized polystyrene microplastics to clams (*Meretrix meretrix*) at three key development stages of life history. Mar Pollut Bull 139:346–354. https://doi.org/10.1016/j.marpolbul.2019.01.003

Luo HW, Xiang YH, He DQ, Li Y, Zhao YY, Wang S, Pan XL (2019) Leaching behavior of fluorescent additives from microplastics and the toxicity of leachate to *Chlorella vulgaris*. Sci Total Environ 678:1–9. https://doi.org/10.1016/j.scitotenv.2019.04.401

Luo H, Zhao Y, Li Y, Xiang Y, He D, Pan X (2020) Aging of microplastics affects their surface properties, thermal decomposition, additives leaching and interactions in simulated fluid. Sci Total Environ 714:136862. https://doi.org/10.1016/j.scitotenv.2020.136862

Lusher AL, McHugh M, Thompson RC (2013) Occurrence of microplastics in the gastrointestinal tract of pelagic and demersal fish from the English Channel. Mar Pollut Bull 67(1):94–99. https://doi.org/10.1016/j.marpolbul.2012.11.028

Lusher AL, Welden NA, Sobral P, Cole M (2017) Sampling, isolating and identifying microplastics ingested by fish and invertebrates. Anal Methods 9(9):1346–1360

Maes T, Jessop R, Wellner N, Haupt K, Mayes AG (2017) A rapid-screening approach to detect and quantify microplastics based on fluorescent tagging with Nile red. Sci Rep 7:44501. https://doi.org/10.1038/srep44501

Mahon AM, O'Connell B, Healy MG, O'Connor I, Officer R, Nash R, Morrison L (2017) Microplastics in sewage sludge: effects of treatment. Environ Sci Technol 51(2):810–818. https://doi.org/10.1021/acs.est.6b04048

Martins A, Guilhermino L (2018) Transgenerational effects and recovery of microplastics exposure in model populations of the freshwater cladoceran *Daphnia magna* Straus. Sci Total Environ 631–632:421–428. https://doi.org/10.1016/j.scitotenv.2018.03.054

Materić D, Kasper-Giebl A, Kau D, Anten M, Greilinger M, Ludewig E, van Sebille E, Röckmann T, Holzinger R (2020) Micro- and nanoplastics in alpine snow: A new method for chemical identification and (semi)quantification in the nanogram range. Environ Sci Technol 54(4):2353–2359. https://doi.org/10.1021/acs.est.9b07540

McGivney E, Cederholm L, Barth A, Hakkarainen M, Hamacher-Barth E, Ogonowski M, Gorokhova E (2020) Rapid physicochemical changes in microplastic induced by biofilm formation. Front Bioeng Biotechnol 8:205. https://doi.org/10.3389/fbioe.2020.00205

Mengistu D, Nilsen V, Heistad A, Kvaal K (2019) Detection and quantification of tire particles in sediments using a combination of simultaneous thermal analysis, Fourier transform infrared, and parallel factor analysis. Int J Environ Res Public Health 16(18):3444. https://doi.org/10.3390/ijerph16183444

Merzel RL, Purser L, Soucy TL, Olszewski M, Colón-Bernal I, Duhaime M, Elgin AK, Banaszak Holl MM (2019) Uptake and retention of nanoplastics in quagga mussels. Global Chall 1800104. https://doi.org/10.1002/gch2.201800104

Michels J, Stippkugel A, Lenz M, Wirtz K, Engel A (2018) Rapid aggregation of biofilm-covered microplastics with marine biogenic particles. Proc R Soc B 285(1885):20181203. https://doi.org/10.1098/rspb.2018.1203

Miller-Chou BA, Koenig JL (2003) A review of polymer dissolution. Prog Polym Sci 28:1223–1270. https://doi.org/10.1016/S0079-6700(03)00045-5

Mintenig SM, Bäuerlein PS, Koelmans AA, Dekker SC, Wezel, A. P. v. (2018) Closing the gap between small and smaller: towards a framework to analyse nano- and microplastics in aqueous environmental samples. Environ Sci Nano 5(7):1640–1649. https://doi.org/10.1039/c8en00186c

Möller JN, Löder MGJ, Laforsch C (2020) Finding microplastics in soils: A review of analytical methods. Environ Sci Technol 54(4):2078–2090. https://doi.org/10.1021/acs.est.9b04618

Munno K, De Frond H, O'Donnell B, Rochman CM (2020) Increasing the accessibility for characterizing microplastics: introducing new application-based and spectral libraries of plastic particles (SLoPP and SLoPP-E). Anal Chem 92(3):2443–2451. https://doi.org/10.1021/acs.analchem.9b03626

Nakajima R, Lindsay DJ, Tsuchiya M, Matsui R, Kitahashi T, Fujikura K, Fukushima T (2019a) A small, stainless-steel sieve optimized for laboratory beaker-based extraction of microplastics from environmental samples. MethodsX 6:1677–1682. https://doi.org/10.1016/j.mex.2019.07.012

Nakajima R, Tsuchiya M, Lindsay DJ, Kitahashi T, Fujikura K, Fukushima T (2019b) A new small device made of glass for separating microplastics from marine and freshwater sediments. PeerJ 7:e7915. https://doi.org/10.7717/peerj.7915

Ng KL, Obbard JP (2006) Prevalence of microplastics in Singapore's coastal marine environment. Marine Pollut Bull 52(7):761–767. https://doi.org/10.1016/j.marpolbul.2005.11.017

Ng EL, Huerta Lwanga E, Eldridge SM, Johnston P, Hu HW, Geissen V, Chen D (2018) An overview of microplastic and nanoplastic pollution in agroecosystems. Sci Total Environ 627:1377–1388. https://doi.org/10.1016/j.scitotenv.2018.01.341

Obbard RW, Sadri S, Wong YQ, Khitun AA, Baker I, Thompson RC (2014) Global warming releases microplastic legacy frozen in Arctic Sea ice. Earth's Future 2(6):315–320. https://doi.org/10.1002/2014ef000240

Ogonowski M, Schür C, Jarsén Å, Gorokhova E (2016) The effects of natural and anthropogenic microparticles on individual fitness in *Daphnia magna*. PLoS One 11(5):e0155063. https://doi.org/10.1371/journal.pone.0155063

Oßmann BE, Sarau G, Schmitt SW, Holtmannspötter H, Christiansen SH, Dicke W (2017) Development of an optimal filter substrate for the identification of small microplastic particles in food by micro-Raman spectroscopy. Anal Bioanal Chem 409(16):4099–4109. https://doi.org/10.1007/s00216-017-0358-y

Pabortsava K, Lampitt RS (2020) High concentrations of plastic hidden beneath the surface of the Atlantic Ocean. Nat Commun 11:4073. https://doi.org/10.1038/s41467-020-17932-9

Pacheco A, Martins A, Guilhermino L (2018) Toxicological interactions induced by chronic exposure to gold nanoparticles and microplastics mixtures in *Daphnia magna*. Sci Total Environ 628–629:474–483. https://doi.org/10.1016/j.scitotenv.2018.02.081

Peñalver R, Arroyo-Manzanares N, López-García I, Hernández-Córdoba M (2020) An overview of microplastics characterization by thermal analysis. Chemosphere 242:125170. https://doi.org/10.1016/j.chemosphere.2019.125170

Peters CA, Hendrickson E, Minor EC, Schreiner K, Halbur J, Bratton SP (2018) Pyr-GC/MS analysis of microplastics extracted from the stomach content of benthivore fish from the Texas Gulf Coast. Mar Pollut Bull 137:91–95. https://doi.org/10.1016/j.marpolbul.2018.09.049

Prata JC, da Costa JP, Duarte AC, Rocha-Santos T (2019) Methods for sampling and detection of microplastics in water and sediment: A critical review. Trends Anal Chem 110:150–159. https://doi.org/10.1016/j.trac.2018.10.029

Prata JC, da Costa JP, Lopes I, Duarte AC, Rocha-Santos T (2020) Environmental exposure to microplastics: an overview on possible human health effects. Sci Total Environ 702:134455. https://doi.org/10.1016/j.scitotenv.2019.134455

Primpke S, Lorenz C, Rascher-Friesenhausen R, Gerdts G (2017) An automated approach for microplastics analysis using focal plane array (FPA) FTIR microscopy and image analysis. Anal Methods 9(9):1499–1511. https://doi.org/10.1039/c6ay02476a

Primpke S, Wirth M, Lorenz C, Gerdts G (2018) Reference database design for the automated analysis of microplastic samples based on Fourier transform infrared (FTIR) spectroscopy. Anal Bioanal Chem 410(21):5131–5141. https://doi.org/10.1007/s00216-018-1156-x

Primpke S, Dias PA, Gerdts G (2019) Automated identification and quantification of microfibres and microplastics. Anal Methods 16:2138–2147. https://doi.org/10.1039/C9AY00126C

Provencher JF, Covernton GA, Moore RC, Horn DA, Conkle JA, Lusher AL (2020) Proceed with caution: the need to raise the publication bar for microplastics research. Sci Total Environ 748:141426. https://doi.org/10.1016/j.scitotenv.2020.141426

Quang TN, He C, Morawska L, Knibbs LD, Falk M (2012) Vertical particle concentration profiles around urban office buildings. Atmos Chem Phys 12(11):5017–5030. https://doi.org/10.5194/acp-12-5017-2012

Reed S, Clark M, Thompson R, Hughes KA (2018) Microplastics in marine sediments near Rothera Research Station, Antarctica. Mar Pollut Bull 133:460–463. https://doi.org/10.1016/j.marpolbul.2018.05.068

Riedel T, Dittmar T (2014) A method detection limit for the analysis of natural organic matter via Fourier transform ion cyclotron resonance mass spectrometry. Anal Chem 86(16):8376–8382. https://doi.org/10.1021/ac501946m

Rivers ML, Gwinnett C, Woodall LC (2019) Quantification is more than counting: actions required to accurately quantify and report isolated marine microplastics. Mar Pollut Bull 139:100–104. https://doi.org/10.1016/j.marpolbul.2018.12.024

Rochman CM et al (2019) Rethinking microplastics as a diverse contaminant suite. Environ Toxicol Chem 38:703–711. https://doi.org/10.1002/etc.4371

Rodrigues SM, Almeida R, C. M., & Ramos, S. (2019) Adaptation of a laboratory protocol to quantity microplastics contamination in estuarine waters. MethodsX 6:740–749. https://doi.org/10.1016/j.mex.2019.03.027

Rummel CD, Jahnke A, Gorokhova E, Kühnel D, Schmitt-Jansen M (2017) Impacts of biofilm formation on the fate and potential effects of microplastic in the aquatic environment. Environ Sci Technol Lett 4(7):258–267. https://doi.org/10.1021/acs.estlett.7b00164

Schirinzi GF, Llorca M, Seró R, Moyano E, Barceló D, Abad E, Farré M (2019) Trace analysis of polystyrene microplastics in natural waters. Chemosphere 236:124321. https://doi.org/10.1016/j.chemosphere.2019.07.052

Schür C, Rist S, Baun A, Mayer P, Hartmann NB, Wagner M (2019) When fluorescence is not a particle: the tissue translocation of microplastics in *Daphnia magna* seems an artifact. Environ Toxicol Chem 38(7):1495–1503. https://doi.org/10.1002/etc.4436

Silva AB, Bastos AS, Justino CIL, da Costa JP, Duarte AC, Rocha-Santos TAP (2018) Microplastics in the environment: challenges in analytical chemistry – A review. Anal Chim Acta 1017:1–19. https://doi.org/10.1016/j.aca.2018.02.043

Simon M, van Alst N, Vollertsen J (2018) Quantification of microplastic mass and removal rates at wastewater treatment plants applying focal plane Array (FPA)- based Fourier transform infrared (FT-IR) imaging. Water Res 142:1–9. https://doi.org/10.1016/j.watres.2018.05.019

Singh N, Duanb H, Tang Y (2020) Toxicity evaluation of E-waste plastics and potential repercussions for human health. Environ Int 137:105559. https://doi.org/10.1016/j.envint.2020.105559

Sommer F, Dietze V, Baum A, Sauer J, Gilge S, Maschowski C, Gieré R (2018) Tire abrasion as a major source of microplastics in the environment. Aerosol Air Qual Res 18:2014–2028. https://doi.org/10.4209/aaqr.2018.03.0099

Song YK, Hong SH, Jang M, Kang J-H, Kwon OY, Han GM, Shim WJ (2014) Large accumulation of micro-sized synthetic polymer particles in the sea surface microlayer. Environ Sci Technol 48(16):9014–9021. https://doi.org/10.1021/es501757s

Song YK, Hong SH, Jang M, Han GM, Jung SW, Shim WJ (2017) Combined effects of UV exposure duration and mechanical abrasion on microplastic fragmentation by polymer type. Environ Sci Technol 51(8):4368–4376. https://doi.org/10.1021/acs.est.6b06155

Spies RB, Andresen BD, Rice DW Jr (1987) Benzthiazoles in estuarine sediments as indicators of street runoff. Nature 327(6124):697–699. https://doi.org/10.1038/327697a0

Su YL, Zhang ZJ, Wu D, Zhan L, Shi HH, Xie B (2019) Occurrence of microplastics in landfill systems and their fate with landfill age. Water Res 164:114968. https://doi.org/10.1016/j.watres.2019.114968

Suaria G, Achtypi A, Perold V, Lee J, Pierucci A, Bornman T, Aliani S, Ryan P (2020) Microfibers in oceanic surface waters: a global characterization. Sci Adv 6(23):8493. https://doi.org/10.1126/sciadv.aay8493

Summers S, Henry T, Gutierrez T (2018) Agglomeration of nano- and microplastic particles in seawater by autochthonous and de novo-produced sources of exopolymeric substances. Mar Pollut Bull 130:258–267. https://doi.org/10.1016/j.marpolbul.2018.03.039

Tagg AS, Sapp M, Harrison JP, Ojeda JJ (2015) Identification and quantification of microplastics in wastewater using focal plane array-based reflectance micro-FT-IR imaging. Anal Chem 87(12):6032–6040. https://doi.org/10.1021/acs.analchem.5b00495

Tagg AS, Harrison JP, Ju-Nam Y, Sapp M, Bradley EL, Sinclair CJ, Ojeda JJ (2017) Fenton's reagent for the rapid and efficient isolation of microplastics from wastewater. Chem Commun 53:372–375. https://doi.org/10.1039/c6cc08798a

Takahashi K (2013) Polymer analysis by supercritical fluid chromatography. J Biosci Bioeng 116(2):133–140. https://doi.org/10.1016/j.jbiosc.2013.02.001

Takahashi CK, Turner A, Millward GE, Glegg GA (2012) Persistence and metallic composition of paint particles in sediments from a tidal inlet. Mar Pollut Bull 64(1):133–137. https://doi.org/10.1016/j.marpolbul.2011.10.010

Tamminga M, Stoewer S-C, Fischer EK (2019) On the representativeness of pump water samples versus manta sampling in microplastic analysis. Environ Pollut 254:112970. https://doi.org/10.1016/j.envpol.2019.112970

Ter Halle A, Jeanneau L, Martignac M, Jardé E, Pedrono B, Brach L, Gigault J (2017) Nanoplastic in the North Atlantic subtropical gyre. Environ Sci Technol 51(23):13689–13697. https://doi.org/10.1021/acs.est.7b03667

Teuten EL, Rowland SJ, Galloway TS, Thompson RC (2007) Potential for plastics to transport hydrophobic contaminants. Environ Sci Technol 41(22):7759–7764. https://doi.org/10.1021/es071737s

Tian LL, Chen QQ, Jiang W, Wang LH, Xie HX, Kalogerakis N, Ma YY, Ji R (2019) A carbon-14 radiotracer-based study on the phototransformation of polystyrene nanoplastics in water versus in air. Environ Sci Nano 6(9):2907–2917. https://doi.org/10.1039/c9en00662a

Turner A (2010) Marine pollution from antifouling paint particles. Mar Pollut Bull 60(2):159–171. https://doi.org/10.1016/j.marpolbul.2009.12.004

Turner A, Wallerstein C, Arnold R (2019) Identification, origin and characteristics of bio-bead microplastics from beaches in western Europe. Sci Total Environ 664:938–947. https://doi.org/10.1016/j.scitotenv.2019.01.281

Unice KM, Kreider ML, Panko JM (2012) Use of a deuterated internal standard with pyrolysis-GC/MS dimeric marker analysis to quantify tire tread particles in the environment. [research support, non-U.S. Gov't]. Int J Environ Res Public Health 9(11):4033–4055. https://doi.org/10.3390/ijerph9114033

Van Cauwenberghe L, Devriese L, Galgani F, Robbens J, Janssen CR (2015) Microplastics in sediments: A review of techniques, occurrence and effects. Marine Environ Res 111:5–17. https://doi.org/10.1016/j.marenvres.2015.06.007

van Sebille E, Wilcox C, Lebreton L, Maximenko N, Hardesty BD, van Franeker JA, Eriksen M, Siegel D, Galgani F, Law KL (2015) A global inventory of small floating plastic debris. Environ Res Lett 10(12):124006. https://doi.org/10.1088/1748-9326/10/12/124006

Vianello A, Jensen RL, Liu L, Vollertsen J (2019) Simulating human exposure to indoor airborne microplastics using a breathing thermal manikin. Sci Rep 9(8670). https://doi.org/10.1038/s41598-019-45054-w

von Moos N, Burkhardt-Holm P, Köhler A (2012) Uptake and effects of microplastics on cells and tissue of the blue mussel *Mytilus edulis* L. after an experimental exposure. Environ Sci Technol 46(20):11327–11335. https://doi.org/10.1021/es302332w

Wagner S, Hüffer T, Klöckner P, Wehrhahn M, Hofmann T, Reemtsma T (2018) Tire wear particles in the aquatic environment – A review on generation, analysis, occurrence, fate and effects. Water Res 139:83–100. https://doi.org/10.1016/j.watres.2018.03.051

Wang ZM, Wagner J, Ghosal S, Bedi G, Wall S (2017) SEM/EDS and optical microscopy analyses of microplastics in ocean trawl and fish guts. Sci Total Environ 603–604:616–626. https://doi.org/10.1016/j.scitotenv.2017.06.047

Wang Z, Taylor SE, Sharma P, Flury M (2018) Poor extraction efficiencies of polystyrene nano- and microplastics from biosolids and soil. PLoS One 13(11):e0208009. https://doi.org/10.1371/journal.pone.0208009

Wang J, Liu XH, Liu GN, Zhang ZX, Wu H, Cui BS, Bai JH, Zhang W (2019) Size effect of polystyrene microplastics on sorption of phenanthrene and nitrobenzene. Ecotoxicol Environ Saf 173:331–338. https://doi.org/10.1016/j.ecoenv.2019.02.037

Ward JE, Zhao S, Holohan BA, Mladinich KM, Griffin TW, Wozniak J, Shumway SE (2019) Selective ingestion and egestion of plastic particles by the blue mussel (*Mytilus edulis*) and eastern oyster (*Crassostrea virginica*): implications for using bivalves as bioindicators of microplastic pollution. Environ Sci Technol 53:8776–8784. https://doi.org/10.1021/acs.est.9b02073

Weidner SM, Trimpin S (2011) Mass spectrometry of synthetic polymers. Anal Chem 82:4811–4829. https://doi.org/10.1002/9783527641826

Welden NA, Cowie PR (2017) Degradation of common polymer ropes in a sublittoral marine environment. Mar Pollut Bull 118(1):248–253. https://doi.org/10.1016/j.marpolbul.2017.02.072

Wesch C, Elert AM, Wörner M, Braun U, Klein R, Paulus M (2017) Assuring quality in microplastic monitoring: about the value of clean-air devices as essentials for verified data. Sci Rep 7(1):5424. https://doi.org/10.1038/s41598-017-05838-4

Whalley J, Zandi S (2016) Particulate matter sampling techniques and data modelling methods. In: Sallis PJ (ed) Air quality – measurement and modeling. https://doi.org/10.5772/62563

Winkler A, Santo N, Ortenzi MA, Bolzoni E, Bacchetta R, Tremolada P (2019) Does mechanical stress cause microplastic release from plastic water bottles? Water Res 166:115082. https://doi.org/10.1016/j.watres.2019.115082

Wolff S, Kerpen J, Prediger J, Barkmann L, Müller L (2019) Determination of the microplastics emission in the effluent of a municipal waste water treatment plant using Raman microspectroscopy. Water Research X 2:100014. https://doi.org/10.1016/j.wroa.2018.100014

Woods MN, Stack ME, Fields DM, Shaw SD, Matrai PA (2018) Microplastic fiber uptake, ingestion, and egestion rates in the blue mussel (*Mytilus edulis*). Mar Pollut Bull 137:638–645. https://doi.org/10.1016/j.marpolbul.2018.10.061

Wright SL, Thompson RC, Galloway TS (2013) The physical impacts of microplastics on marine organisms: A review. Environ Pollut 178:483–492. https://doi.org/10.1016/j.envpol.2013.02.031

Wright SL, Levermore JM, Kelly FJ (2019) Raman spectral imaging for the detection of inhalable microplastics in ambient particulate matter samples. Environ Sci Technol 53(15):8947–8956. https://doi.org/10.1021/acs.est.8b06663

Wu N, Herrmann T, Paepke O, Tickner J, Hale R, Harvey E, La Guardia M, McClean MD, Webster TF (2007) Human exposure to PBDEs: associations of PBDE body burdens with food consumption and house dust concentrations. Environ Sci Technol 41(5):1584–1589. https://doi.org/10.1021/es0620282

Yu XB, Ladewig S, Bao S, Toline CA, Whitmire S, Chow AT (2018) Occurrence and distribution of microplastics at selected coastal sites along the southeastern United States. Sci Total Environ 613–614:298–305. https://doi.org/10.1016/j.scitotenv.2017.09.100

Zettler ER, Mincer TJ, Amaral-Zettler LA (2013) Life in the "plastisphere": microbial communities on plastic marine debris. Environ Sci Technol 47(13):7137–7146. https://doi.org/10.1021/es401288x

Zhang Y, Kang S, Allen S, Allen D, Gao T, Sillanpää M (2020a) Atmospheric microplastics: A review on current status and perspectives. Earth Sci Rev 203. https://doi.org/10.1016/j.earscirev.2020.103118

Zhang XM, Mell A, Li F, Thaysen C, Musselman B, Tice J, Vukovic D, Rochman C, Helm PA, Jobst KJ (2020b) Rapid fingerprinting of source and environmental microplastics using direct analysis in real time-high resolution mass spectrometry. Anal Chim Acta 1100:107–117. https://doi.org/10.1016/j.aca.2019.12.005

Zhang K, Hamidian AH, Tubić A, Zhang Y, Fang JKH, Wu C, Lam PKS (2021) Understanding plastic degradation and microplastic formation in the environment: a review. Environ Pollut 274:116554. https://doi.org/10.1016/j.envpol.2021.116554

3

Aquatic Organisms and Experimental MP Exposure Studies

Puspa L. Adhikari, Wokil Bam, Pamela L. Campbell, Francois Oberhaensli, Marc Metian, Marc Besson, Hugo Jacob and Peter W. Swarzenski

Abstract Environmental microplastic particles (MPs) represent a potential threat to many aquatic animals, and experimental exposure studies, when done well, offer a quantitative approach to assess this stress systematically and reliably. While the scientific literature on MP studies in aquatic environments is rapidly growing, there is still much to learn, and this chapter presents a brief overview of some of the successful methods and pitfalls in experimental MP exposure studies. A short overview of some experimental design types and recommendations are also presented. A proper experimental exposure study will yield useful information on MP-organism impacts and must include the following: a comprehensive MP characterization (e.g., density, buoyancy, type, nature, size, shape, concentration, color, degree of weathering/biofilm formation, an assessment of co-contaminant/surfactant toxicity and behavior, an understanding exposure modes, dose and duration, and the type and life stage of the target species). Finally, more conventional experimental

P. L. Adhikari
Department of Marine and Earth Sciences, The Water School, Florida Gulf Coast University, Fort Myers, FL, USA
e-mail: padhikari@fgcu.edu

W. Bam
International Atomic Energy Agency- Environment Laboratories, Principality of Monaco, Monaco

Department of Oceanography and Coastal Sciences, Louisiana State University, Baton Rouge, LA, USA
e-mail: bamwokil1@gmail.com; wbam1@lsu.edu

P. L. Campbell
US Geological Survey, Pacific Coastal and Marine Science Center, Santa Cruz, CA, USA

F. Oberhaensli · M. Metian · M. Besson · H. Jacob · P. W. Swarzenski (✉)
International Atomic Energy Agency- Environment Laboratories, Principality of Monaco, Monaco
e-mail: f.r.oberhaensli@iaea.org; m.metian@iaea.org; marc.besson@ens-lyon.org; hjacob2012@my.fit.edu; p.swarzenski@iaea.org

considerations, such as time, costs, and access to clean water, specialized instrumentation, and use of appropriate controls, replicate, and robust statistical analyses are also vital. This short review is intended as a necessary first step towards standardization of experimental MP exposure protocols so one can more reliably assess the transport and fate of MP in the aquatic environment as well as their potential impacts on aquatic organisms.

3.1 Introduction

Environmental plastic pollution is a ubiquitous phenomenon, affecting even the most remote environments on Earth, such as the Himalayas, the Arctic, and even the deepest marine trenches (Bergmann et al. 2017; Chiba et al. 2018). In addition to visible, macro-sized plastic litter that adversely may affect megafauna, there is another component of aquatic plastic pollution that remains harder to constrain, the microplastic particles (MP) (GESAMP 2015). MP has been conventionally defined as plastic particles less than 5 mm in size (Hidalgo-Ruz et al. 2012) and is either manufactured (primary MP) or the result of fragmentation and weathering of larger plastics (secondary MP). Some of the principal sources of MP in the aquatic environment are from rivers, wastewater treatment plants, atmospheric deposition (e.g., *municipal dust*), and some marine activities such as fishing and shipping (Cole et al. 2011).

It has been reported that more than 200 marine animal species have already been exposed to MP during some phase of their life cycles (Gall and Thompson 2015), either through direct ingestion or by trophic transfer of plastic-laden food (Lusher et al. 2017; Rochman 2015; Au et al. 2017; Auta et al. 2017; Paul-Pont et al. 2018; Botterell et al. 2019; Nelms et al. 2019). While the ubiquitous nature of MP pollution is an obvious potential threat to many aquatic organisms, we still lack a fundamental understanding of its impacts on biological systems (de Sá et al. 2018; Burns and Boxall 2018; Connors et al. 2017; Bucci et al. 2020). Carefully designed experimental exposure studies will enhance our understanding of the effects and underlying mechanisms of MP toxicity towards aquatic organisms. Such information can then guide policy decisions to strengthen and protect coastal and marine ecosystems.

3.2 MP Parameters

To design and conduct a meaningful MP exposure experiment using aquatic animals, the following parameters must be considered: MP type, chemical form, degree of weathering (or not), size, shape, concentration, color, density, presence of additives, sorbed chemical co-contaminants, exposure pathway and duration, target organism, and life stage (Fig. 3.1).

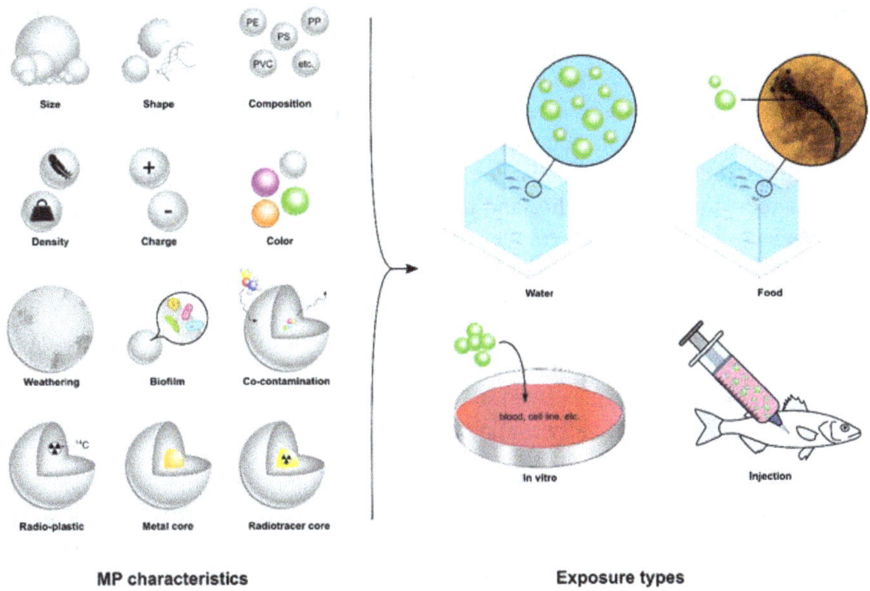

Fig. 3.1 An overview of the characterization of microplastic particles and their potential experimental exposure pathways (food and water) to aquatic organisms

3.2.1 Chemical and Physical Character of MP

In natural aquatic environments, MP are found as complex mixtures with different buoyancies, surface charge, color, composition (e.g., polymer type, presence of adsorbed contaminants and/or chemical additives, presence of biofilm and microorganisms), densities, shapes, and sizes. While some MP characteristics are quite easy to define and control, most require specific considerations. The following section discusses MP characteristics.

There are six plastic polymers that are most widely produced and thus observed in nature: polypropylene (PP), polyethylene (PE) that can occur both as high- and low-density polyethylene (HDPE, LDPE), polyvinyl chloride (PVC), polyurethane (PUR), polyethylene terephthalate (PET), and polystyrene (PS) (Browne et al. 2010; Karapanagioti et al. 2011; Vianello et al. 2013; Isobe et al. 2014; Enders et al. 2015; Frère et al. 2017). Among them, PE, PP, PS, and PET have been found to be the most abundant MP in the marine environment, followed by PVC (Rezania et al. 2018). PS is usually easiest to obtain and thus most widely used in laboratory exposure experiments. For MP fish exposure studies, PE is most utilized, followed by PS and PVC (Phuong et al. 2016; Botterell et al. 2019; Jacob et al. 2020).

MP can also exist in many shapes, such as spheres/beads, pellets, granules, fibers, films, fragments, and foams (Free et al. 2014; Karami 2017). While spheres are most often indicative of a primary MP, fragmentation and weathering will produce secondary MP that can irregularly shape spheres and fibers, films, fragments, and foams (Thompson 2015; Napper and Thompson 2016). Frydkjaer et al. (2017) found that irregular MP fragments were egested at a slower rate than spherical beads in experimental studies using *Daphnia magna*. The shape of a MP is thus an important factor in determining its effects in aquatic organisms (Bucci et al. 2020).

A wide range of MP size classes have been used in experimental exposure studies (Mattsson et al. 2015; Galloway et al. 2017; Ter Halle et al. 2017). According to some studies, the bioavailability and toxicity of MP can be highly size-dependent (Koelmans et al. 2020), with smaller particles generally exhibiting higher toxicity (Betts 2008, Jeong et al. 2016; Wright et al. 2013b; Bucci et al. 2020; Riberio et al. 2019; Wang et al. 2019) due to an increase in bioavailability and potential for translocation across the cell membrane (Browne et al. 2008). Physical blockage in the digestive tract has also been observed with certain MP size classes (Anbumani and Kakkar 2018). Currently, the selection of MP size for exposure experiments is often based on what is commercially available.

MP color can also vary widely, ranging from brightly colored to opaque and clear particles (Shaw and Day 1994; Su et al. 2016; Peters et al. 2017; Wang et al. 2017; Rezania et al. 2018; Zhang et al. 2018). Weathering will fade the original color into a secondary, usually less bright color (Chen et al. 2019). Importantly, the color of some MP may resemble natural food such as phytoplankton, which can affect ingestion rates and/or biological impacts to higher-trophic aquatic organisms (Wright et al. 2013).

The particle surface charge of MP is also an important characteristic that is affected by the ionic strength of natural waters. The shift from freshwater to seawater can dramatically change the aggregation properties and surface charge of particles, including MP. Generally, the physicochemical characterization of MP and its weathering will determine the efficiency of interactions with other particles and/or associated contaminants. The role of the MP surface charge on the toxicity for aquatic organisms is still not well understood (Paul-Pont et al. 2018). However, it has been suggested that the MP charge can play an important role in the transport, fate, and environmental effect of MP in the marine environment (Leslie 2012). The charge and surface properties of MP can play an important role in determining their effects to organisms, primarily due to their interaction with biological membranes (Cole et al. 2013; Rossi et al. 2013).

Polymer density will affect buoyancy and therefore bioavailability to target organisms. For example, high-density particles such as PET quickly sink, increasing bioavailability to benthic dwelling organisms, while pelagic filter/suspension feeders and planktonic feeders will be more readily exposed to low-density MP, such as PE (Wright et al. 2013). Continuous interaction of MP with other marine particles (i.e., ingestion/egestion, adsorption/desorption, aggregation/disaggregation, and biofouling) can also play a role in particle density (Cole et al. 2011, 2016; Kooi et al. 2017; Botterell et al. 2019).

3.2.2 *Primary vs. Weathered MP*

Primary MP consists of various off-the-shelf polymers such as PP, PE, PVC, PUR, PET, and PS, which are most often not directly released into the aquatic environment. Once natural weathering processes occur (e.g., biofouling, organic coatings, or aggregation of MP with other marine particles), a change in the chemical and physical properties will alter the bioavailability and toxicity (White 2006; Cole et al. 2011, 2016; Kooi et al. 2017; Lambert et al. 2017; Botterell et al. 2019; Chen et al. 2019). MP introduced to natural waters for any length of time will develop an organic biofilm that will drastically impact the fate and behavior of MP and associated co-contaminants. The use of weathered MP in exposure studies more closely reflects the natural environment; thus, it is important to account for these weathering changes during an exposure experiment. It is worth noting that most studies to date typically use primary MP for their exposure experiments (Bråte et al. 2018; Paul-Pont et al. 2018; Botterell et al. 2019; Jacob et al. 2020) or have used experimentally weathered MP (e.g., by immersing plastic particles in water for a few weeks or introducing microorganisms to the MP).

3.2.3 *Microplastic Co-contaminants*

Microplastics are complex pollutants consisting of polymer blends, residual monomers, plastic additives, and diverse co-contaminants (Rochman 2015). A large number of chemicals and some persistent organic pollutants (POPs) are added to MP during manufacturing to increase polymerization properties and durability, and these can contribute up to 60% (e.g., PVC: Net et al. 2015) of the plastic polymer mass. The additives most commonly used in the manufacturing process are plasticizers, thermal stabilizers, pigments, lubricants, flame retardants, and acid scavengers. It has been reported that chemicals leached from primary MP pellets may cause more deleterious effects than the ingestion of the MP itself (Botterell et al. 2019). However, studies quantifying the effects of plastic additives on organisms are still rare (Browne et al. 2013; Rochman et al. 2013), and desorption processes of plastic-associated chemicals and their effects on aquatic biota including human health remain poorly understood. Expectedly, organisms with longer gut retention times (i.e., some fish) have the potential for increased exposure and therefore for increased toxicity of MP co-contaminants.

Due to their large surface-to-volume ratio and charged hydrophobic surfaces, MP provide an excellent sorption site to scavenge some particle-reactive, dissolved contaminants (e.g., PBTs, PBDEs, DDT, PAHs, and pharmaceuticals), trace metals (e.g., copper, zinc, lead), and other plastic additives (Teuten et al. 2007, 2009; Beckingham and Ghosh 2017; Ribeiro et al. 2019). Consequently, MP can also become a potential, albeit diffuse source for diverse co-contaminants (Koelmans et al. 2013, 2016; Avio et al. 2015; Brennecke et al. 2016; Nakashima et al. 2016;

Alimi et al. 2018). It has been reported that the transport of HOCs (hydrophobic organic compounds) via MP is insignificant compared to their transport via natural particles (Burns and Boxall 2018; Riberio et al. 2019). Frydkjaer et al. (2017) found that C^{14}-labeled phenanthrene (a three-ring PAH used as a tracer molecule) sorbed more to planktonic organisms than to PE MP in laboratory experiments. Moreover, little is known about the effects of these co-contaminants in the smaller size fractions of microplastics (Velzeboer et al. 2014).

As MP exist as a complex mixture of weathered polymers, additives, organic contaminants, and trace metals, it is very difficult to perform laboratory exposure experiments and differentiate the effects of each component (Galloway et al. 2017; Paul-Pont et al. 2018). Thus, there is a need to carefully characterize the sorbed chemicals and plastic additives when exposing organisms to these MP. As many studies are struggling to accurately characterize the MP itself (Costa et al. 2019), proper quantification of plastic-sorbed chemicals prior to and after an experimental exposure study is even more challenging. Analytically it is often difficult to differentiate the toxicological effects of co-contaminants vs. MP, especially at lower, environmentally relevant exposure concentrations.

3.2.4 *Application of Labelled Microplastics in Experimental Exposure Studies*

Some exposure experiments incorporate labeled MP with either fluorescent or embedded radioisotopes to obtain unique information on transport processes and bioaccumulation kinetics (Cole 2016; Lanctôt et al. 2018). Using fluorescence-labeled MP (i.e., Nile red dye) may enhance imaging (Cole et al. 2016), but one needs to be mindful as MP may also contain an inherent fluorescence which may compromise interpretation. Similarly, stable isotope-labeled MP tracers, using, for example, ^{13}C-labelled MP (Berto et al. 2017), can yield important information on processes such as translocation, cycling, and biological impacts. Gamma- or beta-ray spectrometers are highly sensitive and not readily affected by typical interferences; thus, radiolabeled MP can be accurately quantified, even at trace levels, in complex environmental/biological samples and importantly, even in real time on live target organisms (Lanctôt et al. 2018). Radiolabeled MP can also be used to assess uptake and excretion routes, sorption/desorption kinetics, gut retention time, bioaccumulation, and trophic transfer.

3.3 How to Design a Meaningful Experimental Exposure Study?

Anyone who has worked with MP in controlled exposure studies can attest to the abundant difficulties and challenges. MP introduced to an experimental aquarium will tend to accumulate at the water/air interface and will attach indiscriminately to any surface, including pumps, filters, the exterior of test organisms, and aquaria walls. Thus, MP contact with the target organism must often be facilitated. Experimentalists will almost always have to add a complexing agent/surfactant to the MP to better control the distribution of the MP. The synergistic toxicity of this organic surface-active agent should be carefully evaluated in the context of realistic exposure studies.

An ideal exposure experiment should thus be designed with careful consideration of the physical and chemical properties of MP, the sorbed co-contaminants and additives, as well as the MP concentration, the life cycle of the target organism, and mode and duration of exposure. Environmental parameters such as temperature, salinity, and the pH of ambient aquaria water should be carefully maintained and monitored as these too may have an important effect on the intrinsic chemical properties of the MP. Quantification of MP exposure and retention time, bioaccumulation rates, as well as the concentration of MP are critical for toxicokinetic studies to determine how and where MP is transported in an organism.

3.3.1 Mode of Exposure

Of the four conventional contaminant vectors (food, sediment, water, and parent-to-offspring transfer) commonly traced in experimental exposure studies on aquatic organisms, the two primary pathways of exposure for MP are water and food. For the water pathway, a known concentration of well-characterized MP can be directly introduced into the water column of a controlled aquarium; target organisms can be selected to match the nature of the introduced MP (i.e., bottom- vs. water column-dwelling, life cycle). For the food pathway, target organisms can also be fed prey organisms contaminated with MP so that the target organism ingests the MP with the food (Figs. 3.2 and 3.3). This is a well-proven method to overcome some of the challenges of introducing a toxicant such as MP to living organisms.

3.3.2 Concentration of MP for Exposure Studies

The use of environmentally realistic concentrations of MP in exposure experiments is essential to obtain meaningful information for ecological risk assessments and resource protection (Huvet et al. 2016; Burton 2017; Karami 2017; Nyangoma de

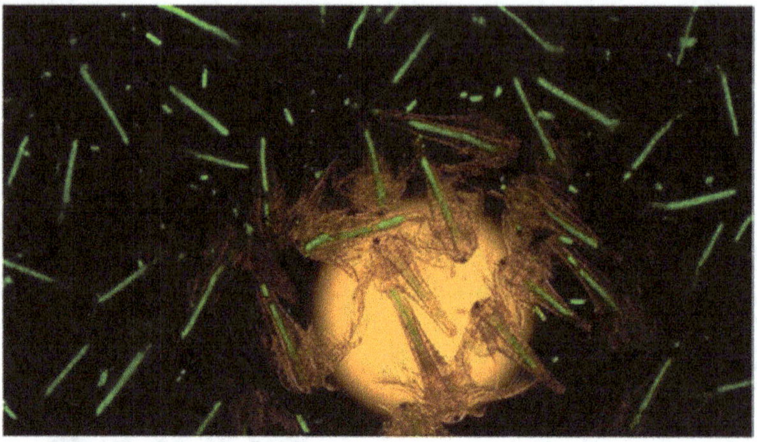

Fig. 3.2 Fluorescent microplastic particles line the stomach of artemia which are used as a microplastic-laden food for experimental exposure studies. (Photo credit: F. Oberhaensli, IAEA, Monaco)

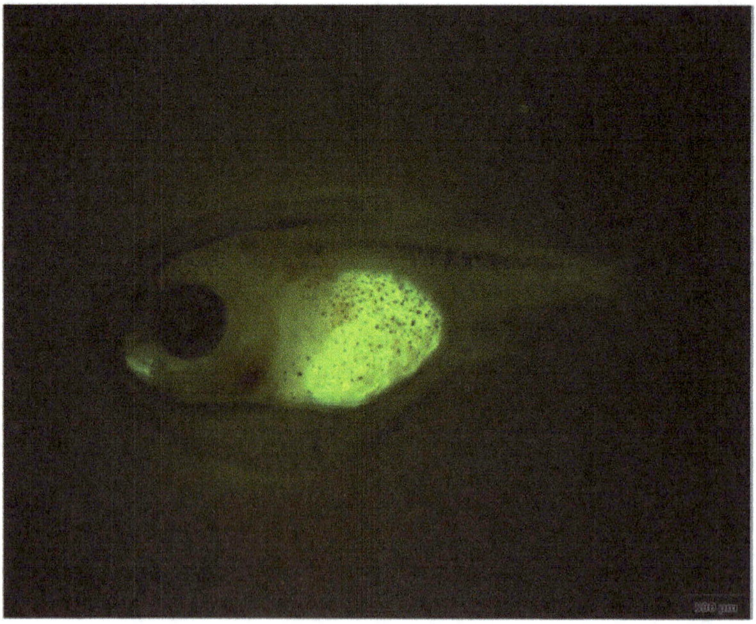

Fig. 3.3 Fluorescent microplastic particles line the stomach of a spiny chromis (*Acanthochromis polyacanthus*) fish. (Photo credit: M Besson, IAEA, Monaco)

Ruijter et al. 2020; Koelmans et al. 2020). Currently, MP concentrations used in laboratory experiments are still often unrealistically elevated (Lenz et al. 2016; Rochman 2016), although we still have a lot to learn about MP abundance in nature (Brandon et al. 2019). Moreover, the deliberate use of elevated concentrations of MP in experiments can be a powerful approach to identify underlying mechanisms and processes that define MP transport and toxicity. The selection of environmentally realistic concentrations of MP for exposure studies is limited mainly by our analytical capabilities (Filella 2015; Lenz et al. 2016; Rochman 2016).

3.3.3 Surfactants

Natural and anthropogenic surfactants are ubiquitous in the aquatic environment, and their inherent toxicity to organisms is generally well-known. Due to the amphiphilic nature of surfactants, the surface tension of the water molecules is decreased which in turn increases the solubility of the HOCs. Surfactants are commonly used in MP exposure experiments to disperse the MP and increase bioavailability. The presence of a surfactant generally increases the formation of homo-agglomerates and promotes adhesion. Indeed, the added presence of a surfactant (MP + surfactant) may increase the toxicity of MP using a surfactant such as Triton X-100 or Tween 20 (Renzi et al. 2019), resulting in higher rates of immobilization. Smaller-sized MP dispersed throughout the water column by surfactants can produce mechanical damage such as impairment of filtration, affecting organism gut residence time, and translocation from the gut into tissues (Cole et al. 2013; Ma et al. 2016; Rehse et al. 2016). Using *Daphnia magna* as a test organism, Renzi et al. (2019) observed the formation of homo-agglomerates of MP, which can adhere to the surfaces of organisms, thereby reducing their motility and increasing energy consumption.

3.3.4 Duration of Exposure

Exposure duration of MP to a target organism is one of the most important parameters that can be easily controlled and one that will directly influence the outcome of an experiment. For example, the residence time and/or retention time of MP within an organism will play a major role in defining its toxicity and will also impact where the MP will eventually reside. The ingestion of MP also depends on the duration of exposure and frequency of feeding which contributes to tissue/organ accumulation and incorporation. Water changes in experimental aquaria must be completed carefully to not remove particles which would change the exposure concentration for the target organisms. Depending upon the duration of the exposure, MP and associated co-contaminants can be leached into the surrounding water column over time with possible additional consequences for aquatic organisms. For example, Pittura et al. (2018) suggested that it might take up to 28 days for a gradual shift in the toxicity of these MP from being mechanical to chemical in nature. Modeling time-series data of chemical toxicity in target organisms can help define acute vs. chronic effects. One of the advantages of using radiolabeled-MP with gamma-emitting radiotracers to study the fate and transport of MP in organisms is that experimental results can be obtained in real time using live target organisms at environmentally relevant concentrations. This permits a real-time assessment of experiment duration to reach an "equilibrium state," and subsequent experimental adjustments can be made to yield the desired outcome. There are few aquarium-based studies that expose test organisms with various concentrations of MP for both short- and long-term in order to determine both the acute and chronic effects of MP simultaneously (Critchell and Hoogenboom 2018; Wang et al. 2019).

3.4 Recommendations

Based on a literature overview (Table 3.1), there exists a need to better standardize MP exposure experiments to be able to provide meaningful and reproducible results. Working with MP in experimental aquaria is challenging, and one needs to keep track of many physicochemical parameters that will affect the experimental outcome, including the chemical form, shape, size and nature (primary vs. weathered, secondary), and the presence of a biofilm and/or co-contaminants (Burns and Boxall 2018; Bucci et al. 2020). Basic experimental exposure study considerations include the following: (i) at which MP concentrations should the experiment be designed, (ii) what are the reporting units, and (iii) what are the QA/QC parameters? An experiment designed with MP concentrations that are close to environmental levels will yield different information than if the experiments are conducted with elevated MP concentrations. Microplastic concentrations are typically expressed in milligram per liter for most toxicity studies although it may be more accurate to report as the number of particles per liter, since different MP types will have variable size ranges. It is therefore important to count the number of particles using a flow cytometer or other suitable counting methods. Surface charge and density considerations are also essential if the MP is to make proper contact with the selected target species (benthic vs. water column species).

Carefully designed experiments can provide useful insight to better understand MP impacts from cellular to organ, organism, and ecosystem levels. Exposure experiments should incorporate a carefully developed approach that includes physical, chemical, and biological factors that have a strong influence on both the target organism. Furthermore, conducting complementary field and/or laboratory-based studies could better define the scientific lacunae in representative sentinel species in single and combined exposure studies. Such complementary field data may provide useful information to better interpret laboratory-based studies to develop realistic assessments of organismal stress to MP (Anbumani and Kakkar 2018; Wright et al. 2019).

Table 3.1 An overview of priorities and recommendations for experimental exposure studies of microplastics on aquatic organisms

Priorities and recommendations	
Microplastics	Use MP with varying physical and chemical properties
	Evaluate the ecotoxicological effects of MP and associated co-contaminants
	Assess the bioaccumulation pathways of MP and co-contaminants through aquatic food webs.
	Use of primary and/or weathered MP to assess the specific organismal impacts
Target organisms	Use multispecies approach with emphasis on early life stages
	Investigate the impacts of MP on less-studied organisms (e.g., echinoderms, cnidarians, and sponges)
	Examine the link between MP, primary producers, and carbon flow
	Assess biological effects on the community, population, and ecosystems
	Investigate the transfer of MP to higher trophic species
Exposures	Use high concentrations to study MP modes of action, kinetics, and processes
	Assess the scavenging potential of natural particles versus MP
	Investigate potential dose rate or threshold responses by using gradient MP concentrations and experiment durations
	Use environmentally relevant MP concentrations to assess potential ecological impacts
	Study MP ingestion and trophic transfer in fish and compare the use of artificial feed or live food
Methods	Develop specific biomonitoring indicators that can track organismal stress including inflammation, intestinal dysbiosis, neurotoxicity and behavioral change, and metabolic alterations
	Develop and use a best practice guide for MP research
	Assess the impacts of MP on various biological functions, e.g., enzymatic, genetic, histological, reproductive, developmental and physiological functions, as well as immune and stress-related responses, cell signaling, energy homeostasis
	Avoid external contamination with MP of experiments to determine accurate impact by a regular monitoring of experimental conditions
	Study the effects of MP at different levels of biological organization (atomic, molecular, cellular, tissue/organ, individual, community, trans-generational)

Previous studies have generally focused on MP effects on target organisms by treating MP as a single pollutant as opposed to a more realistic mixture of pollutants. There is thus the need to conduct experiments on MP and associated co-contaminant mixtures (Burns and Boxall 2018). Because we still have a lot to learn on proper characterization techniques for MP, special emphasis should be placed on the development and standardization of optimized analytical methods. The application of radiolabeled MP exposure experiments can provide better detection limits even at environmental or trace concentrations and can be an excellent method for elucidating the trophic transfer and movement of MP in live organisms.

The ideal experimental setup should be simple in design and should yield reproducible results using realistic MP concentrations, exposure routes, times, and target organisms. Depending on the specific research question, experimental MP exposure studies may first incorporate a simplified experimental design where one indicator species is exposed to a single type of MP. Subsequent studies may then build on these results and more complex experimental designs will yield more precise information on the organismal effects of MP. While the best laboratory exposure experiments currently address the effects of MP on target organisms under a set of environmental conditions, the next generation studies could address synergistic effects of mixed MP and associated co-contaminants on multiple species. This would be a logical extension of current state-of-the-art exposure experiments and would provide information that more closely resembles a natural aquatic ecosystem.

Acknowledgments PWS, MM, and FO of the IAEA are grateful for the support provided to its Environment Laboratories by the Government of the Principality of Monaco. This work was partly funded by generous contributions through the IAEA Peaceful Uses Initiative (PUI).

References

Alimi OS, Budarz JF, Hernandez LM, Tufenkji N (2018) Microplastics and nanoplastics in aquatic environments: aggregation, deposition, and enhanced contaminant transport. Environ Sci Technol 52(4):1704–1724. https://doi.org/10.1021/acs.est.7b05559

Anbumani S, Kakkar P (2018) Ecotoxicological effects of microplastics on biota: a review. Environ Sci Pollut Res 25(15):14373–14396. https://doi.org/10.1007/s11356-018-1999-x

Au SY, Lee CM, Weinstein JE, Hurk PVD, Klaine SJ (2017) Trophic transfer of microplastics in aquatic ecosystems: identifying critical research needs. Integr Environ Assess Manag 13(3):505–509. https://doi.org/10.1002/ieam.1907

Auta H, Emenike C, Fauziah S (2017) Distribution and importance of microplastics in the marine environment: a review of the sources, fate, effects, and potential solutions. Environ Int 102:165–176. https://doi.org/10.1016/j.envint.2017.02.013

Avio CG, Gorbi S, Milan M, Benedetti M, Fattorini D, Derrico G et al (2015) Pollutants bioavailability and toxicological risk from microplastics to marine mussels. Environ Pollut 198:211–222. https://doi.org/10.1016/j.envpol.2014.12.021

Beckingham B, Ghosh U (2017) Differential bioavailability of polychlorinated biphenyls associated with environmental particles: microplastic in comparison to wood, coal and biochar. Environ Pollut 220:150–158. https://doi.org/10.1016/j.envpol.2016.09.033

Bergmann M, Wirzberger V, Krumpen T, Lorenz C, Primpke S, Tekman MB, Gerdts G (2017) High quantities of microplastic in arctic deep-sea sediments from the HAUSGARTEN observatory. Environ Sci Technol 51(19):11000–11010. https://doi.org/10.1021/acs.est.7b03331

Berto D, Rampazzo F, Gion C, Noventa S, Ronchi F, Traldi U et al (2017) Preliminary study to characterize plastic polymers using elemental analyser/isotope ratio mass spectrometry (EA/IRMS). Chemosphere 176:47–56. https://doi.org/10.1016/j.chemosphere.2017.02.090

Betts K (2008) Why small plastic particles may pose a big problem in the oceans. Environ Sci Technol 42(24):8995–8995. https://doi.org/10.1021/es802970v

Botterell ZL, Beaumont N, Dorrington T, Steinke M, Thompson RC, Lindeque PK (2019) Bioavailability and effects of microplastics on marine zooplankton: a review. Environ Pollut 245:98–110. https://doi.org/10.1016/j.envpol.2018.10.065

Brandon JA, Jones W, Ohman MD (2019) Multidecadal increase in plastic particles in coastal ocean sediments. Sci Adv 5(9). https://doi.org/10.1126/sciadv.aax0587

Bråte ILN, Blázquez M, Brooks SJ, Thomas KV (2018) Weathering impacts the uptake of polyethylene microparticles from toothpaste in Mediterranean mussels (*M. galloprovincialis*). Sci Total Environ 626:1310–1318. https://doi.org/10.1016/j.scitotenv.2018.01.141

Brennecke D, Duarte B, Paiva F, Caçador I, Canning-Clode J (2016) Microplastics as vector for heavy metal contamination from the marine environment. Estuar Coast Shelf Sci 178:189–195. https://doi.org/10.1016/j.ecss.2015.12.003

Browne M, Dissanayake A, Galloway T, Lowe D, Thompson R (2008) Ingested microscopic plastic translocates to the circulatory system of the mussel, Mytilus edulis (L.). Environ Sci Technol 42:5026–5031

Browne MA, Galloway TS, Thompson RC (2010) Spatial patterns of plastic debris along estuarine shorelines. Environ Sci Technol 44(9):3404–3409. https://doi.org/10.1021/es903784e

Browne MA, Niven SJ, Galloway TS, Rowland SJ, Thompson RC (2013) Microplastic moves pollutants and additives to worms, reducing functions linked to health and biodiversity. Curr Biol 23(23):2388–2392. https://doi.org/10.1016/j.cub.2013.10.012

Bucci K, Tulio M, Rochman CM (2020) What is known and unknown about the effects of plastic pollution: a meta-analysis and systematic review. Ecol Appl 30(2). https://doi.org/10.1002/eap.2044

Burns EE, Boxall AB (2018) Microplastics in the aquatic environment: evidence for or against adverse impacts and major knowledge gaps. Environ Toxicol Chem 37(11):2776–2796. https://doi.org/10.1002/etc.4268

Burton GA (2017) Stressor exposures determine risk: so, why do fellow scientists continue to focus on superficial microplastics risk? Environ Sci Technol 51(23):13515–13516. https://doi.org/10.1021/acs.est.7b05463

Chen Q, Zhang H, Allgeier A, Zhou Q, Ouellet JD, Crawford SE et al (2019) Marine microplastics bound dioxin-like chemicals: model explanation and risk assessment. J Hazard Mater 364:82–90. https://doi.org/10.1016/j.jhazmat.2018.10.032

Chiba S, Saito H, Fletcher R, Yogi T, Kayo M, Miyagi S et al (2018) Human footprint in the abyss: 30 year records of deep-sea plastic debris. Mar Policy 96:204–212. https://doi.org/10.1016/j.marpol.2018.03.022

Cole M (2016) A novel method for preparing microplastic fibers. Sci Rep 6(1). https://doi.org/10.1038/srep34519

Cole M, Lindeque P, Halsband C, Galloway TS (2011) Microplastics as contaminants in the marine environment: a review. Mar Pollut Bull 62(12):2588–2597. https://doi.org/10.1016/j.marpolbul.2011.09.025

Cole M, Lindeque P, Fileman E, Halsband C, Goodhead R, Moger J, Galloway TS (2013) Microplastic ingestion by zooplankton. Environ Sci Technol 47(12):6646–6655. https://doi.org/10.1021/es400663f

Cole M, Lindeque PK, Fileman E, Clark J, Lewis C, Halsband C, Galloway TS (2016) Microplastics alter the properties and sinking rates of zooplankton faecal pellets. Environ Sci Technol 50(6):3239–3246. https://doi.org/10.1021/acs.est.5b05905

Connors KA, Dyer SD, Belanger SE (2017) Advancing the quality of environmental microplastic research. Environ Toxicol Chem 36(7):1697–1703. https://doi.org/10.1002/etc.3829

Costa JPD, Reis V, Paço A, Costa M, Duarte AC, Rocha-Santos T (2019) Micro(nano)plastics – analytical challenges towards risk evaluation. TrAC Trends Anal Chem 111:173–184. https://doi.org/10.1016/j.trac.2018.12.013

Critchell K, Hoogenboom MO (2018) Effects of microplastic exposure on the body condition and behaviour of planktivorous reef fish (*Acanthochromis polyacanthus*). PLoS One 13(3). https://doi.org/10.1371/journal.pone.0193308

de Sá LC, Oliveira M, Ribeiro F, Rocha TL, Futter MN (2018) Studies of the effects of microplastics on aquatic organisms: what do we know and where should we focus our efforts in the future? Sci Total Environ 645:1029–1039. https://doi.org/10.1016/j.scitotenv.2018.07.207

Enders K, Lenz R, Stedmon CA, Nielsen TG (2015) Abundance, size and polymer composition of marine microplastics ≥ 10 μm in the Atlantic Ocean and their modelled vertical distribution. Mar Pollut Bull 100(1):70–81. https://doi.org/10.1016/j.marpolbul.2015.09.027

Filella M (2015) Questions of size and numbers in environmental research on microplastics: methodological and conceptual aspects. Environ Chem 12(5):527. https://doi.org/10.1071/en15012

Free CM, Jensen OP, Mason SA, Eriksen M, Williamson NJ, Boldgiv B (2014) High levels of microplastic pollution in a large, remote, mountain lake. Mar Pollut Bull 85(1):156–163. https://doi.org/10.1016/j.marpolbul.2014.06.001

Frère L, Paul-Pont I, Rinnert E, Petton S, Jaffré J, Bihannic I et al (2017) Influence of environmental and anthropogenic factors on the composition, concentration and spatial distribution of microplastics: a case study of the bay of Brest (Brittany, France). Environ Pollut 225:211–222. https://doi.org/10.1016/j.envpol.2017.03.023

Frydkjær CK, Iversen N, Roslev P (2017) Ingestion and egestion of microplastics by the cladoceran Daphnia magna: effects of regular and irregular shaped plastic and sorbed phenanthrene. Bull Environ Contam Toxicol 99(6):655–661

Gall S, Thompson R (2015) The impact of debris on marine life. Mar Pollut Bull 92(1–2):170–179. https://doi.org/10.1016/j.marpolbul.2014.12.041

Galloway TS, Cole M, Lewis C (2017) Interactions of microplastic debris throughout the marine ecosystem. Nat Ecol Evol 1(5). https://doi.org/10.1038/s41559-017-0116

GESAMP (2015) Sources, Fate and Effects of Microplastics in the marine Environment: A Global Assessment, Reports and Studies GESAMP. IMO/FAO/UNESCO-ioc/unido/wmo/iaea/un/unep/undp Joint Group of Experts on the Scientific Aspects of Marine Environmental Protection. doi:https://doi.org/10.13140/RG.2.1.3803.7925

Hidalgo-Ruz V, Gutow L, Thompson RC, Thiel M (2012) Microplastics in the marine environment: a review of the methods used for identification and quantification. Environ Sci Technol 46(6):3060–3075. https://doi.org/10.1021/es2031505

Huvet A, Paul-Pont I, Fabioux C, Lambert C, Suquet M, Thomas Y et al (2016) Reply to Lenz et al.: quantifying the smallest microplastics is the challenge for a comprehensive view of their environmental impacts. Proc Natl Acad Sci 113(29). https://doi.org/10.1073/pnas.1607221113

Isobe A, Kubo K, Tamura Y, Kako SI, Nakashima E, Fujii N (2014) Selective transport of microplastics and mesoplastics by drifting in coastal waters. Mar Pollut Bull 89(1–2):324–330. https://doi.org/10.1016/j.marpolbul.2014.09.041

Jacob H, Besson M, Swarzenski PW, Lecchini D, Metian M (2020) Effects of virgin micro- and nanoplastics on fish: trends, meta-analysis, and perspectives. Environ Sci Technol 54(8):4733–4745. https://doi.org/10.1021/acs.est.9b05995

Jeong C-B, Won E-J, Kang H-M, Lee M-C, Hwang D-S, Hwang U-K et al (2016) Microplastic size-dependent toxicity, oxidative stress induction, and p-JNK and p-p38 activation in the monogonont rotifer (Brachionus koreanus). Environ Sci Technol 50(16):8849–8857. https://doi.org/10.1021/acs.est.6b01441

Karami A (2017) Gaps in aquatic toxicological studies of microplastics. Chemosphere 184:841–848. https://doi.org/10.1016/j.chemosphere.2017.06.048

Karapanagioti H, Endo S, Ogata Y, Takada H (2011) Diffuse pollution by persistent organic pollutants as measured in plastic pellets sampled from various beaches in Greece. Mar Pollut Bull 62(2):312–317. https://doi.org/10.1016/j.marpolbul.2010.10.009

Koelmans AA, Redondo-Hasselerharm PE, Nor NHM, Kooi M (2020) Solving the non-alignment of methods and approaches used in microplastic research in order to consistently characterize risk. Environ Sci Technol. https://doi.org/10.1021/acs.est.0c02982

Koelmans AA, Besseling E, Wegner A, Foekema EM (2013) Plastic as a carrier of POPs to aquatic organisms: a model analysis. Environ Sci Technol 47(14):7812–7820. https://doi.org/10.1021/es401169n

Koelmans AA, Bakir A, Burton GA, Janssen CR (2016) Microplastic as a vector for chemicals in the aquatic environment: critical review and model-supported reinterpretation of empirical studies. Environ Sci Technol 50(7):3315–3326. https://doi.org/10.1021/acs.est.5b06069

Koelmans AA, Redondo-Hasselerharm PE, Mohamed Nor NH, Kooi M (2020) Solving the non-alignment of methods and approaches used in microplastic research to consistently characterize risk. Environ Sci Technol 54(19):12307–12315. https://doi.org/10.1021/acs.est.0c02982

Kooi M, Nes EHV, Scheffer M, Koelmans AA (2017) Ups and downs in the ocean: effects of biofouling on vertical transport of microplastics. Environ Sci Technol 51(14):7963–7971. https://doi.org/10.1021/acs.est.6b04702

Lambert S, Scherer C, Wagner M (2017) Ecotoxicity testing of microplastics: considering the heterogeneity of physicochemical properties. Integr Environ Assess Manag 13(3):470–475. https://doi.org/10.1002/ieam.1901

Lanctôt CM, Al-Sid-Cheikh M, Catarino AI, Cresswell T, Danis B, Karapanagioti HK et al (2018) Application of nuclear techniques to environmental plastics research. J Environ Radioact 192:368–375. https://doi.org/10.1016/j.jenvrad.2018.07.019

Lenz R, Enders K, Nielsen TG (2016) Microplastic exposure studies should be environmentally realistic. Proc Natl Acad Sci 113(29). https://doi.org/10.1073/pnas.1606615113

Leslie HA (2012) Microplastic in Noordzee Zwevend stof en Cosmetica. EindrapportageW-12/01, Institute for Environmental Studies, Amsterdam

Lusher AL, Welden NA, Sobral P, Cole M (2017) Sampling, isolating and identifying microplastics ingested by fish and invertebrates. Anal Methods 9(9):1346–1360. https://doi.org/10.1039/c6ay02415g

Ma Y, Huang A, Cao S, Sun F, Wang L, Guo H, Ji R (2016) Effects of nanoplastics and microplastics on toxicity, bioaccumulation, and environmental fate of phenanthrene in fresh water. Environ Pollut 219:166–173. https://doi.org/10.1016/j.envpol.2016.10.061

Mattsson K, Hansson L-A, Cedervall T (2015) Nano-plastics in the aquatic environment. Environ Sci: Processes Impacts 17(10):1712–1721. https://doi.org/10.1039/c5em00227c

Nakashima E, Isobe A, Kako S, Itai T, Takahashi S, Guo X (2016) The potential of oceanic transport and onshore leaching of additive-derived lead by marine macro-plastic debris. Mar Pollut Bull 107(1):333–339. https://doi.org/10.1016/j.marpolbul.2016.03.038

Napper IE, Thompson RC (2016) Release of synthetic microplastic plastic fibres from domestic washing machines: effects of fabric type and washing conditions. Mar Pollut Bull 112(1–2):39–45. https://doi.org/10.1016/j.marpolbul.2016.09.025

Nelms SE, Barnett J, Brownlow A, Davison NJ, Deaville R, Galloway TS et al (2019) Microplastics in marine mammals stranded around the British coast: ubiquitous but transitory? Sci Rep 9(1). https://doi.org/10.1038/s41598-018-37428-3

Net S, Sempéré R, Delmont A, Paluselli A, Ouddane B (2015) Occurrence, fate, behavior and ecotoxicological state of phthalates in different environmental matrices. Environ Sci Technol 49(7):4019–4035. https://doi.org/10.1021/es505233b

Nyangoma de Ruijter V, Redondo-Hasselerharm PE, Gouin T, Koelmans AA (2020) Quality criteria for microplastic effect studies in the context of risk assessment: a critical review. Environ Sci Technol. https://doi.org/10.1021/acs.est.0c03057

Paul-Pont I, Tallec K, Gonzalez-Fernandez C, Lambert C, Vincent D, Mazurais D et al (2018) Constraints and priorities for conducting experimental exposures of marine organisms to microplastics. Front Mar Sci 5. https://doi.org/10.3389/fmars.2018.00252

Peters CA, Thomas PA, Rieper KB, Bratton SP (2017) Foraging preferences influence microplastic ingestion by six marine fish species from the Texas Gulf Coast. Mar Pollut Bull 124(1):82–88. https://doi.org/10.1016/j.marpolbul.2017.06.080

Phuong NN, Zalouk-Vergnoux A, Poirier L, Kamari A, Châtel A, Mouneyrac C, Lagarde F (2016) Is there any consistency between the microplastics found in the field and those used in laboratory experiments? Environ Pollut 211:111–123. https://doi.org/10.1016/j.envpol.2015.12.035

Pittura L, Avio CG, Giuliani ME, Derrico G, Keiter SH, Cormier B et al (2018) Microplastics as vehicles of environmental pahs to marine organisms: combined chemical and physical hazards to the mediterranean mussels, Mytilus galloprovincialis. Front Mar Sci 5. https://doi.org/10.3389/fmars.2018.00103

Rehse S, Kloas W, Zarfl C (2016) Short-term exposure with high concentrations of pristine microplastic particles leads to immobilisation of Daphnia magna. Chemosphere 153:91–99. https://doi.org/10.1016/j.chemosphere.2016.02.133

Renzi M, Grazioli E, Blašković A (2019) Effects of different microplastic types and surfactant-microplastic mixtures under fasting and feeding conditions: a case study on Daphnia magna. Bull Environ Contam Toxicol 103(3):367–373. https://doi.org/10.1007/s00128-019-02678-y

Rezania S, Park J, Din MFM, Taib SM, Talaiekhozani A, Yadav KK, Kamyab H (2018) Microplastics pollution in different aquatic environments and biota: a review of recent studies. Mar Pollut Bull 133:191–208. https://doi.org/10.1016/j.marpolbul.2018.05.022

Ribeiro F, Obrien JW, Galloway T, Thomas KV (2019) Accumulation and fate of nano- and microplastics and associated contaminants in organisms. TrAC Trends Anal Chem 111:139–147. https://doi.org/10.1016/j.trac.2018.12.010

Rochman CM (2015) The complex mixture, fate and toxicity of chemicals associated with plastic debris in the marine environment. In: Marine anthropogenic litter. Springer, Cham, pp 117–140. https://doi.org/10.1007/978-3-319-16510-3_5

Rochman CM (2016) Ecologically relevant data are policy-relevant data. Science 352(6290):1172–1172. https://doi.org/10.1126/science.aaf8697

Rochman CM, Hoh E, Hentschel BT, Kaye S (2013) Long-term field measurement of sorption of organic contaminants to five types of plastic pellets: implications for plastic marine debris. Environ Sci Technol 47(3):1646–1654

Rossi G, Barnoud J, Monticelli L (2013) Polystyrene nanoparticles perturb lipid membranes. J Phys Chem Lett 5(1):241–246. https://doi.org/10.1021/jz402234c

Shaw DG, Day RH (1994) Colour- and form-dependent loss of plastic micro-debris from the North Pacific Ocean. Mar Pollut Bull 28(1):39–43. https://doi.org/10.1016/0025-326x(94)90184-8

Su L, Xue Y, Li L, Yang D, Kolandhasamy P, Li D, Shi H (2016) Microplastics in Taihu Lake, China. Environ Pollut 216:711–719. https://doi.org/10.1016/j.envpol.2016.06.036

Ter Halle A, Jeanneau LT, Martignac MT, Jardé ET, Pedrono BT, Brach LT, Gigault JT (2017) Nanoplastic in the North Atlantic subtropical gyre. Environ Sci Technol 51(23):13689–13697. https://doi.org/10.1021/acs.est.7b03667

Teuten EL, Rowland SJ, Galloway TS, Thompson RC (2007) Potential for plastics to transport hydrophobic contaminants. Environ Sci Technol 41(22):7759–7764. https://doi.org/10.1021/es071737s

Teuten EL, Saquing JM, Knappe DRU, Barlaz MA, Jonsson S, Björn A et al (2009) Transport and release of chemicals from plastics to the environment and to wildlife. Philos Trans R Soc B Biol Sci 364(1526):2027–2045. https://doi.org/10.1098/rstb.2008.0284

Thompson RC (2015) Microplastics in the marine environment: sources, consequences and solutions. In: Marine anthropogenic litter. Springer, Cham, pp 185–200. https://doi.org/10.1007/978-3-319-16510-3_7

Velzeboer I, Kwadijk CJAF, Koelmans AA (2014) Strong sorption of PCBs to nanoplastics, microplastics, carbon nanotubes, and fullerenes. Environ Sci Technol 48(9):4869–4876. https://doi.org/10.1021/es405721v

Vianello A, Boldrin A, Guerriero P, Moschino V, Rella R, Sturaro A, Ros LD (2013) Microplastic particles in sediments of lagoon of Venice, Italy: first observations on occurrence, spatial patterns and identification. Estuar Coast Shelf Sci 130:54–61. https://doi.org/10.1016/j.ecss.2013.03.022

Wang W, Ndungu AW, Li Z, Wang J (2017) Microplastics pollution in inland freshwaters of China: a case study in urban surface waters of Wuhan, China. Sci Total Environ 575:1369–1374. https://doi.org/10.1016/j.scitotenv.2016.09.213

Wang W, Gao H, Jin S, Li R, Na G (2019) The ecotoxicological effects of microplastics on aquatic food web, from primary producer to human: a review. Ecotoxicol Environ Saf 173:110–117. https://doi.org/10.1016/j.ecoenv.2019.01.113

White JR (2006) Polymer ageing: physics, chemistry or engineering? Time to reflect. C R Chim 9(11–12):1396–1408. https://doi.org/10.1016/j.crci.2006.07.008

Wright SL, Rowe D, Thompson RC, Galloway TS (2013) Microplastic ingestion decreases energy reserves in marine worms. Curr Biol 23(23). https://doi.org/10.1016/j.cub.2013.10.068

Wright SL, Thompson RC, Galloway TS (2013b) The physical impacts of microplastics on marine organisms: a review. Environ Pollut 178:483–492. https://doi.org/10.1016/j.envpol.2013.02.031

Wright SL, Levermore JM, Kelly FJ (2019) Raman spectral imaging for the detection of inhalable microplastics in ambient particulate matter samples. Environ Sci Technol 53(15):8947–8956. https://doi.org/10.1021/acs.est.8b06663

Zhang K, Shi H, Peng J, Wang Y, Xiong X, Wu C, Lam PK (2018) Microplastic pollution in Chinas inland water systems: a review of findings, methods, characteristics, effects, and management. Sci Total Environ 630:1641–1653. https://doi.org/10.1016/j.scitotenv.2018.02.300

4

Sources and Pathways of Microplastics to the Environment

Emilie M. F. Kallenbach, Elisabeth S. Rødland, Nina T. Buenaventura and Rachel Hurley

Abstract In recent years, the focus of microplastic research has begun to observe a shift from the marine towards terrestrial and freshwater environments. This is in response to a greater awareness of the predominance of land-based sources in marine microplastic contamination. In this regard, terrestrial and freshwater environments are often perceived as conduits for microplastic particles to the oceans, but this overlooks substantial and important complexities associated with these systems, as well as the need to protect these ecosystems in their own right. This chapter focuses on several critical sources and pathways deemed to be highly important for the release of microplastics to the environment. These include road-associated microplastic particles (RAMP) and emissions related to agriculture that are, thus far, under-researched. Transfers and accumulations of particles within terrestrial and freshwater systems are also reviewed, including the state of knowledge on the occurrence of microplastics in different environmental compartments (air, water, sediments, biota). Methodological constraints are addressed, with particular focus on the need for greater harmonisation along all stages of sampling, analysis, and data handling. Finally, the chapter discusses the ultimate fate of particles released to terrestrial and freshwater environments and highlights critical research gaps that should be addressed to evolve our understanding of microplastic contamination in complex and dynamic environmental systems.

E. M. F. Kallenbach
NIVA Denmark, Copenhagen, Denmark

University of Copenhagen, Copenhagen, Denmark
e-mail: eka@niva-dk.dk

E. S. Rødland
Norwegian Institute for Water Research (NIVA), Oslo, Norway

Norwegian University of Life Science (NMBU), Ås, Norway
e-mail: elisabeth.rodland@niva.no

N. T. Buenaventura · R. Hurley (✉)
Norwegian Institute for Water Research (NIVA), Oslo, Norway
e-mail: nina.buenaventura@niva.no; rachel.hurley@niva.no

4.1 Introduction

Recent research has begun to document widespread and pervasive contamination of terrestrial and freshwater environmental systems by microplastic particles. Several papers have now pointed out a dichotomy that exists: all plastic is produced on land – and the majority of plastic is consumed and disposed of on land – and yet the primary focus for microplastics research still concentrates predominately on the marine environment (e.g. Blettler et al. 2018; Dris et al. 2015; Horton et al. 2017a, b; Lambert and Wagner 2018; Mai et al. 2018). This is where microplastic contamination was first observed (Carpenter et al. 1972; Carpenter and Smith 1972; Shiber 1979) and is highlighted as the eventual recipient for microplastic particles in the environment. Evidence has shown that microplastics are distributed widely across the global ocean and may have negative impacts on the marine ecosystem, particularly in remote and sensitive regions (Avio et al. 2015). Despite this, a focus on marine microplastics misses several important characteristics of their release and geographical distribution that are integral to efforts to reduce environmental contamination. First and foremost, the majority of microplastic particles are released through land-based sources (Rochman 2018). A thorough assessment of these sources is therefore essential to identify actions to effectively reduce microplastic emissions. This is frequently referred to through the 'turning off the tap' analogy (Boucher and Friot 2017; Evans-Pughe 2017); however, this touches upon a second important detail. Many land-based processes, such as fluvial and atmospheric transport, are described as transfers of plastic from land to sea. They should not, however, be considered as pipelines of plastics to the sea: the transport of microplastic particles from their source to the marine environment is expected to be highly complex. Particles released on land likely encounter a range of dynamic environments which can transform particles and may also retain them across a range of timescales, thus acting as a sink of microplastic pollution, and with similar potential impacts as reported for the marine systems. A thorough understanding of these processes is still lacking.

One of the main barriers to establishing this understanding is the paucity of high-quality and comparable data. This is underpinned by methodological complexities associated with the analysis of microplastic contamination in a range of environmental compartments and matrices. Here, a marine-freshwater skew also exists: the majority of harmonisation and standardisation efforts by international bodies and working groups primarily focus on the marine environment (e.g. Frias et al. 2018; GESAMP 2015, 2016, 2019; Isobe et al. 2019). It is important that methods established for the marine environment are not uncritically transferred to terrestrial and freshwater samples. Sampling for microplastics should be closely tied to the specific research questions at hand. There exists a wealth of methodological approaches outside of the plastic research field that may be tailored to include the capture of microplastic particles and which would generate samples that also correspond to a range of relevant hydrological, geomorphological, and aeolian processes. Moreover, in many cases, differences in the type, quantity, and ratios of non-plastic organic and inorganic sample constituents complicate analytical methods and may require the development of new approaches to prepare samples. These new methods require validation to ensure the production of comparable datasets. Recent efforts to crystallise the quality requirements for data reporting have included non-marine sample types (e.g. Koelmans et al. 2019). These describe several good practices which should be applied to all assessments of microplastic contamination: ensuring that sampling is representative, including both blanks and relevant recovery tests, verifying particles as microplastic through chemical analyses, and considering the sources of error in the data.

This chapter will draw together existing research from terrestrial and freshwater environments to address the current state of knowledge and identify important gaps in our understanding of sources and processes related to microplastic contamination across a range of spatial and temporal scales. This will include a review of available data on the occurrence of microplastic particles in selected focused environments: (1) agricultural systems; (2) urban environments, with a particular focus on road-derived microplastics; (3) river systems; (4) lakes; and (5) the atmosphere.

4.2 Microplastics in Terrestrial Environments

The majority of all plastics ever produced – approximately 60% or 4900 Mt. – have been discarded and are now present in either landfills or the environment (Geyer et al. 2017). Establishing the proportions that have been directly (e.g. littering, spills, discharges) or indirectly (e.g. leaching) released to the environment, released to land or the ocean, or released across different spatial and temporal scales is difficult. Plastics used in marine industries (e.g. aquaculture), lost in spills at sea, or directly discarded to the ocean (e.g. littering from ships or at the coast) are likely to represent a small proportion of the total plastics entering the ocean each year. Estimates currently place this at around 5–20%, indicating that the majority of marine plastic waste comes from land-based sources (e.g. Mehlhart and Blepp 2012; Zhou et al. 2011). This chapter addresses microplastic contamination, which is typically associated with more issues due to methodological difficulties, greater heterogeneity, and a lack of clear definitions. The proportion of plastic waste that is released to the environment in the micro-size range is essentially unknown. It is also expected that many plastic items may fragment into micro- or nano-sized particles when exposed to different environmental conditions, but this has not been demonstrated experimentally for many plastic polymer or product types or in a range of relevant environmental settings. Hence, the rates of particle release and associated particle size distributions are not well-understood. The upshot of this is that sources and pathways of microplastic to the terrestrial environment are typically poorly defined.

In recent years, more research has begun to focus on terrestrial environments in regard to microplastic contamination, although the total number of publications remains far below that for the marine environment. This section focuses on agricultural and urban environments, as settings that are likely to be important for the environmental release or impact of microplastic particles. Figure 4.1 presents some

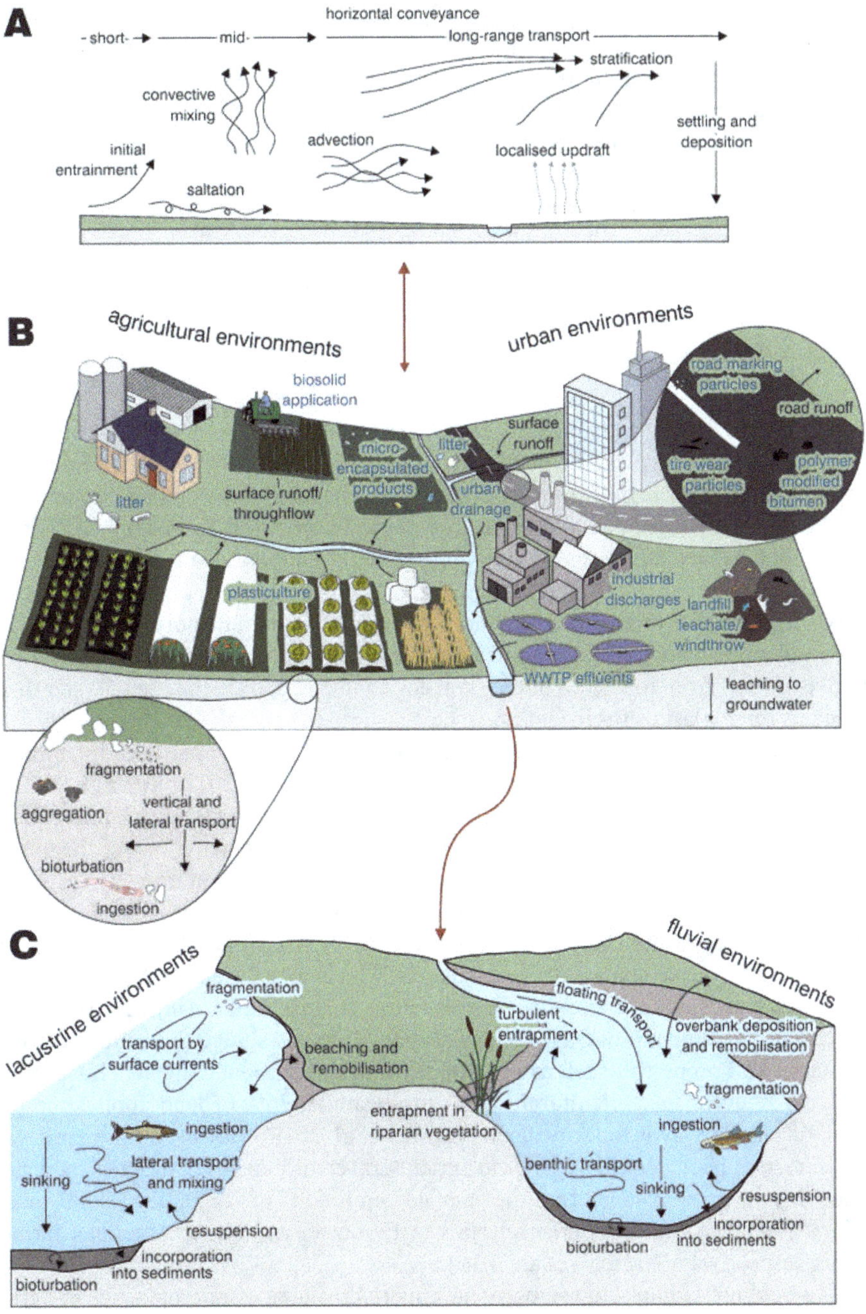

Fig. 4.1 Conceptual diagram showing important sources or release pathways of microplastic (blue text) and processes related to fate and transport of particles (black text) in atmospheric (**a**), terrestrial (**b**), and freshwater (**c**) systems

of the sources and release pathways of microplastics, as well as key processes associated with their fate and transport. In this chapter, sources refer to direct releases of microplastic to the environment – for example, the production of tyre wear particles during vehicle use – whilst release pathways describe processes or practices that release microplastic to the environment but are not the primary source. Land application of sewage sludge is one example of this, where microplastics in sludge are derived from the culmination of a diverse set of sources that occurs prior to environmental release.

4.2.1 Agriculture

Agricultural environments have recently been identified as recipients of considerable plastic debris, typically concentrated into micro- and nanoplastic size fractions (ECHA 2019). This results from the culmination of a wide range of different sources and release pathways of small plastic particles to farmed soil. These include (1) the application of sewage sludge-derived biosolids on land as a soil conditioner and fertiliser; (2) the release of small plastic fragments from an array of plastic products used in agriculture, termed plasticulture; (3) the use of synthetic polymers in micro-encapsulation technologies for agrochemicals and seed coatings; (4) the breakdown of plastic litter from roadsides or the farm environment; and (5) the use of water that contains microplastics for irrigation.

Many studies have reported the enrichment of microplastic particles in sewage sludge (e.g. Li et al. 2018; Lusher et al. 2018; Mahon et al. 2017; Xu et al. 2020). Wastewater treatment plants (WWTPs) receive small plastic particles from a diverse range of sources including households, industry, and stormwater. Many of the treatment processes employed to purify the water are also effective at trapping many of these small particles; reported removal efficiencies range between 64.4 (Liu et al. 2019a, b, c) and 99.9% (Magnusson and Norén 2014; Vollertsen and Hansen 2017). Much of the material that is retained in the WWTP is transferred to the solid sludge phase. One technique to handle the generation of this solid by-product from the wastewater treatment process is the application of treated sludge (biosolids) to land to amend soil properties such as pH, soil texture, and nutrient content. This is particularly relevant for agricultural land – for example, 76% of land application of biosolids in Norway is to farmed soils (Lusher et al. 2018). However, this results in the release of microplastic particles to agricultural environments. Estimates indicate that this is responsible for the annual emission of 63,000–430,000 and 44,000–300,000 tonnes of microplastics to European and North American farmlands, respectively (Nizzetto et al. 2016).

Plasticulture is likely to represent an important source of microplastics to agricultural soils. One of their main applications is mulching: thin films are placed above or below the ground to amend soil conditions, improve water use efficiency, and reduce pests and weeds with the aim of increasing crops yield. Further uses of thin films include greenhouse and tunnel systems and as wrappings for hay bales.

Additionally, plastic netting may be used in plasticulture systems to protect crops from pests. Microplastic debris may be formed during and after use as a result of environmental conditions and agricultural practices that promote degradation or mechanical fragmentation. Typically, it is difficult to remove 100% of these plastic products from the fields following use (Steinmetz et al. 2016). Geographic regions where high-intensity plasticulture converges with stronger environmental degradative forcing (e.g. higher solar insolation, increased variability between day and night temperature, higher humidity) are likely to represent hotspots for microplastic contamination from this source, such as in the Mediterranean agricultural belt and China (Espí et al. 2006; Liu et al. 2014; Scarascia-Mugnozza et al. 2012). Estimates associated with the release of microplastics from plasticulture are largely missing but may be as high as 24% of the total mass of the product when it is removed from the land (Dong et al. 2013). Future projections for Chinese farmlands estimate that mulching-derived plastic contamination in the soil may reach as high as 2000 kg hm^{-2} after 141 years of repeated application (Dong et al. 2013).

Advancements in agricultural technologies include the incorporation of synthetic polymers in the encapsulation of seeds and agrochemicals. This utilises polymerisation, coacervation, coating, and micro- and nano-encapsulation technologies to build a polymer matrix or thin coating, which may include non-biodegradable and insoluble plastic polymers such as polyethylene, polystyrene, polyurethane, or polyesters (França et al. 2019). Through this approach, fertiliser products are coated or encapsulated in a polymer shell which regulates the release of the active ingredient over a period of several months through the process of diffusion. Once the fertiliser is entirely released, the shell remains, representing a direct source of small plastic particles to the environment (Sinha et al. 2019). Seeds may also be coated in a polymeric film that incorporates germination-enhancing products such as fungicides or insecticides (Accinelli et al. 2019). The inclusion of these components into a film reduces the dispersion of agrochemicals that may otherwise be applied in powder form during sowing. The use of these technologies is expected to input between 5400 and 39,700 tonnes of microplastic to agricultural environments in the EU each year (ECHA 2019). Despite this, no published study has observed this release under field conditions.

Additional sources and release pathways for microplastic to agricultural environments include potential inputs derived from plastic litter, irrigation systems, and atmospheric deposition. Plastic litter within farms and from the surrounding environment may fragment due to environmental degradation, leading to the release of particles to soils. Atmospheric deposition, through windthrow of particles from adjacent systems or transport from further distances, may also introduce microplastic to farm environments. This is likely to be particularly relevant for low-density polymer types (Rezaei et al. 2019). Finally, irrigation has been proposed as a potential release pathway including the spreading of microplastic contaminated waters or through degradation of plastic pipe systems (Zhang and Liu 2018). Estimates for the release of microplastics from these sources are entirely unknown.

Only a small number of studies have thus far investigated microplastic loadings in agricultural environments. Table 4.1 presents the plastic exposures and reported

Table 4.1 Published, peer-reviewed studies of microplastics in agricultural systems

Location	Main plastic exposure *and other sources/ pathways*	Reported concentrations	References
China	Mulching film; *plastic litter*	Mean: 78 and 62.5 particles kg^{-1} in shallow and deep soils, respectively	Liu et al. (2018)
China	Mulching film; *irrigation, plastic litter*	Mean: 571 and 263 particles kg^{-1} in mulched and non-mulched fields, respectively	Zhou et al. (2019)
China	Mulching film	Mean: 80.3 ± 49.3, 308 ± 138.1, and 1075.6 ± 346.8 particles kg^{-1} in fields with 5, 15, and 24 years of continuous mulching, respectively	Huang et al. (2020)
China	Plastic films	Mean: 10.3 ± 2.2 particles kg^{-1}	Lv et al. (2019a)
China	Greenhouse system; *sewage sludge, irrigation*	7100–42,960 particles kg^{-1} (mean: 18,760)	Zhang and Liu (2018)
Chile	Sewage sludge	0.6–10.4 particles g^{-1} (approximately equivalent to 600–10,400 particles kg^{-1})	Corradini et al. (2019)
Spain	Sewage sludge	Mean: 5190 and 2030 particles kg^{-1} in fields with and without sludge application, respectively	van den Berg et al. (2020)
China	*Various: household sewage, textiles, plastic netting, plastic bags, roads*	320–12,560 particles kg^{-1} (mean: 2020)	Chen et al. (2020)
Germany	None; *windblown litter*	Mean: 0.34 ± 0.36 particles kg^{-1}	Piehl et al. (2018)

concentrations of these studies. The highest values are associated with fields undergoing multiple plastic exposure routes, greenhouse systems, sewage, irrigation, plastic litter, and proximity to roads, and are located in China (Chen et al. 2020; Zhang and Liu 2018). Fields that have undergone multiple applications of sewage sludge also present high soil microplastic concentrations (Corradini et al. 2019). Fields undergoing mulching with plastic films exhibit variable concentrations across two orders of magnitude (Huang et al. 2020; Liu et al. 2018; Lv et al. 2019a; Zhou et al. 2019), but this range may be linked to both the intensity and temporal frame of the plastic-cropping systems (Huang et al. 2020). Piehl et al. (2018) studied a farm in Germany that had no history of plasticulture or sludge application to soils. Despite this, low levels of microplastic contamination were observed, potentially derived from atmospheric deposition. The concentrations reported by Piehl et al. (2018) were several orders of magnitude below those reported by other studies. This suggests that agricultural practices involving plastics can significantly increase microplastic contamination in soils; however, several methodological disparities also exist between studies which may explain some of the observed variance. Differences in agricultural practices resulting from regional, seasonal, or crop-type variability may also result in large variations in soil microplastic concentrations. For example, different crops and environmental settings benefit from different forms of plasticulture and different countries implement a range of restrictions on the use of sewage sludge-derived biosolids.

Soil represents a complex matrix from which to isolate microplastic particles. Methods for analysing small plastic particle sizes are costly and time-consuming and require additional processing steps to clean up soil samples. For this reason, few studies examine the smallest microplastic size fractions, and, therefore, current assessments may represent an underestimate. Methods for analysing environmental *nanoplastic* contamination are largely non-existent, and complex soil matrices likely present an additional analytical challenge. Hence, there are no studies reporting nanoplastic contamination in soil environments, and so estimations of the contributions from agricultural products that contain or are expected to generate nanoplastics have not yet been possible. This is despite the possibility that nanoplastic particles could negatively influence soil functioning (Benckiser 2019).

The distribution of microplastic particles within agricultural environments is expected to be driven by a complex range of processes. Agricultural practices are likely to be highly relevant; in particular, the intensity and spatial scales associated with the use of plastic and plastic-containing products, the efficiency of plasticulture removal and waste handling, and the extent of ploughing or tilling of the land. This will govern the initial spread of particles across land and within soil profiles. Beyond this, processes related to wind erosion (Rezaei et al. 2019), bioturbation (Huerta Lwanga et al. 2017; Maaß et al. 2017; Yu et al. 2019), and water-mediated transport (Keller et al. 2020; O'Connor et al. 2019) are expected to transfer particles both within and from agricultural environments. Further research is necessary to document and quantify these processes under relevant field conditions.

4.2.2 Urban Environments

Urban environments can be expected to represent important domains for the release and cycling of plastic debris, based on the concentration of plastic production, consumption, and waste generation activities in these areas. Urban zones are characterised by higher population densities and may also comprise industrial areas that are involved in the production of plastics or manufacturing of plastic products. Releases of microplastics may include emissions from industry via air or water, the breakdown of larger plastic items (such as litter) due to environmental degradation or mechanical stress, and shedding from textiles such as clothing and home furnishings. Despite this, very few studies have thus far reported microplastic concentrations in samples from urban terrestrial environments. Fuller and Gautam (2016) identified poly(vinyl) chloride, polyethylene, and polystyrene microplastic in soils from an industrial area in Sydney, Australia, as part of a method development case study. Plastic debris was also noted in urban soil profiles from Stuttgart, Germany, but the size (macro-, meso-, micro-) of these particles was not described in detail (Lorenz and Kandeler 2005). Three studies have documented microplastics in urban dust samples from sites across Iran (Abbasi et al. 2017, 2019; Dehghani et al. 2017). Microplastic fibres and fragments were also observed at concentrations of approximately 2.9–166 particles g^{-1}, and spatial patterns of microplastic abundance were correlated with factors such as the location of commercial or industrial districts, population density, and traffic load. Particles exhibited a range of colours and morphologies, representing a heterogeneous mix of potential sources. Importantly, particles potentially derived from tyre rubber and other road sources dominated the samples. Several additional studies have pointed towards roads as important sources of microplastics (e.g. Kole et al. 2017; Sommer et al. 2018). Hence, this section will focus on road environments as a critical component of urban systems regarding microplastic contamination and releases.

4.2.2.1 Roads

Roads represent complex anthropogenic environments comprising artificial ground, a high degree of mechanical abrasion from vehicle tyres on the road surface, and emissions of a range of contaminants from exhaust, tyres, the road surface, and other debris. Runoff from road environments is typically characterised by high levels of particulates and may be contaminated by a range of heavy metals (e.g. zinc, copper, cadmium, nickel) and organic pollutants (e.g. polycyclic aromatic hydrocarbons, organophosphates, octylphenols, phthalates) (Grung et al. 2017; Hallberg et al. 2014; Meland et al. 2010a, b; Meland and Rødland 2018). Road runoff has received renewed interest in recent years due to the presence of particles with polymer components; it has been identified as one of the largest sources of microplastic particles in the environment (Baensch-Baltruschat et al. 2020; Kole et al. 2017; Wagner et al. 2018). In particular, particles created by the wear and tear of car tyres are estimated to be the single largest source of microplastics in several countries, such as Norway, Sweden, and Denmark (Lassen et al. 2015; Magnusson et al. 2016; Sundt et al. 2014; Sundt et al. 2016; Vogelsang et al. 2019); although, these estimations are based on emission factors and need to be supported by peer-reviewed experimental or environmental evidence. Similar estimations using emission factors have been conducted in China, estimating that close to 55% of all primary microplastic emissions are derived from tyres (Wang et al. 2019a, b). The authors also compared their emissions to Norway and Denmark and calculated that the release in China is 85 times higher than in Norway and 400 times higher compared to Denmark. For this review we use the term road-associated microplastic particles (RAMP), first introduced in Vogelsang et al. (2019). RAMP comprises several categories of particle types: tyre-wear particles (TWP), road-wear particles from polymer-modified bitumen (RWP_{PMB}), and road-wear particles from road marking (RWP_{RM}) (Vogelsang et al. 2019). The RAMP terminology differs from the tire and road wear particle (TRWP) terminology, used in several other studies (Baensch-Baltruschat et al. 2020; Klöckner et al. 2019), by only including particles with plastic components whilst TRWP may also include particles without plastic components.

In general, very little research has thus far been conducted on RAMP. It is expected that large particles from road runoff will accumulate at the roadside or be captured in gully-pots, whilst smaller particles have the potential to spread further and be transported with the runoff (Vogelsang et al. 2019). However, these assumptions are based on the behaviour of road runoff particles in general, and further research on the emission of RAMP from road environments is needed.

A small number of studies have attempted to measure concentrations of TWP in the terrestrial environment. Fig. 4.2 presents the range of concentrations that have thus far been reported (Baensch-Baltruschat et al. 2020; Bye and Johnsen 2019; Rødland et al. 2020; Wik and Dave 2009). Tyres are composed of a complex mix of ingredients including natural and synthetic rubbers, mineral oils, fillers, antioxidants, and antiozonants (Wik and Dave 2009). Hence, they are difficult to quantify using chemical analyses. Nearly all the current studies have used tracers to estimate the concentration of TWP, such as different benzothiazoles and zinc that are present in tyres, whilst some others have instead measured concentrations of tyre-related polymers. The most commonly studied matrices include road dust from road

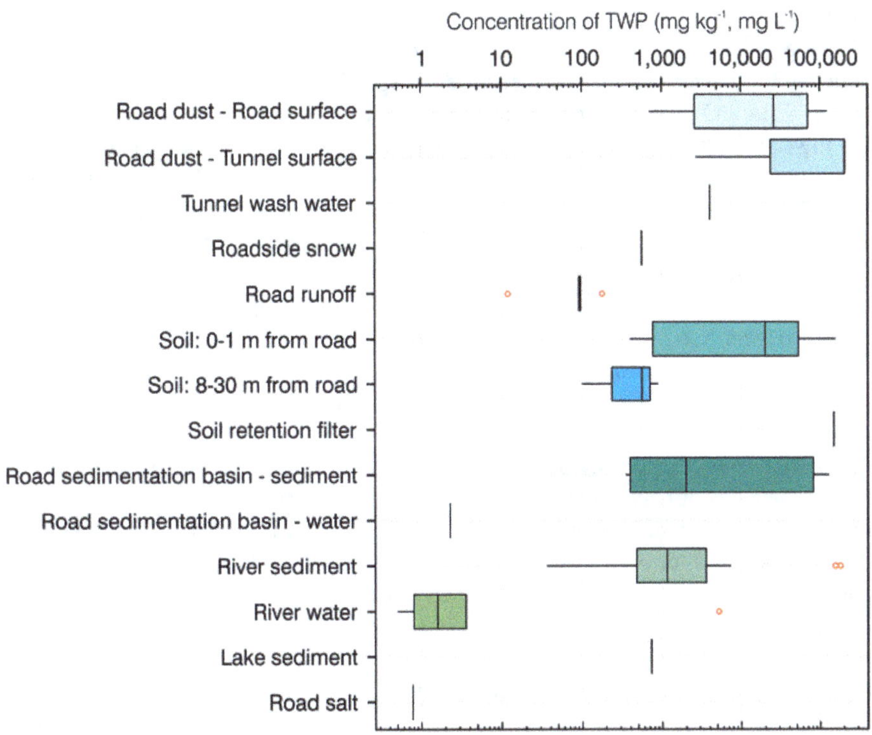

Fig. 4.2 Reported concentrations of TWP in terrestrial and freshwater environments. Each data entry is a mean value and red dots represent outliers. The figure summarizes a number of different studies from 1997 to 2020. (These are further described in Baensch-Baltruschat et al. 2020; Bye and Johnsen 2019; Rødland et al. 2020; Wik and Dave 2009)

surfaces, roadside soils, sediments from sedimentation basins, and river sediments. The studies have a wide geographic spread and represent different traffic volumes. Moreover, they employ a wide range of different analytical approaches, so comparisons between studies should be approached with caution (Rødland 2019).

Concentrations of TWP in the environment vary by several orders of magnitude (Fig. 4.2). This variability has a spatial component, related to proximity to the road environment and different environmental matrices. For example, concentrations of TWP in road dust differed between road surfaces outside tunnels (700–124, 000 mg kg^{-1}) and inside tunnels (2700–210,000 mg kg^{-1}) (Wik and Dave 2009). Moreover, Bye and Johnsen (2019) measured TWP in tunnel wash water from Smestad tunnel in Oslo (annual average daily traffic (AADT): 66,322) and found 4038 mg kg^{-1} of TWP. This corresponded to the accumulation of TWP since the previous tunnel wash – a period of 60 days – and a production of nearly 3 kg of TWP per day. Concentrations for road runoff material also ranged between an order of magnitude: 12–179 mg kg^{-1} (Wik and Dave 2009). In roadside soil, the highest concentrations were found closest to the road, ranging between 400 and 158,000 mg kg^{-1} at 0 m, with considerably lower concentrations from 1 to 30 m from the road (0–900 mg kg^{-1}) (Baensch-Baltruschat et al. 2020; Wik and Dave 2009). One study also demonstrated the accumulation of TWP in roadside snow (563 mg L^{-1}; Bauman and Ismeyer 1998). In road sedimentation basins, the highest concentrations were found in the sediments, 350–130,000 mg kg^{-1} (Klöckner et al. 2019; Wik and Dave 2009), and lower concentrations were found in water, 2.3 mg L^{-1} (Wik and Dave 2009). One study has looked at the retention of TWP in a soil retention filters and reported a concentration of 150,000 mg kg^{-1} (Klöckner et al. 2019). A study of microplastic particles in road de-icing salt (Rødland et al. 2020), used in areas with cold winter climate to ensure traffic safety, also found TWP in the salt, coming from the production sites and/or roads nearby the salt collection sites. However, the concentrations are very low compared to the contribution from tire wear itself, only 0.77 mg kg $^{-1}$ salt.

These data include some discrepancies in the analytical approach. For example, Eisentraut et al. (2018) used thermal extraction desorption gas chromatography-mass spectrometry (TED-GC-MS) to measure the amount of styrene-butadiene rubber (SBR), a synthetic polymer, from sediments in a road runoff treatment. They recorded values of between 3.9 and 9.3 mg g^{-1} SBR in their samples. The approximate concentration of SBR in common tyres is 11.3% (Eisentraut et al. 2018). This probably varies a lot between different tyre brands and types of tyres (e.g. summer and winter tyres, studded and non-studded tyres); however, it can be used to calculate the concentration of TWP. This gives a result of 34.5–82.0 mg kg^{-1} for runoff sediments.

In addition to these studies, rubber particles potentially derived from tyre wear have been reported for snow samples from several sites in the Swiss Alps, Bremen, and Svalbard and from ice floes in the Fram Strait (Bergmann et al. 2019). This implies that TWP may be mobilised by atmospheric transport processes. However, the methods used in the study could not provide confirmation of TWP occurrence, and suspected particles were reported based upon their morphology.

Other components of RAMP are wholly under-researched. A single peer-reviewed publication reported RWP_{RM} in sediments from the River Thames, UK (Horton et al. 2017a). No studies have recorded their occurrence in terrestrial samples. Furthermore, no studies have thus far measured the concentrations of RWP_{PMB} alone in the environment. More research on the occurrence of RAMP in the terrestrial environment is needed to establish the relative contributions from different components and identify the scale of emissions from road environments.

4.2.3 Occurrence of Microplastics in Terrestrial Organisms

Very few studies have thus far reported the occurrence of microplastic particles in terrestrial organisms. Entanglement and incorporation of plastic into nests have been reported for both urban and agricultural crows in California, USA (Townsend and Barker 2014). Anthropogenic material, most commonly composed of plastic, was observed in 85.2% of nests, but this was typically in the size range of meso- or macro-plastic. Ingestion of microplastics by terrestrial birds in Shanghai, China, has been reported by Zhao et al. (2016), where plastic fibres and fragments accounted for 62.6% of litter items identified in digestive tracts. Carlin et al. (2020) observed an average of 6.22 microplastic particles present in the gastrointestinal tracts of birds of prey from central Florida, although many of these were identified to be rayon, which is sometimes excluded from microplastic counts as it is not a true synthetic polymer. It has been noted that microplastics now appear to be ubiquitous in the gut contents of bird species (Holland et al. 2016). Only a single study has documented the occurrence of microplastics in a terrestrial macroinvertebrate under field conditions. Panebianco et al. (2019) observed concentrations of 0.07 ± 0.01 particles g^{-1} tissue in three species of edible snails (*H. aperta*, *H. aspersa*, and *H. pomatia*) from Italy. Despite the current paucity of data on the uptake of microplastics by terrestrial organisms, numerous laboratory studies have demonstrated ingestion of microplastics by a range of species and have investigated related impacts. This is addressed in more detail in Chap. 8: *Ecotoxicology of Plastic Pollution*.

4.3 Pathways to Freshwater Environments

Most inputs of microplastics to freshwater systems can be characterised as *release pathways* – emissions are typically not direct *sources* and have instead travelled through other systems first. There are a small number of sources of microplastic to freshwater environments, such as the in situ fragmentation of plastic litter, point discharges from plastic industries, and the generation of micro-sized particles of polymeric paint or plastics from boats or other aquatic infrastructures. This section describes five key pathways to freshwater environments: spread from agricultural environments, releases from littering and landfill leachate, discharges from urban drainage systems, road runoff, and WWTP effluents.

4.3.1 *Transfers from Agricultural Environments*

Agricultural soils have been highlighted as potentially highly significant reservoirs of microplastics, which may actually exceed loadings currently observed in the global ocean (Nizzetto et al. 2016). The potential for agricultural microplastic contamination to propagate across wider spatial scales is, therefore, of particular interest in terms of global microplastic patterns and cycling. Processes such as windthrow, surface runoff, throughflow, and leaching are likely to be relevant for the transfer of plastic particles from soil systems, dependent upon factors such as particle size, morphology, and surface charge (Hurley and Nizzetto 2018). A small number of studies have demonstrated some of these processes experimentally within soil profiles (e.g. Keller et al. 2020; O'Connor et al. 2019), but, thus far, no published, peer-reviewed study has quantified the release of microplastics from soils and, especially, agricultural environments. A report from Ranneklev et al. (2019) presented preliminary data of microplastics in water discharged from a field amended with sewage sludge into a sedimentation pond connected to a stream in Norway. Approximately 2 particles L^{-1} were observed in the discharge water; however, the flow of discharge water from the field was not quantified. Nevertheless, this indicates that agricultural soils represent a release pathway for microplastic to freshwater systems. Based on the potential scale of microplastic contamination associated with agriculture, further research is urgently required to quantify transfers to the wider environment.

4.3.2 *Transfers from Urban Environments*

4.3.2.1 Littering and Leaching of Plastic Waste

Estimates for the mismanagement of plastic waste have been used as a means of assessing emissions of plastic to the marine environment (e.g. Jambeck et al. 2015). The transport pathways connecting this land-based release of plastic to the oceans are described as inland waterways, wastewater outflows, and wind action. Hence, the transfer of litter from populated or industrial areas to freshwater systems is often assumed. The generation of litter can be from littering practices or accidental releases during stages of waste handling, such as municipal waste management (Kum et al. 2005; Muñoz-Cadena et al. 2012). This litter may already be in the size range of microplastic particles or may act as a source of microplastics through the breakdown of larger plastics into micro-sized fragments. This fragmentation can be caused by weathering processes that chemically alter and weaken plastic polymers or through mechanical abrasion. Movement via water is likely to be an important process for the transport of litter from urban terrestrial environments to nearby freshwater systems and is addressed in more detail in Sect. 4.3.2.2. Wind action has also been identified as an important process distributing microplastics around the environment and is described in Sect. 4.5.

An additional release pathway associated with this form of (micro)plastic contamination is landfill leachate. Thus far, two published, peer-reviewed studies have reported concentrations of between 0.42 and 24.6 particles L^{-1} in leachate from municipal solid waste landfills in China (He et al. 2019; Su et al. 2019). Fewer particles were detected in older landfill systems, which has been linked to increased consumption and disposal of plastic in recent years (Su et al. 2019). Leachate discharges may emit microplastics to nearby soils or to freshwater systems.

4.3.2.2 Urban Drainage

Urban drainage systems designed to handle surface water runoff during precipitation events represent a key pathway linking urban and freshwater environments. Larger plastic items, typically litter, are often captured by drainage systems, and a body of research exists around documenting this process and engineering solutions to reduce blockages or prevent release into waterways (Armitage 2007; Armitage et al. 2001; Armitage and Rooseboom 2000; Marais et al. 2001, 2004). As described above, urban environments are expected to represent hotspots for microplastic contamination. Urban drainage is likely to act as a conduit for these particles to enter freshwater systems.

Several studies identify combined sewer overflows (CSOs) as a potentially important source of microplastics to freshwater systems (Ballent et al. 2016; Eriksen et al. 2013; Hurley et al. 2018), but very little work has attempted to quantify the scale of release or investigate the composition of microplastics. UNEP (2009) specifically identify sewage treatment and CSOs as one of the eight key land-based sources of marine litter, highlighting the important role that they are expected to play as a pathway for particles from urban environments. CSOs are a feature in many urban drainage systems; they allow for the direct release of untreated wastewater during periods of increased precipitation to prevent the system from backing up. Dris et al. (2018) sampled three CSOs in Paris, France, during a storm event. Very high levels of synthetic fibres (up to 190,000 fibres L^{-1}) and fragments (up to 3100 fragments L^{-1}) were reported. These results were higher than those observed for wastewater and stormwater alone, and it was suggested that this could be due to the accumulation of particles within the system during dry weather periods, which may then be resuspended once the CSOs are activated (Dris et al. 2018). This would represent a pulse of very high concentrations of microplastic released into rivers during storm conditions.

Microplastics have also been observed in stormwater ponds (Liu et al. 2019a, 2019b; Olesen et al. 2019). These receive runoff from a range of urban environments and aim to retain particles. The role of these systems in conveying microplastics to recipient water bodies – often freshwater systems – has not yet been quantified.

These ponds do show potential for accumulating microplastic particles in their sediments, which may act as a temporary sink (Liu et al. 2019b). However, a diverse range of polymer and particle types has been reported including many which might not be expected to settle out to sediments.

4.3.2.3 Road Runoff

As discussed in Sect. 4.2.2.1, there are a limited number of studies that have investigated RAMP, and this includes the pathways from land to freshwater systems. The highest concentrations of TWP in environmental samples are found close to the road environment: from road dust and soil at the roadside. Lower concentrations have been reported from environments further from the road (Fig. 4.2). Values for freshwater environment vary considerably: between 36 and 179,000 mg kg^{-1} for river sediments and 1.6 and 36 mg L^{-1} for surface water (Baensch-Baltruschat et al. 2020; Unice et al. 2013; Wik and Dave 2009). However, there may be large variations in river water as well, especially due to the input during rainfall, as seen in the study by Kamata et al. (2000), where they reported TWP concentrations of 2200–5200 mg kg^{-1} during a storm flow. However, this demonstrates that transfers to freshwater systems do occur. Comparing the concentrations found in river water and water from sedimentation ponds to river sediments and sediments from sedimentation ponds, the current data provide clear indications that TWP will accumulate in the sediments. Additionally, transport of TWP from the road to freshwater is expected to be limited in areas where there is soil, sediment, and vegetation between the road and the recipient water body, as TWP is more likely to be retained.

Direct releases from the road environment to freshwaters are likely to occur. For example, particles may be released from bridges passing over freshwater via splashing or direct water outlets. Additionally, some larger roads have in-built drainage systems that collect and transport road runoff material and release it directly into a freshwater recipient. Many of these systems employ gully-pots which are expected to trap larger particles, but they rarely include sedimentation ponds or treatment systems to remove particulate and contaminants from runoff waters. The retention of TWP and other RAMP constituents in gully-pots is expected to be limited (Vogelsang et al. 2019); hence, direct discharges to freshwater environments are likely to represent an important pathway for RAMP release. Tunnel wash water represents a third direct discharge pathway. All road tunnels are washed regularly in order to maintain traffic safety and to avoid damage to technical instruments. The frequency of these tunnel wash events differs, usually determined by the number of cars passing through per day (AADT). In between these wash events, the tunnel is typically dry, and, therefore, there is a limited release of runoff from the tunnel. During a tunnel wash, large volumes of water are used, and this is collected by the tunnel drainage system. In some cases, the water passes through sedimentation and filtration treatment systems, but in other instances, it may be directly discharged to receiving water bodies. For example, Norway has over 1200 road tunnels, the third highest value globally (Vegkart, 2020). Only a small fraction of these tunnels receives any kind of treatment of the tunnel drainage water, and most release the runoff directly into a water recipient (Rødland and Helgadottir 2018). Several studies have documented high levels of particulate matter in tunnel wash water (Hallberg et al. 2014; Meland et al. 2010a, b; Meland and Rødland 2018); however, there are very few that investigate concentrations of TWP or other microplastics in this matrix. It is expected that tunnel wash waters may be highly enriched in RAMP. This is an area of research that should be addressed in future studies.

4.3.2.4 Wastewater Treatment Plant (WWTP) Effluents

Wastewater treatment plants (WWTPs) typically discharge treated effluents directly into recipient water bodies. This has the potential to represent an important release pathway for microplastics to freshwater environments. As discussed in Sect. 4.2.1, WWTPs receive microplastics from a diverse range of sources. Many of these are captured by wastewater treatment processes and transferred to the sewage sludge phase (64.4–99.9%). Despite this, total discharges from WWTPs are significant, and so this remaining percentage is expected to represent a significant release pathway for microplastics across temporal scales (Carr et al. 2016; Lv et al. 2019b; Sun et al. 2019).

Several studies have now provided an estimation of this release. The average concentration of microplastic in treated effluents generally falls below 1 particle L^{-1}; however, large WWTPs can process several million litres of wastewater each day, resulting in daily emissions that are significant (Gatidou et al. 2019; Sun et al. 2019). For example, Mason et al. 2016a, b estimated that on average approximately four million microplastic particles were released each day from a single plant, in a study of 17 WWTPs in the USA. This study had a lower size limit of detection of 125 µm. Studies that go below this report far higher concentrations in the lowest size fraction (<20 µm), so the number of particles that are released can be expected to be far higher (e.g. Simon et al. 2018). The release of microplastics by WWTPs is further supported by studies that have documented significant increases in fluvial microplastic concentrations downstream from WWTP effluent releases (e.g. Estahbanati and Fahrenfield 2016; Kay et al. 2018; Vermaire et al. 2017). The release of microplastics from WWTPs can be expected to vary through time and space. Smaller, rural WWTPs process far smaller volumes of wastewater per day but may also have low trapping efficiencies for microplastic particles (e.g. Wei et al. 2020). It is also of interest to capture the influence of precipitation events on microplastic release by WWTPs, where plants may release pulses of untreated wastewater into recipient waterbodies. Moreover, approximately 80% of the world's wastewater is emitted without sufficient treatment (UNESCO 2017), for which the potential microplastic release is unknown.

Fibres are commonly reported to be the dominant particle type present in WWTP effluents (Mason et al. 2016a, b; Ruan et al. 2019; Yang et al. 2019). A single garment can shed more than 1900 fibres per domestic laundry wash, resulting in >100 fibres L^{-1} of laundry effluent (Browne et al. 2011). Many different treatment steps are employed by WWTPs globally, with capture rates ranging from 0 to 99.9% (Zhang and Chen 2020). The efficacy of these clean-up steps is expected to differ across the spectrum of particle sizes and morphologies (Carr et al. 2016; Lusher et al. 2019; Sun et al. 2019). This is particularly relevant for fibrous particles, which have been noted as the most challenging microplastic type for removal in wastewater due to their high length to width ratio and potential to curve and bend (Ngo et al. 2019). In some cases, the presence of microplastics can actually exhibit a negative impact on the efficiency of wastewater treatment processes (Zhang and Chen 2020), further complicating their removal.

It is important to note that comparing studies of microplastic releases from WWTP is challenging, as the methodologies used in each study, from sample collection, sample processing, and the size range of particles may differ substantially, and no harmonised methodology has yet emerged. There are also different approaches to sampling microplastic particles in streams receiving WWTP effluents, many of which may not capture very small microplastic particles. Excluding the smaller fraction of microplastics from studies may result in underestimating the microplastics released from WWTPs as they may be less likely to be captured by treatment processes (Lusher et al. 2019). Further research is required to quantify the scale of microplastic release by WWTP effluents.

4.4 Microplastics in Freshwater Systems

4.4.1 Microplastics in Rivers

Fluvial systems comprise running bodies of water that connect terrestrial, lacustrine, glacial, and marine environments. They represent important long-range transport pathways and act as conduits for suspended sediments and contamination through the landscape. Rivers and streams are expected to be highly complex and dynamic regarding the accumulation and transfer of microplastic particles. As has been established in the previous section, fluvial environments are connected, with many sources and release pathways for microplastic particles, including both point and diffuse releases across different spatial and temporal scales. Numerous studies have now documented microplastic contamination in rivers or streams (Scherer et al. 2020).

Variation is observed in the microplastic contamination reported within river systems. This spatial and temporal heterogeneity is not common across all studied catchments. These differences point towards some of the complexity associated with river systems. For example, some studies report a common longitudinal pattern of increasing microplastic concentrations with distance downstream (e.g. Jiang et al. 2019; Shruti et al. 2019), which likely represents a culmination of microplastic sources and pathways. In contrast, other studies show a less clear-cut pattern of microplastic abundance, especially in highly urbanised systems (e.g. Hurley et al.

2018). River sediments, on average, present higher concentrations than overlying waters (Li et al. 2020) and may act as a temporary store for microplastic particles (Castañeda et al. 2014). However, relating microplastic abundance in moving waters to that in more static sediments is complicated.

The different methodological approaches to sampling fluvial environments make data comparability challenging. Many variations exist between samples; for example, methods of sampling (sediment grabs, sediment cores, nets, pumps, etc.), sample matrices (surface waters, water column, sediments, etc.), particle size fractions, laboratory analytical methods (sample purification, density separation, pore sizes of filtration approaches, etc.), and reporting units (particles m^{-2}, m^{-3}, L^{-1}, kg^{-1}, etc.) (Blettler et al. 2018). Table 4.2 presents a selection of studies reporting fluvial microplastic contamination that utilise a range of different approaches. Based on this degree of discrepancy between methodologies, it is difficult to partition the observed differences in reported concentrations between methodological and environmental factors. This is further hampered by the wide range of potential controls

Table 4.2 Selected studies of microplastic contamination in river systems that utilise a range of different sampling and analytical approaches.

Location	Matrix	Sampling method	Reported concentrations	Sample volume	Particle sizes	References
Rivers in Tibetan Plateau, China	Surface water	Bulk water	483–967 particles m^{-3}	30 L	>45 μm	Jiang et al. (2019)
Pear River, China	Surface water	Plankton net	0.57–0.71 particles L^{-1}	18,860–138,134 L	160 μm–5 mm	Fan et al. (2019)
Rhine River, Europe	Surface water	Manta trawl	892,777 particles km^{-2}	4634 m^3	300 μm–5 mm	Mani et al. (2015)
Antua River, Portugal	Surface water	Surface water pump	58–193 particles m^{-3}	*n.r.*	55 μm–5 mm	Rodrigues et al. (2018)
Marne River, France	Surface water	Manta trawl	5.7–398 particles m^{-3}	*n.r.*	80 μm–5 mm	Dris et al. (2018)
Pearl River, China	Sediment	Grasp bucket	685 particles kg^{-1}	*n.r.*	>100 μm	Fan et al. (2019)
Beijiang River, China	Sediment	Stainless steel shovel	178 ± 69–544 ± 107 particles kg^{-1}	Triplicates of 30 g per site, 8 sites	1 μm–5 mm	Wang et al. (2017)
Thames River, UK	Sediment	Stainless steel scoop	18.5 to 66 particles 100 g^{-1}	*n.r.*	1–4 mm	Horton et al. (2017a, b)
Antua River, Portugal	Sediment	Van Veen Grab	100 to 629 kg^{-1}	*n.r.*	55 μm–5 mm	Rodrigues et al. (2018)

on microplastic release and distribution in river systems including varying hydrological and geomorphological conditions, density and proximity of sources and release pathways, catchment characteristics and land use, anthropogenic modifications such as dams, and seasonal variability in microplastic releases and river characteristics (Blettler et al. 2018; Mai et al. 2019; McCormick et al. 2016). Furthermore, it has been suggested that existing assessments of riverine microplastic contamination may miss significant variability due to their selected spatial and temporal scales (Stanton et al. 2020).

In dynamic systems such as rivers, there is a need to examine how microplastic distributions change across spatial and temporal scales and in response to different controls. Figure 4.1 presents some of the processes likely to be relevant for microplastic transport in river systems. For example, microplastics can settle in riverbeds but may be resuspended during high energy events, such as floods, and transported further along the river (Hurley et al. 2018). The extent and controls of this remobilisation are essentially unknown (Alimi et al. 2018), and only a small number of studies have begun to investigate important hydrological controls on particle retention and transport (e.g. Ockelford et al. 2020). It is assumed that the smaller the microplastic particle, the lower its retention in river systems based upon the lower flow velocities required for entrainment (Besseling et al. 2017). Connectivity between river channels and the overbank zone during flood events may lead to deposition or mobilisation of microplastic particles. Seasonal variability is also likely to play an important role in some systems. Watkins et al. (2019) identified that hydrological differences between spring high flow and summer low flow were the dominant factor determining microplastic concentrations in two streams in New York, USA. These studies suggest that the hydrodynamics of the river strongly impact microplastic distributions and emissions to the marine environment (Besseling et al. 2017; McCormick et al. 2016). Furthermore, although more attention is typically directed to larger river systems, smaller streams should also be investigated as in many cases they are the primary interface between land, usage of plastics, and drainage networks (Dikareva and Simon 2019). Microplastic contamination of headwater streams has been reported (Hurley et al. 2018), demonstrating the pervasive nature of fluvial microplastic contamination.

4.4.2 Microplastics in Lakes

Microplastics were first recorded in a lake environment in 2012 (Faure et al. 2012). Since then, 36 additional published studies have investigated the occurrence of microplastics in the waters or sediments of lakes, globally. The majority of these studies can be broadly grouped into three key locations: Great Lakes system, European lakes, and Chinese lakes (Fig. 4.3). Lakes may receive microplastic particles from a wide range of potential sources or release pathways, including WWTP effluents (Uurasjärvi et al. 2020), industrial discharges (Eriksen et al. 2013), fisheries (Wang et al. 2018; Yuan et al. 2019), and inflowing rivers (Ballent et al. 2016;

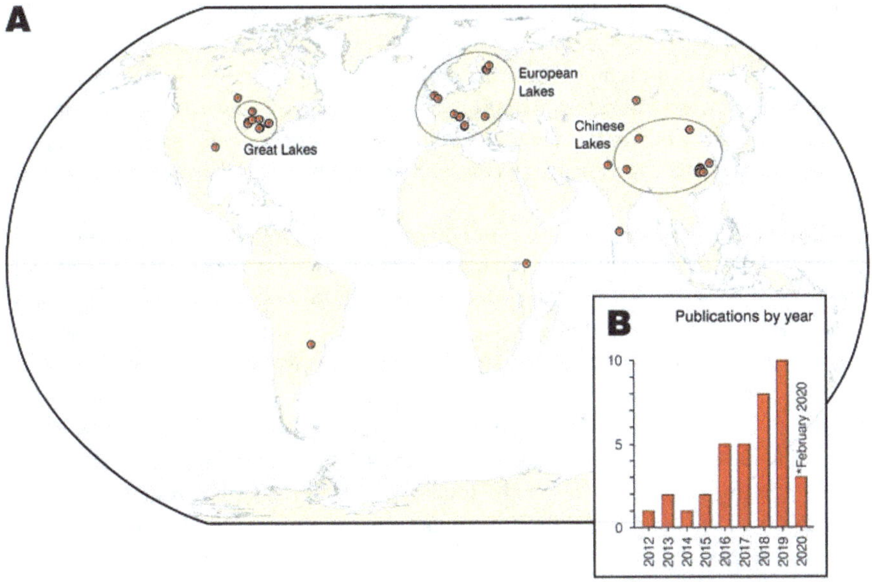

Fig. 4.3 Map of published, peer-reviewed studies of microplastic occurrence in lake waters or sediments (**a**) includes the broad geographic grouping of the three key areas for lake microplastic research. The total number of studies published each year is shown in **b**, showing results up to February 2020

Corcoran et al. 2015). They represent complex environmental systems that have the potential to transport, disperse, or accumulate particles according to an array of different processes (Fig. 4.1).

Methodological disparities complicate efforts to compare findings from different studies. This includes the field sampling procedure, which has been shown to result in significant differences between reported microplastic concentrations for different sampling apparatus types (Uurasjärvi et al. 2020). Further variation in the particle sizes classes analysed introduces additional uncertainty. The lakes investigated also represent a spectrum of lake and catchment sizes and types. Hence, it is difficult to determine whether differences between studies are mainly derived from the methodological approach or to a higher extent relate to environmental factors. Despite this, variability between spatial or temporal concentrations within single studies that apply one methodology indicates that environmental factors are important in governing levels of microplastic contamination (Nan et al. 2020; Scherer et al. 2020). It is notable that despite possible methodological difference, the presence of microplastics has been reported in all lakes studied thus far, even in remote locations (Free et al. 2014; Zhang et al. 2016a, b).

Assessments of surface waters report concentrations ranging from 0.21 (Fischer et al. 2016) to 34,000 particles m^{-3} (Yuan et al. 2019). Low-density plastic types such as polyethylene and polypropylene are commonly reported as the dominant microplastic types (Sighicelli et al. 2018; Wang et al. 2018; Xiong et al. 2018). This concurs with the expected buoyant properties of these polymer types. Fibres are also commonly reported (Anderson et al. 2017; Wang et al. 2018), despite typically being composed of polymer types that are denser than water. Several studies show high concentrations close to population centres and point towards the surface currents generated by prevailing winds as an important process governing microplastic distributions at the water surface (Fischer et al. 2016; Free et al. 2014; Migwi et al. 2020).

Lake sediment concentrations vary between studies. In smaller lakes, concentrations of microplastics in lake sediments have been shown to reflect processes influencing surface water distributions, such as prevailing wind or proximity to inputs, suggesting that denser polymer types undergo transport through the lake prior to sedimentation (Vaughan et al. 2017). In larger lake systems, a lack of correlation between surface water and sediment concentrations is often reported (Yuan et al. 2019), but this may be due to the spatial resolution of sampling campaigns. River tributaries have been identified as a depositional environment for microplastics as energy conditions change during the transition into the lake environment (Ballent et al. 2016). Lenaker et al. (2019) demonstrated that partitioning between surface waters, sub-surface waters, and sediments occurred at a density threshold of 1.1 g cm^{-3} in a North American freshwater lake system. Despite this, low-density polymer types are sometimes observed in lake sediments (e.g. Sruthy and Ramasamy 2017). This is contrary to the expected buoyancy of these particles but may be explained by processes such as biofouling that increases particle bulk density (Chen et al. 2019). Lake sediments also have the potential to accumulate and preserve microplastic particles through processes of sedimentation and burial. This has been reported for Lake Ontario (Corcoran et al. 2015); Hampstead Pond, London, UK (Turner et al. 2019); and Donghu Lake, Wuhan, China (Dong et al. 2020).

4.4.3 Occurrence of Microplastic in Freshwater Organisms

To assess the status of freshwater systems, there has been a long tradition to use macroinvertebrates as indicator species. They represent a diverse group of organisms that show tolerance and sensitivities towards different stressors, present different feeding strategies, inhabit different environments, and have a range of lifespans (including long life cycles allowing for accumulation of contaminants). Research has now begun to investigate macroinvertebrates as a measure for microplastic contamination.

The majority of studies on freshwater macroinvertebrates has been conducted in the laboratory to measure the ecotoxicity of different polymers and particles types. Only a comparatively small number of studies have documented the occurrence of microplastics in macroinvertebrates in environmental samples (Akindele et al. 2019; Hurley et al. 2017; Nan et al. 2020; Nel et al. 2018; Su et al. 2018; Windsor et al. 2019a). These investigate a range of organisms including mayflies, caddisflies, gastropods, clams, and shrimp. All of the 20 different investigated species of six classes: Insecta, Clitellata, Bivalvia, Amphibia, Malacostraca, and Gastropoda were found to contain microplastic particles. The investigated sites covered lakes, river deltas, urban rivers, and small streams. Concentrations vary from 0.07 to 5 particles individual^{-1}, 0.01 to 0.042 particles mg^{-1} d.w., and 0.0003 to 1.12 particles mg^{-1} w.w (Akindele et al. 2019; Hurley et al. 2017; Nan et al. 2020; Nel et al. 2018; Su et al. 2018; Windsor et al. 2019a). At present, there is insufficient data to draw conclusions regarding differences in microplastic uptake based on different feeding traits, trophic interactions, or microplastic particle size. However, it has been suggested that generalist species are more likely to ingest microplastic than predators (Scherer et al. 2018) and that non-selective feeders are more likely to ingest microplastic particles than selective feeders (Scherer et al. 2017). Fibres are the most commonly reported particle type (e.g. Akindele et al. 2019; Hurley et al. 2017; Nan et al. 2020; Su et al. 2018), but it is not known whether this reflects the feeding behaviour of macroinvertebrate species or the dominant particle type present in the local environment.

These studies have thus far applied several different methods for separating microplastic from organisms. These include alkaline hydrolysis (KOH or NaOH), digestion with hydrogen peroxide, acid digestion (HNO_3), and combined approaches (KOH + H_2O_2). Some of the studies rely on visual assessment of microplastic particles only – two thirds verify a subsample of particles using chemical analytical techniques. Since most macroinvertebrates are small, and the majority of their food items are thereby also small, a visual analysis may not be sufficient to capture the full-size spectrum of plastic particles that may be ingested. There is also an absence of quality assurance and quality control measures such as including both blank and spiked samples in many of the studies, emphasising the need for methodological improvements.

A larger number of studies have reported the occurrence of microplastic in freshwater fish (Andrade et al. 2019; Biginagwa et al. 2016; Horton et al. 2018; Jabeen et al. 2017; Phillips and Bonner 2015; Sanchez et al. 2014; Silva-Cavalcanti et al. 2017). Over 50 species of fish have been analysed for microplastic ingestion under field conditions thus far. This has mostly been documented for riverine specimens, but studies have also been conducted in lakes (Biginagwa et al. 2016) and a stormwater pond (Olesen et al. 2019). Concentrations range from 0 to 65 microplastic particles individual^{-1}. It is important to note, however, that most studies only investigate the gut content so total concentrations are not known. Although, it is expected that only small microplastic particles have the potential to pass gut membranes. A single study found no plastic particles in the gut contents of fish: northern pikes, roach, and bream from Lake Geneva (Faure et al. 2012). The majority of studies perform a visual examination of the gastrointestinal tract or digest the gut contents using KOH or H_2O_2; however, small fish may be freeze-dried and digested (e.g. Olesen et al. 2019), and one study has also documented the occurrence of polyethylene and polystyrene in liver samples digested using sodium hypochlorite (e.g. Collard et al. 2018). Some trends have been reported. For example, McGoran et al. (2017) found that benthic-feeding fish ingested more microplastics (75%) than pelagic-feeding fish (20%) in the Thames River. Moreover, Horton et al. (2018)

observed that microplastics in gut content was positively correlated with fish size, which is in turn typically associated with sex. The reported microplastic burden on freshwater fish species may also represent an underestimate due to the lower size limit of the studies; Roch et al. (2019) reported that the majority of microplastic ingested may be below 40 μm. This is supported by the high concentrations reported when using high-resolution μFT-IR imaging methodologies (e.g. Olesen et al. 2019).

Thus far, there are no studies that document the occurrence of RAMP in organisms under field conditions. Several studies have demonstrated the uptake of hazardous compounds associated with tyres during laboratory toxicity testing, which are summarised in Table 4.3. However, only a single study was able to confirm uptake of tyre particles by any of the organisms (Redondo-Hasselerharm et al. 2018); on average between 2.5 and 4 tyre tread particles were ingested by freshwater benthic macroinvertebrates. No published, peer-reviewed studies, in the field or laboratory, have yet looked for ingestion of RWP_{RM} or RWP_{PMB}.

Table 4.3 Published, peer-reviewed studies confirming uptake of hazardous compounds due to TWP in freshwater organisms based on laboratory exposures. Due to difficulties in finding the TWP in the environment, many studies have used lab-made tyre particles (TP) in their toxicity tests. These can be made in different ways, ground tyres or tyre scrap (granulates: TP_{GR}), cryo-fractured particles (TP_{CF}), particles abraded from the tyres with different rasps or steel files (TP_{AB}), or road simulators (TP_{RS})

Type of tyre material	Particle size	Concentration	Organisms	References
TP_{GR}	10–586 μm	0, 0.1, 0.3, 1, 3, 10% sediment d.w.	*Asellus aquaticus* *Gammarus pulex* *Tubifex* spp.	Redondo-Hasselerharm et al. (2018)
Tyre leachate, TP_{RS}	10–80 μm	50,000–100,000 mg L^{-1} 10% dilution 50,000–100,000 mg L^{-1} 100% dilution	*Daphnia magna* *Xenopus laevis*	Gualtieri et al. (2005a)
Tyre leachate, TP_{GR}	<590 μm	100,000 mg L-1 0.1–100% dilution	*Aedes albopictus* *Aedes triseriatus*	Villena et al. (2017)
Tyre leachate, TP_{CF}	*n.r.*	50,000–100,000 mg L^{-1} 0–100% dilution	*Xenopus laevis*	Gualtieri et al. (2005b)
Tyre leachate, TP_{CF}	*n.r.*	50–1400 mg L^{-1}	*Xenopus laevis*	Mantecca et al. (2007)
Tyre leachate, TP_{AB}	*n.r.*	250–16,000 mg L^{-1}	*Daphnia magna*	Wik and Dave (2005)
Tyre leachate, TP_{AB}	*n.r.*	900 mg 900 ml^{-1} 44 °C, 72 hours	*Daphnia magna*	Wik and Dave (2006)
Tyre leachate, TP_{AB}	*n.r.*	10, 100, 1000, 10,000 mg L^{-1} Leaching 5–11 days	*Daphnia magna* *Ceriodaphnia dubia* *Danio rerio* *Pseudokirchneriella subcapitata*	Wik et al. (2009)

(continued)

Table 4.3 (continued)

Type of tyre material	Particle size	Concentration	Organisms	References
Spiked sediments, TP_{GR}	*n.r.*	83,800 mg kg^{-1}	*Rana sylvatica*	Camponelli et al. (2009)
Spiked sediments, TP_{RS}	<150 μm	10,000 mg kg^{-1}	*Chironomus dilutes* *Hyalella azteca*	Panko et al. (2013)
Direct exposure, TP_{AB} Tyre leachate, TP_{AB}	<500 μm <500 μm	0–15,000 particles ml^{-1} 0.125,000 particles ml^{-1}	*Hyalella azteca* *Hyalella azteca*	Khan et al. (2019)

4.5 Microplastics in the Atmosphere

The potential for microplastic occurrence in the atmosphere above both land and sea is as yet largely unexplored. Processes of initial entrainment, localised updraft, convective mixing and advection, horizontal conveyance, and settling are expected to be relevant for the suspension, dispersion, transport, and deposition of particles across spatial scales (Fig. 4.1). These processes are likely affected by the size, morphology, and density of plastic particles. For example, particle size influences the movement of particles by wind at the land-air interface. Larger particles may move in a rolling motion, known as 'creep', whilst smaller particles may be transported through saltation motion, hopping along the land surface, or by suspension, based upon thresholds for particle motion and entrainment (Raupach and Lu 2004). Obstacles in the landscape may also represent temporary stores for microplastic particles, as has been demonstrated for terrestrial plants (Liu et al. 2020a).

Atmospheric deposition represents a pathway to terrestrial and freshwater environments. Several studies have reported deposition rates of between 0 and 11,130 particles m^{-2} day^{-1} (Allen et al. 2019; Cai et al. 2017; Dris et al. 2016, 2017; Klein and Fischer 2019; Liu et al. 2019a, b, c, d; Wright et al. 2020; Zhou et al. 2017); however, methodological differences, including discrepancies in the particle size classes analysed, hinder comparisons between datasets. For most studies, deposition appears to be higher in urban areas, which is likely associated with the quantity and proximity of sources. However, Klein and Fischer (2019) report higher concentrations in rural areas, which they attribute to the influence of forest canopy textures in combing out suspended particles. Rayon, polyamides, and polyesters are the dominant polymer types associated with fibrous microplastic, whilst polyethylene, polypropylene, and polystyrene are regularly reported for other particles types. For particle types such as fragments and films, studies typically report a higher deposition of microplastics concentrated in the smallest size categories and associated with lower-density polymer types. This demonstrates the influence of particle characteristics on atmospheric transport. For fibres, however, larger particles are commonly observed, and particles are composed of higher-density polymer types. Here, shape likely plays a dominant role in initial entrainment, transport, and deposition, where the irregular form of fibres encourages continued suspension. This has been demonstrated by Abbasi et al. (2019) who analysed urban dust from Asaluyeh county in Iran and found that, whilst deposited dusts were composed of a diverse range of particles types, suspended dusts contained only fine or fibrous particles. Furthermore, several studies of atmospheric deposition of microplastics report fibres as the dominant particle shape (e.g. Wright et al. 2020; Zhou et al. 2017; Dris et al. 2017). It has now been estimated that between 7.64 and 33.76 tonnes of fibrous atmospheric microplastics were generated globally during the year 2018 (Liu et al. 2020b).

It is difficult to quantify the role of atmospheric deposition as a pathway for microplastics to freshwater systems, due to the influence of multiple potential sources of microplastic contamination which complicates the assessment of individual inputs (Free et al. 2014). However, a recent study of atmospheric contamination of glacial ice shed some light on the potential contribution of atmospheric deposition. Ambrosini et al. (2019) found 74 ± 28 microplastics kg^{-1} of supraglacial sediments found on the glacier surface, which indicate a baseline level of contamination for that region. Further research is required to quantify the rates of deposition across different spatial and temporal scales. Moreover, the cryosphere represents a vastly understudied environmental compartment, which may yield insights into baseline atmospheric deposition rates and the dynamics of long-range transport (Windsor et al. 2019b).

Factors such as precipitation and wind speed have been positively correlated with microplastic deposition in a remote catchment in the Pyrenees (Allen et al. 2019). Both rain and snowfall events led to increased deposition of particles, where event occurrence and intensity were found to be more important than the duration of precipitation. This is supported by a recent study identifying high concentrations of microplastics in snow samples from Europe and the Arctic (Bergmann et al. 2019).

Tracking air mass trajectory through atmospheric modelling has successfully demonstrated medium-range transport of microplastic particles over an extended sampling duration (Allen et al. 2019); however, transport over longer distances within a regional context was also likely to have occurred. This is further studied by an assessment of microplastics in the sea air, which used the same backward trajectory modelling approach to identify the terrestrial-to-marine transfer of microplastics in the west Pacific Ocean (Liu et al. 2019a, b, c, d). The same study reported that trajectory modelling indicates that suspended microplastic particles from that region could be transported to the Arctic through the movement of air masses. Microplastics have been observed in several remote regions that are typically considered ‘pristine’ due to the very low levels of anthropogenic influence in the vicinity (Allen et al. 2019; Free et al. 2014; Zhang et al. 2016b, 2019). Medium- and long-range transport of particles is considered to be a key mechanism delivering microplastic contamination to these locations. Questions remain regarding the potential for long-range atmospheric transport of microplastic particles, atmospheric residence times, and transformation (e.g. degradation, fragmentation) of microplastic within the atmosphere.

4.6 Microplastics: Where Do They End Up?

4.6.1 Export to the Marine Environment

Due to catchment dynamics, erosion, and transport processes, many of the microplastic particles released on land are expected to eventually end up in the marine environment (Hale et al. 2020). Several seminal studies have estimated significant fluxes of plastic from land to the ocean (Jambeck et al. 2015; Lebreton et al. 2017; Schmidt et al. 2017); however, these deal with mass estimates, which predominately illustrate flows of *macroplastic*. Some studies have instead modelled microplastic release, demonstrating increases in microplastic export over the next several decades (Siegfried et al. 2017; van Wijnen et al. 2019). These studies highlight some geographical hotspots for release, such as South East Asia, and highlight wastewater treatment and TWP as important origins for microplastic that reaches the oceans. Improvements in sewage treatment were identified as a potential solution to significantly reduce future marine export of microplastics from land-based sources.

As previously discussed, microplastic transport in freshwater systems may vary in regard to seasonal or episodic changes in hydrological conditions. This is likely to influence the flux of microplastic to the oceans. Flood events are important for the transport of suspended sediments; over 90% of the annual suspended sediment flux of a river may be associated with storm events (Walling et al. 1992). Hurley et al. (2018) reported an export of 0.85 tonnes of microplastic particles from bed sediments in a medium-sized catchment in the UK. This was associated with a high-magnitude flood event that scoured accumulated microplastics from riverbeds and transferred them downstream and potentially out to the ocean. This is supported by evidence for significant increases in coastal microplastic contamination in the vicinity of river outlets following flood events (e.g. Gündoğdu et al. 2018; Lee et al. 2013; Veerasingam et al. 2016).

This transfer from land to sea may not always be unidirectional. Rivers influenced by tidal changes see a reversal in flow direction for some, or all, of the cross section during high-tide conditions, which may transport plastics upstream (van Emmerik et al. 2019). Moreover, coastal flooding may return marine microplastics to the land through deposition during the inundation of land. The impact of these factors should be considered when establishing robust flux estimations and assessing the fate of microplastic particles.

4.6.2 Microplastic Sinks in Terrestrial and Freshwater Environments

In some cases, land-based sources of microplastic and the terrestrial and freshwater systems involved in their dispersal and transformation are perceived as vectors for marine microplastic contamination. In regard to hydrological and geological cycles, across long timescales the majority of waters and sediments can be expected to reach the marine environment. Hence, microplastics are also likely to end up at this ultimate destination. Despite this, it is important not to overlook the need to better understand terrestrial and freshwater contamination dynamics, the risks posed to these ecosystems, and measures to limit or remediate contamination in these settings. Microplastics in the environment may pass through several terrestrial and marine cycles related to continuous and complex movement between both biotic and abiotic environmental compartments (Bank and Hansson 2019). Without a thorough understanding of the transfer of particles from the source to the ocean, efforts to reduce or remediate microplastic contamination will be hindered.

In addition, within terrestrial and freshwater systems, there are several candidate environments that may act as environmental sinks for microplastic particles, interrupting their ultimate transport to the coast. These may represent temporary or permanent sinks across different temporal and spatial scales. For example, lake sediments, where microplastics may accumulate and become buried by sediment deposits, have been identified as sites of plastic preservation and storage (Corcoran et al. 2015; Dong et al. 2020; Turner et al. 2019). At the bottom of a lake and beneath sediment layers, plastic particles are isolated from many of the degradative forces that initiate weathering, such as photodegradation (Corcoran et al. 2015). Microplastic particles have been identified in sediment layers as deep as 75 cm and dated to have been deposited during the early twentieth century, at the onset of plastic production (Turner et al. 2019). Once particles are buried to that depth, a significant disturbance event is required to remobilise sediments – such as dredging activities or a very high-magnitude storm. In the absence of such disturbance, lake sediments can be considered permanent or very long-term sinks for microplastic particles.

Other environments that may represent environmental sinks – but have not yet been studied regarding this specific question – include a range of sedimentary landscapes. These comprise settings that have been identified as environmental sinks for other contaminant types. For example, alluvial environments act as stores for many sediment-bound contaminants (e.g. Lecce and Pavlowsky 1997; Walling et al. 2003; Winter et al. 2001). Floodplain soils have already been shown to contain microplastic particles (Scheurer and Bigalke 2018). Depending on the geomorphological conditions of the environmental setting, floodplains may represent long-term stores of microplastic particles. Additionally, they may constitute future diffuse sources of microplastic particles as sediments with connectivity to freshwater systems may be reworked into active channels.

Environmental sinks can be defined by their temporal frame. From this perspective, not all sinks may be sedimentary. For example, residence times of waters in large lake systems, such as the Great Lakes, can reach close to 100 years (Mason et al. 2016a, b). If particles are also retained in these water masses, surface and subsurface waters in lacustrine environments may represent a short- to medium-term sink for microplastic particles. This is particularly relevant for lakes that are not consistently connected with fluvial systems, such as floodplain lakes. Additionally, entrapment in low-energy zones in fluvial systems – such as in dense riparian vegetation – may constitute short-term storage of buoyant microplastics. The residence times associated with these stores for microplastics particles, as well as the thresholds required to transition these environments to 'sources', require further investigation.

4.7 Future Research Agendas

Microplastic contamination is globally pervasive across terrestrial and freshwater environments. This review has drawn together research on several important sources and release pathways for microplastics including roads, agriculture, and wastewater treatment. Many environmental settings are expected to receive significant microplastic loadings, which likely represents a greater annual release than that estimate for the marine environment. Terrestrial and freshwater systems can be characterised by considerable complexity, whereby a range of dynamic processes are expected to influence the distribution, transport, and fate of microplastic particles.

Through this review, a set of specific directives for future research have been identified:

i. Harmonisation of Methods and Reporting, Including Improved Quality Assurance and Control (QA/QC) Practices, to Ensure Sufficient Data Quality and Permit Comparability Between Datasets

The review of studies of microplastic occurrence in terrestrial and freshwater environments, including samples of water, sediment, and biota, is characterised by the wide range of methodological approaches undertaken. This includes discrepancies between the sampling techniques, sample treatment, analytical technologies employed, and particle size classes analysed. In addition, many studies do not employ a similar set of QA/QC measures, so it is not possible to assess the quality of reported data. Findings are also often reported in different ways, for example, using different units or publishing only summary statistics that also differ (e.g. minimum/maximum, mean, median). The culmination of this variability is the lack of comparability between different studies. Harmonisation of analytical methods and reporting formats, and the publication of data in appropriate repositories, will help to reduce uncertainties in a holistic, global overview of the status of contamination, as well as providing meaningful baselines from which to track the impact of reduction or remediation measures.

ii. A Thorough Assessment of Microplastic Sources, Fate, and Impacts in Agricultural Environments

Agricultural environments represent the convergence of several sources and release pathways of microplastic particles. Particles may also be associated with higher chemical burdens from plastic additives (such as for decreasing photodegradation of mulching films) or sorbed contaminants (such as from WWTPs), although the extent and significance of this are relatively unknown. The status of contamination of agricultural settings is relatively unknown, and the fate of particles in agricultural soils remains under-researched, which makes it difficult to assess the relative contributions of different sources under relevant environmental conditions or the accumulation of particles over time. A small body of research on the impacts of microplastic contamination in agricultural environments is just now emerging. More research is required to gain a holistic perspective on the risks posed by microplastics across spatial and temporal scales. This is particularly important given the potential for any identified negative effects to impact upon soil health and food security.

iii. Quantification of Road-Associated Microplastic Particles (RAMP) as a Source of Microplastic to the Environment

For road-associated microplastic particles (RAMP), there are major knowledge gaps concerning environmental loadings, transport from the road to different matrices, and retention in gully-pots and water treatment systems. More research is urgently needed on RAMP in order to accurately assess how much is released into the environment, including the relative contributions from different road-related sources (TWP, $RAMP_{PMB}$, $RAMP_{RM}$). This is important given the spotlight that has now been placed on RAMP in several assessments of globally significant sources of microplastic to the environment. Quantification should be achieved through a new and optimised approach to analysing RAMP in environmental samples.

iv. Measures and Technologies to Reduce Microplastic Emissions to Wastewater or to Separate Particles Within WWTP Systems

This review highlighted the role of wastewater systems, such as WWTPs and CSOs, as a release pathway for microplastic particles to both terrestrial and freshwater environments. A diverse range of sources input microplastic particles to wastewater, and efforts should be made to reduce these at the source to reduce the burden on WWTPs and limit releases from untreated discharges such as CSOs. Much of the world's wastewater is not connected to a WWTP and is instead released untreated. Improvements in the global capacity of wastewater treatment would limit the environmental release of microplastics in many countries. Land application of sewage sludge has been identified as a primary release pathway for many microplastic types (ECHA 2019). Technologies to capture and remove plastic particles in WWTPs may help to reduce the burden on global soil environments.

v. A Better Understanding of the Controls Underpinning the Retention and Transport of Microplastic Particles in Freshwater Systems, Including More Accurate Flux Estimates to the Marine Environment

Microplastic particles in freshwater systems are likely to follow a complex pathway from their release to their ultimate fate. This may include several processes that interrupt downstream transport. These dynamics require further investigation to establish thresholds and controls on microplastic transport in freshwater environments. The majority of particles are expected to eventually end up in the marine environment, via fluvial systems. Estimates for this flux need to draw upon process-based research to incorporate appropriate complexity and identify the relevant controls on microplastic release to the marine environment. Moreover, further research on the dynamics of microplastic transport and spatial patterns of contamination will identify zones of microplastic accumulation in freshwater systems and can highlight the areas at greatest risk to potential negative impacts of contamination. This will help to focus efforts to protect freshwater ecosystems.

vi. Further Investigation of the Occurrence of Microplastics in Terrestrial and Freshwater Organisms, with a Specific Focus on Particle Types Such as RAMP

Many ecotoxicological studies are determining the effects associated with different microplastic particles and loadings, but evidence for uptake under field conditions is still scarce. Exposure represents half of the equation to evaluate risk, and so a more detailed investigation of the uptake of particles in real environment conditions is essential to contextualise ecotoxicological studies and inform risk assessments.

vii. Assessment of the Spatial and Temporal Scales of Environment Sinks for Microplastic Particles

This review highlighted several candidate environments that may act as temporary, long-term, or even permanent sinks for microplastic particles in freshwater and terrestrial settings. Some initial studies have investigated particle accumulation and potential residence times for some of these, but further research is required to establish the spatial and temporal scales upon which these environments act as stores, including the potential for them to become future sources of microplastic contamination through reworking and remobilisation. This is necessary to gain a better long-term perspective of environmental contamination and build more appropriate and better-targeted approaches to remediation instead of short-term fixes (Table 4.4).

Table 4.4 References for Fig. 4.3

Location	Study references
Lake Geneva, Switzerland	Faure et al. (2012)
Laurentian Great Lakes	Eriksen et al. (2013)
Lake Garda, Italy	Imhof et al. (2013)
Lake Hovsgol, Mongolia	Free et al. (2014)
Lake Ontario, Canada	Corcoran et al. (2015)
Lake Bolsena and Chiusi, Italy	Fischer et al. (2016)
Lake Michigan, USA	Mason et al. (2016a, b)
Taihu Lake, China	Su et al. (2016)
Tibet plateau lakes, China	Zhang et al. (2016a, b)
Lake Winnipeg, Canada	Anderson et al. (2017)
Paraná lakes, South America	Blettler et al. (2017)
Vembanad Lake, India	Sruthy and Ramasamy (2017)
Edgbaston Pool, UK	Vaughan et al. (2017)
Wuhan lakes, China	Wang et al. (2017)
Lake Erie, Canada	Dean et al. (2018)
Lake Superior, USA	Hendrickson et al. (2018)
Lake Garda, Italy	Imhof et al. (2018)

(continued)

Table 4.4 (continued)

Location	Study references
Dongting Lake, China	Jiang et al. (2019)
Italian lakes	Sighicelli et al. (2018)
Dongting and Hong Lakes, China	Wang et al. (2018)
Changsha lakes, China	Wen et al. (2018)
Qinghai Lake, China	Xiong et al. (2018)
Carpathian basin, Hungary	Bordós et al. (2019)
Lake Michigan, USA	Lenaker et al. (2019)
Poyang Lake, China	Liu et al. (2019a, b, c, d)
Lake Ulansuhai, China	Qin et al. (2019)
Vesijärvi lake and Pikku Vesijärvi pond, Finland	Scopetani et al. (2019)
Hampstead Pool 1, UK	Turner et al. (2019)
Lake Kallavesi, Finland	Uurasjärvi et al. (2020)
Lake Ulansuhai, China	Wang et al. (2019a, b)
Changsha lakes, China	Yin et al. (2019)
Poyang Lake, China	Yuan et al. (2019)
Donghu Lake, China	Dong et al. (2020)
Lake Naivasha, Kenya	Migwi et al. (2020)

Acknowledgements EK was supported by the Innovation Fund Denmark under the project *Udvikling af nye metoder til at detektere mikroplastik i vandmiljøet.* ER was supported by both the MicroROAD project, funded by both the NordFoU-project REHIRUP, consisting of the Norwegian Public Roads Administration, the Swedish Transport Administration and the Danish Road Directorate, and the Norwegian Institute for Water Research (NIVA). RH was supported by the Research Council of Norway (grant number 271825/E50), in the frame of the collaborative international consortium (IMPASSE) financed under the ERA-NET WaterWorks2015 co-funded call. This ERA-NET is an integral part of the 2016 Joint Activities developed by the Water Challenges for a Changing World Joint Programme Initiative (Water JPI).

References

Abbasi S, Keshavarzi B, Moore F, Delshab H, Soltani N, Sorooshian A (2017) Investigation of microrubbers, microplastics and heavy metals in street dust: a study in Bushehr city, Iran. Environ Earth Sci 76(23):798. https://doi.org/10.1007/s12665-017-7137-0

Abbasi S, Keshavarzi B, Moore F, Turner A, Kelly FJ, Dominguez AO, Jaafarzadeh N (2019) Distribution and potential health impacts of microplastics and microrubbers in air and street dusts from Asaluyeh County, Iran. Environ Pollut 244:153–164. https://doi.org/10.1016/j.envpol.2018.10.039

Accinelli C, Abbas HK, Shier WT, Vicari A, Little NS, Aloise MR, Giacomini S (2019) Degradation of microplastic seed film-coating fragments in soil. Chemosphere 226:645–650. https://doi.org/10.1016/j.chemosphere.2019.03.161

Akindele EO, Ehlers SM, Koop JHE (2019) First empirical study of freshwater microplastics in West Africa using gastropods from Nigeria as bioindicators. Limnologica 78:125708. https://doi.org/10.1016/j.limno.2019.125708

Alimi OS, Farner Budarz J, Hernandez LM, Tufenkji N (2018) Microplastics and nanoplastics in aquatic environments: aggregation, deposition, and enhanced contaminant transport. Environ Sci Technol 52(4):1704–1724. https://doi.org/10.1021/acs.est.7b05559

Allen S, Allen D, Phoenix VR, Roux GL, Jiménez PD, Simonneau A et al (2019) Atmospheric transport and deposition of microplastics in a remote mountain catchment. Nat Geosci 12(5):339–344. https://doi.org/10.1038/s41561-019-0335-5

Ambrosini R, Azzoni RS, Pittino F, Diolaiuti G, Franzetti A, Parolini M (2019) First evidence of microplastic contamination in the supraglacial debris of an alpine glacier. Environ Pollut 253:297–301. https://doi.org/10.1016/j.envpol.2019.07.005

Anderson PJ, Warrack S, Langen V, Challis JK, Hanson ML, Rennie MD (2017) Microplastic contamination in Lake Winnipeg, Canada. Environ Pollut 225:223–231. https://doi.org/10.1016/j.envpol.2017.02.072

Andrade MC, Winemiller KO, Barbosa PS, Fortunati A, Chelazzi D, Cincinelli A, Giarrizzo T (2019) First account of plastic pollution impacting freshwater fishes in the Amazon: ingestion of plastic debris by piranhas and other serrasalmids with diverse feeding habits. Environ Pollut 244:766–773. https://doi.org/10.1016/j.envpol.2018.10.088

Armitage N (2007) The reduction of urban litter in the stormwater drains of South Africa. Urban Water J 4(3):151–172. https://doi.org/10.1080/15730620701464117

Armitage N, Rooseboom A (2000) The removal of urban litter from stormwater conduits and streams: paper 1-The quantities involved and catchment litter management options. Water SA 26(2):181–188. Retrieved from https://journals.co.za/content/waters/26/2/AJA03784738_2350

Armitage N, Marais M, Pithey S (2001) Reducing urban litter in South Africa through catchment based litter management plans. Models and Applications to Urban Water Systems, Monograph 9:37. Retrieved from https://www.chijournal.org/R207-03

Avio CG, Gorbi S, Regoli F (2015) Experimental development of a new protocol for extraction and characterization of microplastics in fish tissues: first observations in commercial species from Adriatic Sea. Mar Environ Res 111(Supplement C):18–26. https://doi.org/10.1016/j.marenvres.2015.06.014

Baensch-Baltruschat B, Kocher B, Stock F, Reifferscheid G (2020) Tyre and road wear particles (TRWP) – a review of generation, properties, emissions, human health risk, ecotoxicity, and fate in the environment. Sci Total Environ 733:137823. https://doi.org/10.1016/j.scitotenv.2020.137823

Ballent A, Corcoran PL, Madden O, Helm PA, Longstaffe FJ (2016) Sources and sinks of microplastics in Canadian Lake Ontario nearshore, tributary and beach sediments. Mar Pollut Bull 110(1):383–395. https://doi.org/10.1016/j.marpolbul.2016.06.037

Bank MS, Hansson SV (2019) The plastic cycle: a novel and holistic paradigm for the anthropocene. Environ Sci Technol 53(13):7177–7179. https://doi.org/10.1021/acs.est.9b02942

Baumann M, Ismeier M (1998) Emissionen bei bestimmungsgemäßem Gebrauch von Reifen. In: KGK Kautschuk Gummi Kunststoffe, 51st year, No.3/98, 182-186 (in German)

Benckiser G (2019) Plastics, micro- and nanomaterials, and virus-soil microbe-plant interactions in the environment. In: Prasad R (ed) Plant nanobionics. Springer, pp 83–101. https://doi.org/10.1007/978-3-030-12496-0

Bergmann M, Mützel S, Primpke S, Tekman MB, Trachsel J, Gerdts G (2019) White and wonderful? Microplastics prevail in snow from the Alps to the Arctic. Sci Adv 5(8):eaax1157. https://doi.org/10.1126/sciadv.aax1157

Besseling E, Quik JTK, Sun M, Koelmans AA (2017) Fate of nano- and microplastic in freshwater systems: a modeling study. Environ Pollut 220:540–548. https://doi.org/10.1016/j.envpol.2016.10.001

Biginagwa FJ, Mayoma BS, Shashoua Y, Syberg K, Khan FR (2016) First evidence of microplastics in the African Great Lakes: recovery from Lake Victoria Nile perch and Nile tilapia. J Great Lakes Res 42(1):146–149. https://doi.org/10.1016/j.jglr.2015.10.012

Blettler MCM, Ulla MA, Rabuffetti AP, Garello N (2017) Plastic pollution in freshwater ecosystems: macro-, meso-, and microplastic debris in a floodplain lake. Environ Monit Assess 189(11):581. https://doi.org/10.1007/s10661-017-6305-8

Blettler MC, Abrial E, Khan FR, Sivri N, Espinola LA (2018) Freshwater plastic pollution: Recognizing research biases and identifying knowledge gaps. Water Res 143:416–424. https://doi.org/10.1016/j.watres.2018.06.015

Bordós G, Urbányi B, Micsinai A, Kriszt B, Palotai Z, Szabó I, Hantosi Z, Szoboszlay S (2019) Identification of microplastics in fish ponds and natural freshwater environments of the Carpathian basin, Europe. Chemosphere 216:110–116. https://doi.org/10.1016/j.chemosphere.2018.10.110

Boucher J, Friot D (2017) Primary microplastics in the oceans: A global evaluation of sources. IUCN International Union for Conservation of Nature. https://doi.org/10.2305/IUCN.CH.2017.01.en

Browne MA, Crump P, Niven SJ, Teuten E, Tonkin A, Galloway T, Thompson R (2011) Accumulation of microplastic on shorelines worldwide: sources and sinks. Environ Sci Technol 45(21):9175–9179. https://doi.org/10.1021/es201811s

Cai L, Wang J, Peng J, Tan Z, Zhan Z, Tan X, Chen Q (2017) Characteristic of microplastics in the atmospheric fallout from Dongguan city, China: preliminary research and first evidence. Environ Sci Pollut Res:1–8. https://doi.org/10.1007/s11356-017-0116-x

Camponelli KM, Casey RE, Snodgrass JW, Lev SM, Landa ER (2009) Impacts of weathered tire debris on the development of Rana sylvatica larvae. Chemosphere 74(5):717–722. https://doi.org/10.1016/j.chemosphere.2008.09.056

Carlin J, Craig C, Little S, Donnelly M, Fox D, Zhai L, Walters L (2020) Microplastic accumulation in the gastrointestinal tracts in birds of prey in central Florida, USA. Environ Pollut 264:114633. https://doi.org/10.1016/j.envpol.2020.114633

Carpenter EJ, Smith KL (1972) Plastics on the Sargasso Sea Surface. Science 175(4027):1240–1241. https://doi.org/10.1126/science.175.4027.1240

Carpenter EJ, Anderson SJ, Harvey GR, Miklas HP, Peck BB (1972) Polystyrene spherules in coastal waters. Science 178(4062):749–750. https://doi.org/10.1126/science.178.4062.749

Carr SA, Liu J, Tesoro AG (2016) Transport and fate of microplastic particles in wastewater treatment plants. Water Res 91:174–182. https://doi.org/10.1016/j.watres.2016.01.002

Castañeda RA, Avlijas S, Simard MA, Ricciardi A (2014) Microplastic pollution in St. Lawrence River sediments. Can J Fish Aquat Sci 71(12):1767–1771. https://doi.org/10.1139/cjfas-2014-0281

Chen X, Xiong X, Jiang X, Shi H, Wu C (2019) Sinking of floating plastic debris caused by biofilm development in a freshwater lake. Chemosphere 222:856–864. https://doi.org/10.1016/j.chemosphere.2019.02.015

Chen Y, Leng Y, Liu X, Wang J (2020) Microplastic pollution in vegetable farmlands of suburb Wuhan, central China. Environ Pollut 257:113449. https://doi.org/10.1016/j.envpol.2019.113449

Collard F, Gasperi J, Gilbert B, Eppe G, Azimi S, Rocher V, Tassin B (2018) Anthropogenic particles in the stomach contents and liver of the freshwater fish Squalius cephalus. Sci Total Environ 643:1257–1264. https://doi.org/10.1016/j.scitotenv.2018.06.313

Corcoran PL, Norris T, Ceccanese T, Walzak MJ, Helm PA, Marvin CH (2015) Hidden plastics of Lake Ontario, Canada and their potential preservation in the sediment record. Environ Pollut 204:17–25. https://doi.org/10.1016/j.envpol.2015.04.009

Corradini F, Meza P, Eguiluz R, Casado F, Huerta-Lwanga E, Geissen V (2019) Evidence of microplastic accumulation in agricultural soils from sewage sludge disposal. Sci Total Environ 671:411–420. https://doi.org/10.1016/j.scitotenv.2019.03.368

Dean BY, Corcoran PL, Helm PA (2018) Factors influencing microplastic abundances in nearshore, tributary and beach sediments along the Ontario shoreline of Lake Erie. J Great Lakes Res 44(5):1002–1009. https://doi.org/10.1016/j.jglr.2018.07.014

Dehghani S, Moore F, Akhbarizadeh R (2017) Microplastic pollution in deposited urban dust, Tehran metropolis, Iran. Environ Sci Pollut Res 24(25):20360–20371. https://doi.org/10.1007/s11356-017-9674-1

Dikareva N, Simon KS (2019) Microplastic pollution in streams spanning an urbanisation gradient. Environ Pollut 250:292–299. https://doi.org/10.1016/j.envpol.2019.03.105

Dong H, Liu T, Li Y, Liu H, Wang D (2013) Effects of plastic film residue on cotton yield and soil physical and chemical properties in Xinjiang. Trans Chinese Soc Agric Eng 29(8):91–99. https://doi.org/10.3969/j.issn.1002-6819.2013.08.011

Dong M, Luo Z, Jiang Q, Xing X, Zhang Q, Sun Y (2020) The rapid increases in microplastics in urban lake sediments. Sci Rep 10(1):1–10. https://doi.org/10.1038/s41598-020-57933-8

Dris R, Imhof H, Sanchez W, Gasperi J, Galgani F, Tassin B, Laforsch C (2015) Beyond the ocean: contamination of freshwater ecosystems with (micro-)plastic particles. Environ Chem 12(5):539–550. https://doi.org/10.1071/EN14172

Dris R, Gasperi J, Saad M, Mirande C, Tassin B (2016) Synthetic fibers in atmospheric fallout: a source of microplastics in the environment? Mar Pollut Bull 104(1):290–293. https://doi.org/10.1016/j.marpolbul.2016.01.006

Dris R, Gasperi J, Mirande C, Mandin C, Guerrouache M, Langlois V, Tassin B (2017) A first overview of textile fibers, including microplastics, in indoor and outdoor environments. Environ Pollut (Barking, Essex: 1987) 221:453–458. https://doi.org/10.1016/j.envpol.2016.12.013

Dris R, Gasperi J, Tassin B (2018) Sources and fate of microplastics in urban areas: a focus on Paris megacity. In: Freshwater microplastics. Springer, Cham, pp 69–83. https://doi.org/10.1007/978-3-319-61615-5_4

ECHA (2019) ANNEX XV RESTRICTION REPORT, PROPOSAL FOR A RESTRICTION: intentionally added microplastics, p 146. Retrieved from https://echa.europa.eu/documents/10162/05bd96e3-b969-0a7c-c6d0-441182893720

Eisentraut P, Dümichen E, Ruhl AS, Jekel M, Albrecht M, Gehde M, Braun U (2018) Two birds with one stone—fast and simultaneous analysis of microplastics: microparticles derived from thermoplastics and tire wear. Environ Sci Technol Lett 5(10):608–613. https://doi.org/10.1021/acs.estlett.8b00446

Eriksen M, Mason S, Wilson S, Box C, Zellers A, Edwards W et al (2013) Microplastic pollution in the surface waters of the Laurentian Great Lakes. Mar Pollut Bull 77(1):177–182. https://doi.org/10.1016/j.marpolbul.2013.10.007

Espí E, Salmerón A, Fontecha A, García Y, Real AI (2006) Plastic Films for Agricultural Applications. J Plastic Film Sheeting 22(2):85–102. https://doi.org/10.1177/8756087906064220

Estahbanati S, Fahrenfield NL (2016) Influence of wastewater treatment plant discharges on microplastic concentrations in surface water. Chemosphere 162:277–284. https://doi.org/10.1016/j.chemosphere.2016.07.083

Evans-Pughe C (2017) All at sea: cleaning up the great Pacific garbage patch. Eng Technol 12(1):52–55. https://doi.org/10.1049/et.2017.0105

Fan Y, Zheng K, Zhu Z, Chen G, Peng X (2019) Distribution, sedimentary record, and persistence of microplastics in the Pearl River catchment, China. Environ Pollut 251:862–870. https://doi.org/10.1016/j.envpol.2019.05.056

Faure F, Corbaz M, Baecher H, Alencastro D, Felippe L (2012) Pollution due to plastics and microplastics in Lake Geneva and in the Mediterranean Sea. Arch Sci 65:157–164. Retrieved from https://pdfs.semanticscholar.org/5744/79e4e81161ff2d93615a2af7a3e83432035e.pdf

Fischer EK, Paglialonga L, Czech E, Tamminga M (2016) Microplastic pollution in lakes and lake shoreline sediments–a case study on Lake Bolsena and Lake Chiusi (central Italy). Environ Pollut 213:648–657. https://doi.org/10.1016/j.envpol.2016.03.012

França D, Messa LL, Souza CF, Faez R (2019) Nano and microencapsulated nutrients for enhanced efficiency fertilizer. In: Gutiérrez TJ (ed) Polymers for agri-food applications. Springer, Cham, pp 29–44. https://doi.org/10.1007/978-3-030-19416-1_3

Free CM, Jensen OP, Mason SA, Eriksen M, Williamson NJ, Boldgiv B (2014) High-levels of microplastic pollution in a large, remote, mountain lake. Mar Pollut Bull 85(1):156–163. https://doi.org/10.1016/j.marpolbul.2014.06.001

Frias JPGL, Pagter E, Nash R, O'Connor I, Carretero O, Filgueiras A et al (2018) Standardised protocol for monitoring microplastics in sediments. https://doi.org/10.13140/RG.2.2.36256.89601/1

Fuller S, Gautam A (2016) A procedure for measuring microplastics using pressurized fluid extraction. Environ Sci Technol 50(11):5774–5780. https://doi.org/10.1021/acs.est.6b00816

Gatidou G, Arvanti OS, Stasinakis AS (2019) Review on the occurrence and fate of microplastics in Sewage Treatment Plants. J Hazard Mater 367:504–512. https://doi.org/10.1016/j.jhazmat.2018.12.081

GESAMP (2015) Sources, fate and effects of microplastics in the marine environment (part 1), No. 2015 #90. IMO, p 98. Retrieved from: http://www.gesamp.org/publications/reports-and-studies-no-90

GESAMP (2016) Sources, fate and effects of microplastics in the marine environment (part 2), No. 2016 #93. IMO, p 221. Retrieved from: http://www.gesamp.org/publications/microplastics-in-the-marine-environment-part-2

GESAMP (2019) Guidelines for the monitoring and assessment of plastic litter in the ocean, No. 2019 #99. IMO, p 123. Retrieved from: http://www.gesamp.org/publications/guidelines-for-the-monitoring-and-assessment-of-plastic-litter-in-the-ocean

Geyer R, Jambeck JR, Law KL (2017) Production, use, and fate of all plastics ever made. Sci Adv 3(7):e1700782. https://doi.org/10.1126/sciadv.1700782

Grung M, Kringstad A, Bæk K, Allan IJ, Thomas KV, Meland S, Ranneklev SB (2017) Identification of non-regulated polycyclic aromatic compounds and other markers of urban pollution in road tunnel particulate matter. J Hazard Mater 323:36–44. https://doi.org/10.1016/j.jhazmat.2016.05.036

Gualtieri M, Andrioletti M, Vismara C, Milani M, Camatini M (2005a) Toxicity of tire debris leachates. Environ Int 31(5):723–730. https://doi.org/10.1016/j.envint.2005.02.001

Gualtieri M, Andrioletti M, Mantecca P, Vismara C, Camatini M (2005b) Impact of tire debris on in vitro and in vivo systems. Part Fibre Toxicol 2(1):1. https://doi.org/10.1186/1743-8977-2-1

Gündoğdu S, Cevik C, Ayat B, Aydoğan B, Karaca S (2018) How microplastics quantities increase with flood events? An example from Mersin Bay NE Levantine coast of Turkey. Environ Pollut 239:342–350. https://doi.org/10.1016/j.envpol.2018.04.042

Hale RC, Seeley ME, La Guardia MJ, Mai L, Zeng EY (2020) A global perspective on microplastics. J Geophys Res Oceans 125(1):5, e2018JC014719. https://doi.org/10.1029/2018JC014719

Hallberg M, Renman G, Byman L, Svenstam G, Norling M (2014) Treatment of tunnel wash water and implications for its disposal. Water Sci Technol J Int Assoc Water Pollut Res 69(10):2029–2035. https://doi.org/10.2166/wst.2014.113

He P, Chen L, Shao L, Zhang H, Lü F (2019) Municipal solid waste (MSW) landfill: a source of microplastics? -Evidence of microplastics in landfill leachate. Water Res 159:38–45. https://doi.org/10.1016/j.watres.2019.04.060

Hendrickson E, Minor EC, Schreiner K (2018) Microplastic abundance and composition in western Lake Superior as determined via microscopy, Pyr-GC/MS, and FTIR. Environ Sci Technol 52(4):1787–1796. https://doi.org/10.1021/acs.est.7b05829

Holland ER, Mallory ML, Shutler D (2016) Plastics and other anthropogenic debris in freshwater birds from Canada. Sci Total Environ 571:251–258. https://doi.org/10.1016/j.scitotenv.2016.07.158

Horton AA, Svendsen C, Williams RJ, Spurgeon DJ, Lahive E (2017a) Large microplastic particles in sediments of tributaries of the River Thames, UK – Abundance, sources and methods for effective quantification. Mar Pollut Bull 114(1):218–226. https://doi.org/10.1016/j.marpolbul.2016.09.004

Horton AA, Walton A, Spurgeon DJ, Lahive E, Svendsen C (2017b) Microplastics in freshwater and terrestrial environments: evaluating the current understanding to identify the knowledge gaps and future research priorities. Sci Total Environ 586:127–141. https://doi.org/10.1016/j.scitotenv.2017.01.190

Horton AA, Jürgens MD, Lahive E, van Bodegom PM, Vijver MG (2018) The influence of exposure and physiology on microplastic ingestion by the freshwater fish Rutilus rutilus (roach) in the River Thames, UK. Environ Pollut 236:188–194. https://doi.org/10.1016/j.envpol.2018.01.044

Huang Y, Liu Q, Jia W, Yan C, Wang J (2020) Agricultural plastic mulching as a source of microplastics in the terrestrial environment. Environ Pollut 114096. https://doi.org/10.1016/j.envpol.2020.114096

Huerta Lwanga E, Gertsen H, Gooren H, Peters P, Salánki T, van der Ploeg M et al (2017) Incorporation of microplastics from litter into burrows of Lumbricus terrestris. Environ Pollut 220:523–531. https://doi.org/10.1016/j.envpol.2016.09.096

Hurley RR, Nizzetto L (2018) Fate and occurrence of micro(nano)plastics in soils: knowledge gaps and possible risks. Curr Opin Environ Sci Health 1:6–11. https://doi.org/10.1016/j.coesh.2017.10.006

Hurley RR, Woodward JC, Rothwell JJ (2017) Ingestion of microplastics by freshwater Tubifex worms. Environ Sci Technol 51(21):12844–12851. https://doi.org/10.1021/acs.est.7b03567

Hurley RR, Woodward JC, Rothwell JJ (2018) Microplastic contamination of river beds significantly reduced by catchment-wide flooding. Nat Geosci. https://doi.org/10.1038/s41561-018-0080-1

Imhof HK, Ivleva NP, Schmid J, Niessner R, Laforsch C (2013) Contamination of beach sediments of a subalpine lake with microplastic particles. Curr Biol 23(19):R867–R868. https://doi.org/10.1016/j.cub.2013.09.001

Imhof HK, Wiesheu AC, Anger PM, Niessner R, Ivleva NP, Laforsch C (2018) Variation in plastic abundance at different lake beach zones-A case study. Sci Total Environ 613:530–537

Isobe A, Buenaventura NT, Chastain S, Chavanich S, Cózar A, DeLorenzo M et al (2019) An interlaboratory comparison exercise for the determination of microplastics in standard sample bottles. Mar Pollut Bull 146:831–837. https://doi.org/10.1016/j.marpolbul.2019.07.033

Jabeen K, Su L, Li J, Yang D, Tong C, Mu J, Shi H (2017) Microplastics and mesoplastics in fish from coastal and fresh waters of China. Environ Pollut 221:141–149. https://doi.org/10.1016/j.envpol.2016.11.055

Jambeck JR, Geyer R, Wilcox C, Siegler TR, Perryman M, Andrady A et al (2015) Plastic waste inputs from land into the ocean. Science 347(6223):768–771. https://doi.org/10.1126/science.1260352

Jiang C, Yin L, Li Z, Wen X, Luo X, Hu S et al (2019) Microplastic pollution in the rivers of the Tibet Plateau. Environ Pollut 249:91–98. https://doi.org/10.1016/j.envpol.2019.03.022

Johnsen JP, Bye NH (2019) Assessment of tire wear emission in a road tunnel, using benzothiazoles as tracer in tunnel wash water 79. Retrieved from https://nmbu.brage.unit.no/nmbu-xmlui/handle/11250/2612289

Kamata H, Masuda K, Yamada J, Takada H (2000) Water-Particle distribution of hydrophobic micro pollutants in storm water runoff. Polycycl Aromat Compd 20(1–4):39–54. https://doi.org/10.1080/10406630008034774

Kay P, Hiscoe R, Moberley I, Bajic L, McKenna N (2018) Wastewater treatment plants as a source of microplastics in river catchments. Environ Sci Pollut Res 25(20):20264–20267. https://doi.org/10.1007/s11356-018-2070-7

Keller AS, Jimenez-Martinez J, Mitrano DM (2020) Transport of nano- and Microplastic through unsaturated porous media from sewage sludge application. Environ Sci Technol 54(2):911–920. https://doi.org/10.1021/acs.est.9b06483

Khan FR, Halle LL, Palmqvist A (2019) Acute and long-term toxicity of micronized car tire wear particles to Hyalella azteca. Aquat Toxicol 213:105216. https://doi.org/10.1016/j.aquatox.2019.05.018

Klein M, Fischer EK (2019) Microplastic abundance in atmospheric deposition within the Metropolitan area of Hamburg, Germany. Sci Total Environ 685:96–103. https://doi.org/10.1016/j.scitotenv.2019.05.405

Klöckner P, Reemtsma T, Eisentraut P, Braun U, Ruhl AS, Wagner S (2019) Tire and road wear particles in road environment – Quantification and assessment of particle dynamics by Zn determination after density separation. Chemosphere 222:714–721. https://doi.org/10.1016/j.chemosphere.2019.01.176

Koelmans AA, Mohamed Nor NH, Hermsen E, Kooi M, Mintenig SM, De France J (2019) Microplastics in freshwaters and drinking water: critical review and assessment of data quality. Water Res 155:410–422. https://doi.org/10.1016/j.watres.2019.02.054

Kole PJ, Löhr AJ, Van Belleghem FGAJ, Ragas AMJ (2017) Wear and tear of tyres: a stealthy source of microplastics in the environment. Int J Environ Res Public Health 14(10):1265. https://doi.org/10.3390/ijerph14101265

Kum V, Sharp A, Harnpornchai N (2005) Improving the solid waste management in Phnom Penh city: a strategic approach. Waste Manag 25(1):101–109. https://doi.org/10.1016/j.wasman.2004.09.004

Lambert S, Wagner M (2018) Microplastics are contaminants of emerging concern in freshwater environments: an overview. In: Freshwater microplastics. Springer, Cham, pp 1–23. https://doi.org/10.1007/978-3-319-61615-5_1

Lassen, C., Hansen, S. F., Magnusson, K., Hartmann, N. B., Jensen, P. R., Nielsen, T. G., & Brinch, A. (2015). Microplastics: occurrence, effects and sources of releases to the environment in Denmark. Danish Environmental Protection Agency. Retrieved from https://orbit.dtu.dk/en/publications/microplastics-occurrence-effects-and-sources-of-releases-to-the-e

Lebreton LCM, van der Zwet J, Damsteeg J-W, Slat B, Andrady A, Reisser J (2017) River plastic emissions to the world's oceans. Nat Commun 8:15611. https://doi.org/10.1038/ncomms15611

Lecce SA, Pavlowsky RT (1997) Storage of mining-related zinc in floodplain sediments, Blue River, Wisconsin. Phys Geogr 18(5):424–439. https://doi.org/10.1080/02723646.1997.10642628

Lee J, Hong S, Song YK, Hong SH, Jang YC, Jang M et al (2013) Relationships among the abundances of plastic debris in different size classes on beaches in South Korea. Mar Pollut Bull 77(1–2):349–354. https://doi.org/10.1016/j.marpolbul.2013.08.013

Lenaker PL, Baldwin AK, Corsi SR, Mason SA, Reneau PC, Scott JW (2019) Vertical distribution of microplastics in the water column and surficial sediment from the Milwaukee River Basin to Lake Michigan. Environ Sci Technol. https://doi.org/10.1021/acs.est.9b03850

Li X, Chen L, Mei Q, Dong B, Dai X, Ding G, Zeng EY (2018) Microplastics in sewage sludge from the wastewater treatment plants in China. Water Res 142:75–85. https://doi.org/10.1016/j.watres.2018.05.034

Li C, Busquets R, Campos LC (2020) Assessment of microplastics in freshwater systems: a review. Sci Total Environ 707:135578. https://doi.org/10.1016/j.scitotenv.2019.135578

Liu EK, He WQ, Yan CR (2014) 'White revolution' to 'white pollution'—agricultural plastic film mulch in China. Environ Res Lett 9(9):091001. https://doi.org/10.1088/1748-9326/9/9/091001

Liu M, Lu S, Song Y, Lei L, Hu J, Lv W et al (2018) Microplastic and mesoplastic pollution in farmland soils in suburbs of Shanghai, China. Environ Pollut 242:855–862. https://doi.org/10.1016/j.envpol.2018.07.051

Liu F, Vianello A, Vollertsen J (2019a) Retention of microplastics in sediments of urban and highway stormwater retention ponds. Environ Pollut 113335. https://doi.org/10.1016/j.envpol.2019.113335

Liu K, Wu T, Wang X, Song Z, Zong C, Wei N, Li D (2019b) Consistent transport of terrestrial microplastics to the ocean through atmosphere. Environ Sci Technol 53(18):10612–10619. https://doi.org/10.1021/acs.est.9b03427

Liu S, Jian M, Zhou L, Li W (2019c) Distribution and characteristics of microplastics in the sediments of Poyang Lake, China. Water Sci Technol 79(10):1868–1877. https://doi.org/10.2166/wst.2019.185

Liu X, Yuan W, Di M, Li Z, Wang J (2019d) Transfer and fate of microplastics during the conventional activated sludge process in one wastewater treatment plant of China. Chem Eng J 362:176–182. https://doi.org/10.1016/j.cej.2019.01.033

Liu K, Wang X, Song Z, Wei N, Li D (2020a) Terrestrial plants as a potential temporary sink of atmospheric microplastics during transport. Sci Total Environ 742:140523. https://doi.org/10.1016/j.scitotenv.2020.140523

Liu K, Wang X, Song Z, Wei N, Ye H, Cong, X. … Li, D. (2020b) Global inventory of atmospheric fibrous microplastics input into the ocean: an implication from the indoor origin. J Hazard Mater 400:123223. https://doi.org/10.1016/j.jhazmat.2020.123223

Lorenz K, Kandeler E (2005) Biochemical characterization of urban soil profiles from Stuttgart, Germany. Soil Biol Biochem 37(7):1373–1385. https://doi.org/10.1016/j.soilbio.2004.12.009

Lusher AL, Hurley RR, Vogelsang C, Nizzetto L, Olsen M (2018) Mapping microplastics in sludge, No. M907, p 55. Retrieved from http://www.miljodirektoratet.no/Documents/publikasjoner/M907/M907.pdf

Lusher AL, Hurley RR, Vogelsang C (2019) Microplastics in sewage sludge: captured but released? In: Karapanagioti HK, Kalavrouziotis IK (eds) Microplastics in water and wastewater. IWA Publishing, London, pp 85–100

Lv W, Zhou W, Lu S, Huang W, Yuan Q, Tian M et al (2019a) Microplastic pollution in rice-fish co-culture system: a report of three farmland stations in Shanghai, China. Sci Total Environ 652:1209–1218. https://doi.org/10.1016/j.scitotenv.2018.10.321

Lv X, Dong Q, Zuo Z, Liu Y, Huang X, Wu W-M (2019b) Microplastics in a municipal wastewater treatment plant: fate, dynamic distribution, removal efficiencies, and control strategies. Retrieved from https://pubag.nal.usda.gov/catalog/6367168

Maaß S, Daphi D, Lehmann A, Rillig MC (2017) Transport of microplastics by two collembolan species. Environ Pollut 225:456–459. https://doi.org/10.1016/j.envpol.2017.03.009

Magnusson K, Norén F (2014) Screening of microplastic particles in and down-stream a wastewater treatment plant, No. C 55, p 22. Retrieved from http://www.diva-portal.org/smash/record.jsf?pid=diva2:773505

Magnusson K, Eliasson K, Fråne A, Haikonen K, Hultén J, Olshammar M et al (2016) Swedish sources and pathways for microplastics to the marine environment. A Review of Existing Data IVL, C, 183. Retrieved from https://www.naturvardsverket.se/upload/miljoarbete-i-samhallet/miljoarbete-i-sverige/regeringsuppdrag/2016/mikroplaster/swedish-sources-and-pathways-for-microplastics-to-marine%20environment-ivl-c183.pdf

Mahon AM, O'Connell B, Healy MG, O'Connor I, Officer R, Nash R, Morrison L (2017) Microplastics in sewage sludge: effects of treatment. Environ Sci Technol 51(2):810–818. https://doi.org/10.1021/acs.est.6b04048

Mai L, Bao L-J, Wong CS, Zeng EY (2018) Microplastics in the terrestrial environment. In: Zeng EY (ed) Microplastic contamination in aquatic environments; an emerging matter of environmental urgency. Elsevier, pp 365–378. https://doi.org/10.1016/B978-0-12-813747-5.00012-6

Mai L, You S-N, He H, Bao L-J, Liu L, Zeng EY (2019) Riverine microplastic pollution in the Pearl River Delta, China: are modeled estimates accurate? Environ Sci Technol. https://doi.org/10.1021/acs.est.9b04838

Mani T, Hauk A, Walter U, Burkhardt-Holm P (2015) Microplastics profile along the Rhine River. Sci Rep 5:17988. https://doi.org/10.1038/srep17988

Mantecca P, Gualtieri M, Andrioletti M, Bacchetta R, Vismara C, Vailati G, Camatini M (2007) Tire debris organic extract affects Xenopus development. Environ Int 33(5):642–648. https://doi.org/10.1016/j.envint.2007.01.007

Marais M, Armitage N, Pithey S (2001) A study of the litter loadings in urban drainage systems – methodology and objectives. Water Sci Technol 44(6):99–108. https://doi.org/10.2166/wst.2001.0350

Marais M, Armitage N, Wise C (2004) The measurement and reduction of urban litter entering stormwater drainage systems: paper 1 – quantifying the problem using the City of Cape Town as a case study [Text]. Retrieved February 24, 2020, from https://www.ingentaconnect.com/content/sabinet/waters/2004/00000030/00000004/art00006

Mason SA, Garneau D, Sutton R, Chu Y, Ehmann K, Barnes J et al (2016a) Microplastic pollution is widely detected in US municipal wastewater treatment plant effluent. Environ Pollut 218:1045–1054. https://doi.org/10.1016/j.envpol.2016.08.056

Mason SA, Kammin L, Eriksen M, Aleid G, Wilson S, Box C, Williamson N, Riley A (2016b) Pelagic plastic pollution within the surface waters of Lake Michigan, USA. J Great Lakes Res 42(4):753–759. https://doi.org/10.1016/j.jglr.2016.05.009

McCormick AR, Hoellein TJ, London MG, Hittie J, Scott JW, Kelly JJ (2016) Microplastic in surface waters of urban rivers: concentration, sources, and associated bacterial assemblages. Ecosphere 7(11):e01556. https://doi.org/10.1002/ecs2.1556

McGoran AR, Clark PF, Morritt D (2017) Presence of microplastic in the digestive tracts of European flounder, Platichthys flesus, and European smelt, Osmerus eperlanus, from the River Thames. Environ Pollut 220:744–751. https://doi.org/10.1016/j.envpol.2016.09.078

Mehlhart G, Blepp M (2012) Study on land-sourced litter (LSL) in the marine environment: review of sources and literature. Öko-Institut. https://www.oeko.de/oekodoc/1487/2012-058-en.pdf

Meland S, Rødland E (2018) Forurensning i tunnelvaskevann –en studie av 34 veitunneler i Norge. VANN 1:54–65. Retrieved from: https://vannforeningen.no/wp-content/uploads/2018/07/Meland.pdf

Meland S, Borgstrøm R, Heier LS, Rosseland BO, Lindholm O, Salbu B (2010a) Chemical and ecological effects of contaminated tunnel wash water runoff to a small Norwegian stream. Sci Total Environ 408(19):4107–4117. https://doi.org/10.1016/j.scitotenv.2010.05.034

Meland S, Heier LS, Salbu B, Tollefsen KE, Farmen E, Rosseland BO (2010b) Exposure of brown trout (Salmo trutta L.) to tunnel wash water runoff—chemical characterisation and biological impact. Sci Total Environ 408(13):2646–2656. https://doi.org/10.1016/j.scitotenv.2010.03.025

Migwi FK, Ogunah JA, Kiratu JM (2020) Occurrence and spatial distribution of microplastics in the surface waters of Lake Naivasha, Kenya. Environ Toxicol Chem n/a(n/a). https://doi.org/10.1002/etc.4677

Muñoz-Cadena CE, Lina-Manjarrez P, Estrada-Izquierdo I, Ramón-Gallegos E (2012) An approach to litter generation and littering practices in a Mexico City neighborhood. Sustainability 4(8):1733–1754. https://doi.org/10.3390/su4081733

Nan B, Su L, Kellar C, Craig NJ, Keough MJ, Pettigrove V (2020) Identification of microplastics in surface water and Australian freshwater shrimp Paratya australiensis in Victoria, Australia. Environ Pollut 259:113865. https://doi.org/10.1016/j.envpol.2019.113865

Nel HA, Dalu T, Wasserman RJ (2018) Sinks and sources: assessing microplastic abundance in river sediment and deposit feeders in an Austral temperate urban river system. Sci Total Environ 612:950–956. https://doi.org/10.1016/j.scitotenv.2017.08.298

Ngo PL, Pramanik BK, Shah K, Roychand R (2019) Pathway, classification and removal efficiency of microplastics in wastewater treatment plants. Environ Pollut 255(2):113326. https://doi.org/10.1016/j.envpol.2019.113326

Nizzetto L, Futter M, Langaas S (2016) Are agricultural Soils dumps for microplastics of urban origin? Environ Sci Technol 50(20):10777–10779. https://doi.org/10.1021/acs.est.6b04140

O'Connor D, Pan S, Shen Z, Song Y, Jin Y, Wu W-M, Hou D (2019) Microplastics undergo accelerated vertical migration in sand soil due to small size and wet-dry cycles. Environ Pollut. https://doi.org/10.1016/j.envpol.2019.03.092

Ockelford A, Cundy A, Ebdon JE (2020) Storm response of fluvial sedimentary microplastics. Sci Rep 10(1):1–10. https://doi.org/10.1038/s41598-020-58765-2

Olesen KB, Stephansen DA, van Alst N, Vollertsen J (2019) Microplastics in a stormwater pond. Water 11(7):1466. https://doi.org/10.3390/w11071466

Panebianco A, Nalbone L, Giarratana F, Ziino G (2019) First discoveries of microplastics in terrestrial snails. Food Control 106:106722. https://doi.org/10.1016/j.foodcont.2019.106722

Panko JM, Kreider ML, McAtee BL, Marwood C (2013) Chronic toxicity of tire and road wear particles to water- and sediment-dwelling organisms. Ecotoxicology 22(1):13–21. https://doi.org/10.1007/s10646-012-0998-9

Phillips MB, Bonner TH (2015) Occurrence and amount of microplastic ingested by fishes in watersheds of the Gulf of Mexico. Mar Pollut Bull 100(1):264–269. https://doi.org/10.1016/j.marpolbul.2015.08.041

Piehl S, Leibner A, Löder MGJ, Dris R, Bogner C, Laforsch C (2018) Identification and quantification of macro- and microplastics on an agricultural farmland. Sci Rep 8(1):17950. https://doi.org/10.1038/s41598-018-36172-y

Qin Y, Wang Z, Li W, Chang X, Yang J, Yang F (2019) Microplastics in the sediment of Lake Ulansuhai of Yellow River Basin, China. Water Environ Res 92(6). https://doi.org/10.1002/wer.1275

Ranneklev SB, Hurley R, Bråte ILN, Vogelsang C (2019) Plast i landbruket: kilder, massebalanse og spredning til lokale vannforekomster (Plastland). Norsk insitutt for vannforskning. Retrieved from https://niva.brage.unit.no/niva-xmlui/handle/11250/2632595

Raupach MR, Lu H (2004) Representation of land-surface processes in aeolian transport models. Environ Model Softw 19(2):93–112. https://doi.org/10.1016/S1364-8152(03)00113-0

Redondo-Hasselerharm PE, de Ruijter VN, Mintenig SM, Verschoor A, Koelmans AA (2018) Ingestion and chronic effects of car tire tread particles on freshwater benthic macroinvertebrates. Environ Sci Technol 52(23):13986–13994. https://doi.org/10.1021/acs.est.8b05035

Rezaei M, Riksen MJ, Sirjani E, Sameni A, Geissen V (2019) Wind erosion as a driver for transport of light density microplastics. Sci Total Environ. https://doi.org/10.1016/j.scitotenv.2019.02.382

Roch S, Walter T, Ittner LD, Friedrich C, Brinker A (2019) A systematic study of the microplastic burden in freshwater fishes of south-western Germany – Are we searching at the right scale? Sci Total Environ 689:1001–1011. https://doi.org/10.1016/j.scitotenv.2019.06.404

Rochman CM (2018) Microplastics research—from sink to source. Science 360(6384):28–29. https://doi.org/10.1126/science.aar7734

Rødland E (2019) Ecotoxic potential of road-associated microplastic particles (RAMP). VANN 3:166–183. Retrieved from: https://vannforeningen.no/wp-content/uploads/2019/12/R%C3%B8dland.pdf

Rødland E, Helgadottir D (2018) Prioriteringsverktøy for vurdering av forurensingsbelastning fra urenset tunnelvaskevann. VANN 4:367–376. Retrieved from: https://vannforeningen.no/wp-content/uploads/2019/04/R%C3%B8dland.pdf

Rødland E, Okoffo ED, Rauert C, Heier LS, Lind OC, Reid M, Thomas KV, Meland S (2020) Road de-icing salt: assessment of a potential new source and pathway of microplastics particles from roads. Sci Total Environ 738:139352. https://doi.org/10.1016/j.scitotenv.2020.139352

Rodrigues MO, Abrantes N, Gonçalves FJM, Nogueira H, Marques JC, Gonçalves AMM (2018) Spatial and temporal distribution of microplastics in water and sediments of a freshwater system (Antuã River, Portugal). Sci Total Environ 633:1549–1559. https://doi.org/10.1016/j.scitotenv.2018.03.233

Ruan Y, Zhang K, Wu C, Wu R, Lam PKS (2019) A preliminary screening of HBCD enantiomers transported by microplastics in wastewater treatment plants. Sci Total Environ 674:171–178. https://doi.org/10.1016/j.scitotenv.2019.04.007

Sanchez W, Bender C, Porcher J-M (2014) Wild gudgeons (Gobio gobio) from French rivers are contaminated by microplastics: Preliminary study and first evidence. Environ Res 128:98–100. https://doi.org/10.1016/j.envres.2013.11.004

Scarascia-Mugnozza G, Sica C, Russo G (2012) Plastic materials in European agriculture: actual use and perspectives. J Agric Eng 42(3):15–28. https://doi.org/10.4081/jae.2011.3.15

Scherer C, Brennholt N, Reifferscheid G, Wagner M (2017) Feeding type and development drive the ingestion of microplastics by freshwater invertebrates. Sci Rep 7(1):17006. https://doi.org/10.1038/s41598-017-17191-7

Scherer C, Weber A, Lambert S, Wagner M (2018) Interactions of microplastics with freshwater biota. In: Freshwater microplastics. Springer, Cham, pp 153–180. https://doi.org/10.1007/978-3-319-61615-5_8

Scherer C, Weber A, Stock F, Vurusic S, Egerci H, Kochleus, C. … Reifferscheid, G. (2020) Comparative assessment of microplastics in water and sediment of a large European river. Sci Total Environ 738:139866. https://doi.org/10.1016/j.scitotenv.2020.139866

Scheurer M, Bigalke M (2018) Microplastics in Swiss floodplain soils. Environ Sci Technol 52(6):3591–3598. https://doi.org/10.1021/acs.est.7b06003

Schmidt C, Krauth T, Wagner S (2017) Export of plastic debris by rivers into the sea. Environ Sci Technol 51(21):12246–12253. https://doi.org/10.1021/acs.est.7b02368

Scopetani C, Chelazzi D, Cincinelli A, Esterhuizen-Londt M (2019) Assessment of microplastic pollution: occurrence and characterisation in Vesijärvi lake and Pikku Vesijärvi pond, Finland. Environ Monit Assess 191(11):652. https://doi.org/10.1007/s10661-019-7843-z

Shiber JG (1979) Plastic pellets on the coast of Lebanon. Mar Pollut Bull 10(1):28–30. https://doi.org/10.1016/0025-326X(79)90321-7

Shruti VC, Jonathan MP, Rodriguez-Espinosa PF, Rodríguez-González F (2019) Microplastics in freshwater sediments of Atoyac River basin, Puebla City, Mexico. Sci Total Environ 654:154–163. https://doi.org/10.1016/j.scitotenv.2018.11.054

Siegfried M, Koelmans AA, Besseling E, Kroeze C (2017) Export of microplastics from land to sea. A modelling approach. Water Res 127(Supplement C):249–257. https://doi.org/10.1016/j.watres.2017.10.011

Sighicelli M, Pietrelli L, Lecce F, Iannilli V, Falconieri M, Coscia L et al (2018) Microplastic pollution in the surface waters of Italian Subalpine Lakes. Environ Pollut 236:645–651. https://doi.org/10.1016/j.envpol.2018.02.008

Silva-Cavalcanti JS, Silva JDB, de França EJ, de Araújo MCB, Gusmão F (2017) Microplastics ingestion by a common tropical freshwater fishing resource. Environ Pollut 221:218–226. https://doi.org/10.1016/j.envpol.2016.11.068

Simon M, van Alst N, Vollertsen J (2018) Quantification of microplastic mass and removal rates at wastewater treatment plants applying Focal Plane Array (FPA)-based Fourier Transform Infrared (FT-IR) imaging. Water Res 142:1–9. https://doi.org/10.1016/j.watres.2018.05.019

Sinha T, Bhagwatwar P, Krishnamoorthy C, Chidambaram R (2019) Polymer based micro- and nanoencapsulation of agrochemicals. In: Gutiérrez TJ (ed) Polymers for agri-food applications. Springer, Cham, pp 5–28. https://doi.org/10.1007/978-3-030-19416-1_2

Sommer F, Dietze V, Baum A, Sauer J, Gilge S, Maschowski C, Gieré R (2018) Tire abrasion as a major source of microplastics in the environment. Aerosol Air Qual Res 18(8):2014–2028. https://doi.org/10.4209/aaqr.2018.03.0099

Sruthy S, Ramasamy EV (2017) Microplastic pollution in Vembanad Lake, Kerala, India: the first report of microplastics in lake and estuarine sediments in India. Environ Pollut 222:315–322. https://doi.org/10.1016/j.envpol.2016.12.038

Stanton T, Johnson M, Nathanail P, MacNaughton M, Gomes RL (2020) Freshwater microplastic concentrations vary through both space and time. Environ Pollut 263(B):114481. https://doi.org/10.1016/j.envpol.2020.114481

Steinmetz Z, Wollmann C, Schaefer M, Buchmann C, David J, Tröger J et al (2016) Plastic mulching in agriculture. Trading short-term agronomic benefits for long-term soil degradation? Sci Total Environ 550:690–705. https://doi.org/10.1016/j.scitotenv.2016.01.153

Su L, Xue Y, Li L, Yang D, Kolandhasamy P, Li D, Shi H (2016) Microplastics in Taihu Lake, China. Environ Pollut 216(Supplement C):711–719. https://doi.org/10.1016/j.envpol.2016.06.036

Su L, Cai H, Kolandhasamy P, Wu C, Rochman CM, Shi H (2018) Using the Asian clam as an indicator of microplastic pollution in freshwater ecosystems. Environ Pollut 234:347–355. https://doi.org/10.1016/j.envpol.2017.11.075

Su Y, Zhang Z, Wu D, Zhan L, Shi H, Xie B (2019) Occurrence of microplastics in landfill systems and their fate with landfill age. Water Res 164:114968. https://doi.org/10.1016/j.watres.2019.114968

Sun J, Dai X, Wang Q, van Loosdrecht MCM, Ni B-J (2019) Microplastics in wastewater treatment plants: detection, occurrence and removal. Water Res 152:21–37. https://doi.org/10.1016/j.watres.2018.12.050

Sundt P, Schulze P-E, Syversen F (2014) Sources of microplastics pollution to the marine environment, No. M-321 | 2015. MEPEX fpr the Norwegian Environment Agency, p 86. Retrieved from http://www.miljodirektoratet.no/Documents/publikasjoner/M321/M321.pdf

Sundt P, Syversen F, Skogesal O, Schulze P-E (2016) Primary microplasticpollution: measures and reduction potentials in Norway, No. M-545. Norwegian Environment Agency, p 117. Retrieved from http://www.miljodirektoratet.no/Documents/publikasjoner/M545/M545.pdf

Townsend AK, Barker CM (2014) Plastic and the nest entanglement of urban and agricultural crows. PLoS One 9(1):e88006. https://doi.org/10.1371/journal.pone.0088006

Turner S, Horton AA, Rose NL, Hall C (2019) A temporal sediment record of microplastics in an urban lake, London. UK J Paleolimnol:1–14. https://doi.org/10.1007/s10933-019-00071-7

UNEP (2009) Marine litter: a global challenge. Regional Seas, United Nations Environment Programme, Nairobi. Retrieved from: http://wedocs.unep.org/handle/20.500.11822/7787

UNESCO (2017) Waste water: the untapped resources. The United Nations World Water Development Report. Retrieved from: https://www.unido.org/sites/default/files/2017-03/UN_World_Water_Development_Report_-_Full_0.pdf

Unice KM, Kreider ML, Panko JM (2013) Comparison of tire and road wear particle concentrations in sediment for watersheds in France, Japan, and the United States by quantitative pyrolysis GC/MS Analysis. Environ Sci Technol 47(15):8138–8147. https://doi.org/10.1021/es400871j

Uurasjärvi E, Hartikainen S, Setälä O, Lehtiniemi M, Koistinen A (2020) Microplastic concentrations, size distribution, and polymer types in the surface waters of a northern European lake. Water Environ Res 92(1):149–156. https://doi.org/10.1002/wer.1229

van den Berg P, Huerta-Lwanga E, Corradini F, Geissen V (2020) Sewage sludge application as a vehicle for microplastics in eastern Spanish agricultural soils. Environ Pollut 261:114198. https://doi.org/10.1016/j.envpol.2020.114198

van Emmerik T, Loozen M, van Oeveren K, Buschman F, Prinsen G (2019) Riverine plastic emission from Jakarta into the ocean. Environ Res Lett 14(8):084033. https://doi.org/10.1088/1748-9326/ab30e8

van Wijnen J, Ragas AMJ, Kroeze C (2019) Modelling global river export of microplastics to the marine environment: Sources and future trends. Sci Total Environ 673:392–401. https://doi.org/10.1016/j.scitotenv.2019.04.078

Vaughan R, Turner SD, Rose NL (2017) Microplastics in the sediments of a UK urban lake. Environ Pollut 229:10–18. https://doi.org/10.1016/j.envpol.2017.05.057

Veerasingam S, Mugilarasan M, Venkatachalapathy R, Vethamony P (2016) Influence of 2015 flood on the distribution and occurrence of microplastic pellets along the Chennai coast, India. Mar Pollut Bull 109(1):196–204. https://doi.org/10.1016/j.marpolbul.2016.05.082

Vegkart (2020) Road statistics from the road map applications Vegkart https://vegkart.atlas.vegvesen.no/. Accessed 8th May 2020

Vermaire JC, Pomeroy C, Herczegh SM, Haggart O, Murphy M (2017) Microplastic abundance and distribution in the open water and sediment of the Ottawa River, Canada, and its tributaries. FACETS 2:201–314. https://doi.org/10.1139/facets-2016-0070

Villena OC, Terry I, Iwata K, Landa ER, LaDeau SL, Leisnham PT (2017) Effects of tire leachate on the invasive mosquito Aedes albopictus and the native congener Aedes triseriatus. PeerJ 5:e3756. https://doi.org/10.7717/peerj.3756

Vogelsang C, Lusher AL, Dadkhah ME, Sundvor I, Umar M, Ranneklev SB et al (2019) Microplastics in road dust – characteristics, pathways and measures. Norsk institutt for vannforskning. Retrieved from https://brage.bibsys.no/xmlui/handle/11250/2493537

Vollertsen J, Hansen AA (2017) Microplastic in Danish wastewater: Sources, occurrences and fate, Environmental Project No. 1906. Danish Environmental Protection Agency, pp 1–55. Retrieved from https://www2.mst.dk/Udgiv/publications/2017/03/978-87-93529-44-1.pdf

Wagner S, Hüffer T, Klöckner P, Wehrhahn M, Hofmann T, Reemtsma T (2018) Tire wear particles in the aquatic environment – a review on generation, analysis, occurrence, fate and effects. Water Res 139:83–100. https://doi.org/10.1016/j.watres.2018.03.051

Walling DE, Webb BW, Woodward JC (1992) Some sampling considerations in the design of effective strategies for monitoring sediment-associated transport. Erosion and Sediment Transport Monitoring Programmes in River Basins (Proceedings of the Oslo Symposium, August 1992), IAHS Publication no. 210, pp 279–288. Retrieved from: https://pdfs.semanticscholar.org/c3db/f67dbe41df9f7194ebe55d3748015fd816ad.pdf?_ga=2.165832254.18612053.1583661455-2018662021.1583059758

Walling DE, Owens P, Carter J, Leeks GJ, Lewis S, Meharg A, Wright J (2003) Storage of sediment-associated nutrients and contaminants in river channel and floodplain systems. Appl Geochem 18(2):195–220. https://doi.org/10.1016/S0883-2927(02)00121-X

Wang W, Ndungu AW, Li Z, Wang J (2017) Microplastics pollution in inland freshwaters of China: a case study in urban surface waters of Wuhan, China. Sci Total Environ 575:1369–1374. https://doi.org/10.1016/j.scitotenv.2016.09.213

Wang W, Yuan W, Chen Y, Wang J (2018) Microplastics in surface waters of Dongting Lake and Hong Lake, China. Sci Total Environ 633:539–545. https://doi.org/10.1016/j.scitotenv.2018.03.211

Wang T, Li B, Zou X, Wang Y, Li Y, Xu Y et al (2019a) Emission of primary microplastics in mainland China: Invisible but not negligible. Water Res 162:214–224. https://doi.org/10.1016/j.watres.2019.06.042

Wang Z, Qin Y, Li W, Yang W, Meng Q, Yang J (2019b) Microplastic contamination in freshwater: first observation in Lake Ulansuhai, Yellow River Basin, China. Environ Chem Lett 17(4):1821–1830. https://doi.org/10.1007/s10311-019-00888-8

Watkins L, Sullivan PJ, Walter MT (2019) A case study investigating temporal factors that influence microplastic concentration in streams under different treatment regimes. Environ Sci Pollut Res 26(21):21797–21807. https://doi.org/10.1007/s11356-019-04663-8

Wei S, Luo H, Zou J, Chen J, Pan X, Rousseau DPL, Li J (2020) Characteristics and removal of microplastics in rural domestic wastewater treatment facilities of China. Sci Total Environ 739:139935. https://doi.org/10.1016/j.scitotenv.2020.139935

Wen X, Du C, Xu P, Zeng G, Huang D, Yin L, Yin Q, Hu L, Wan J, Zhang J, Tan S, Deng R (2018) Microplastic pollution in surface sediments of urban water areas in Changsha, China: Abundance, composition, surface textures. Mar Pollut Bull 136:414–423. https://doi.org/10.1016/j.marpolbul.2018.09.043

Wik A, Dave G (2005) Environmental labeling of car tires—toxicity to Daphnia magna can be used as a screening method. Chemosphere 58(5):645–651. https://doi.org/10.1016/j.chemosphere.2004.08.103

Wik A, Dave G (2006) Acute toxicity of leachates of tire wear material to Daphnia magna—variability and toxic components. Chemosphere 64(10):1777–1784. https://doi.org/10.1016/j.chemosphere.2005.12.045

Wik A, Dave G (2009) Occurrence and effects of tire wear particles in the environment – a critical review and an initial risk assessment. Environ Pollut 157(1):1–11. https://doi.org/10.1016/j.envpol.2008.09.028

Wik A, Nilsson E, Källqvist T, Tobiesen A, Dave G (2009) Toxicity assessment of sequential leachates of tire powder using a battery of toxicity tests and toxicity identification evaluations. Chemosphere 77(7):922–927. https://doi.org/10.1016/j.chemosphere.2009.08.034

Windsor FM, Tilley RM, Tyler CR, Ormerod SJ (2019a) Microplastic ingestion by riverine macroinvertebrates. Sci Total Environ 646:68–74. https://doi.org/10.1016/j.scitotenv.2018.07.271

Windsor FM, Durance I, Horton AA, Thompson RC, Tyler CR, Ormerod SJ (2019b) A catchment-scale perspective of plastic pollution. Glob Chang Biol 25:1207–1221

Winter LT, Foster IDL, Charlesworth SM, Lees JA (2001) Floodplain lakes as sinks for sediment-associated contaminants—a new source of proxy hydrological data? Sci Total Environ 266(1–3):187–194. https://doi.org/10.1016/s0048-9697(00)00745-2

Wright SL, Ulke J, Font A, Chan KLA, Kelly FJ (2020) Atmospheric microplastic deposition in an urban environment and an evaluation of transport. Environ Int 136:105411. https://doi.org/10.1016/j.envint.2019.105411

Xiong X, Zhang K, Chen X, Shi H, Luo Z, Wu C (2018) Sources and distribution of microplastics in China's largest inland lake – Qinghai Lake. Environ Pollut 235:899–906. https://doi.org/10.1016/j.envpol.2017.12.081

Xu Q, Gao Y, Xu L, Shi W, Wang F, LeBlanc GA et al (2020) Investigation of the microplastics profile in sludge from China's largest Water reclamation plant using a feasible isolation device. J Hazard Mater 388:122067. https://doi.org/10.1016/j.jhazmat.2020.122067

Yang L, Li K, Cui S, Kang Y, An L, Lei K (2019) Removal of microplastics in municipal sewage from China's largest water reclamation plant. Water Res 155:175–181. https://doi.org/10.1016/j.watres.2019.02.046

Yin L, Jiang C, Wen X, Du C, Zhong W, Feng Z, Long Y, Ma Y (2019) Microplastic pollution in surface water of urban lakes in Changsha, China. Int J Environ Res Public Health 16(9):1650. https://doi.org/10.3390/ijerph16091650

Yu M, van der Ploeg M, Lwanga EH, Yang X, Zhang S, Ma X et al (2019) Leaching of microplastics by preferential flow in earthworm (Lumbricus terrestris) burrows. Environ Chem 16(1):31–40. https://doi.org/10.1071/EN18161

Yuan W, Liu X, Wang W, Di M, Wang J (2019) Microplastic abundance, distribution and composition in water, sediments, and wild fish from Poyang Lake, China. Ecotoxicol Environ Saf 170:180–187. https://doi.org/10.1016/j.ecoenv.2018.11.126

Zhang Z, Chen Y (2020) Effects of microplastics on wastewater and sewage sludge treatment and their removal: a review. Chem Eng J 382:122955. https://doi.org/10.1016/j.cej.2019.122955

Zhang GS, Liu YF (2018) The distribution of microplastics in soil aggregate fractions in southwestern China. Sci Total Environ 642:12–20. https://doi.org/10.1016/j.scitotenv.2018.06.004

Zhang K, Su J, Xiong X, Wu X, Wu C, Liu J (2016a) Microplastic pollution of lakeshore sediments from remote lakes in Tibet plateau, China. Environ Pollut 219:450–455. https://doi.org/10.1016/j.envpol.2016.05.048

Zhang K, Su J, Xiong X, Wu X, Wu C, Liu J (2016b) Microplastic pollution of lakeshore sediments from remote lakes in Tibet plateau, China. Environ Pollut 219:450–455. https://doi.org/10.1016/j.envpol.2016.05.048

Zhang Y, Gao T, Kang S, Sillanpää M (2019) Importance of atmospheric transport for microplastics deposited in remote areas. Environ Pollut 254:112953. https://doi.org/10.1016/j.envpol.2019.07.121

Zhao S, Zhu L, Li D (2016) Microscopic anthropogenic litter in terrestrial birds from Shanghai, China: Not only plastics but also natural fibers. Sci Total Environ 550:1110–1115. https://doi.org/10.1016/j.scitotenv.2016.01.112

Zhou P, Huang C, Fang H, Cai W, Li D, Li X, Yu H (2011) The abundance, composition and sources of marine debris in coastal seawaters or beaches around the northern South China Sea (China). Mar Pollut Bull 62(9):1998–2007. https://doi.org/10.1016/j.marpolbul.2011.06.018

Zhou Q, Tian C, Luo Y (2017) Various forms and deposition fluxes of microplastics identified in the coastal urban atmosphere. Chin Sci Bull 62(33):3902–3909. https://doi.org/10.1360/N972017-00956

Zhou B, Wang J, Zhang H, Shi H, Fei Y, Huang S et al (2019) Microplastics in agricultural soils on the coastal plain of Hangzhou Bay, east China: multiple sources other than plastic mulching film. J Hazard Mater 121814. https://doi.org/10.1016/j.jhazmat.2019.121814

5

Microplastics in Seafood and the Impact on Human Health

Anne-Katrine Lundebye, Amy L. Lusher and Michael S. Bank

Abstract Seafood is an important food source, and this chapter addresses the food safety concerns related to plastic particles in different seafood. Here we focus on those species which are commonly consumed by humans, such as bivalves, gastropods, cephalopods, echinoderms, crustaceans, and finfish. The objectives of this chapter are to (1) outline the major sources, fate, and transport dynamics of microplastics in marine ecosystems, (2) provide a critical assessment and synthesis of microplastics in seafood taxa commonly consumed by humans, (3) discuss the implications of microplastics with regard to human health risk assessments, and (4) suggest future research priorities and recommendations for assessing microplastics in marine ecosystems in the context of global food security and ocean and human health.

5.1 Introduction

Seafood is an important food source – with fisheries and aquaculture production predicted to increase by about 17.5% from 171 million tonnes in 2016 to approximately 201 million tonnes in 2030 (FAO 2018). It is a necessity that these marine-based foods are carefully managed and are safe for human consumption. Food

A.-K. Lundebye (✉)
Institute of Marine Research, Bergen, Norway
e-mail: Anne-Katrine.Lundebye@hi.no

A. L. Lusher
Norwegian Institute for Water Research, Oslo, Norway

University of Bergen, Bergen, Norway
e-mail: Amy.Lusher@niva.no

M. S. Bank
Institute of Marine Research, Bergen, Norway

University of Massachusetts Amherst, Amherst, MA, USA
e-mail: Michael.Bank@hi.no; mbank@eco.umass.edu

security is defined by the Food and Agriculture Organization (FAO) as "a situation that exists when all people, at all times, have physical, social and economic access to sufficient, safe and nutritious food that meets their dietary needs and food preferences for an active and healthy life" (FAO 2017). In this chapter, the food safety concerns related to plastic particles in seafood will be addressed.

Global annual production of plastics is estimated to be approximately 300 million tonnes (Galloway 2015) and is still increasing steadily. Most plastic polymers are resistant to complete degradation and pose a potential risk to both human and environmental health. Of particular concern are microplastics, which are defined as particles <5 mm (GESAMP 2019) and which are the focus of this chapter.

Microplastics occur in different shapes and sizes and are formed from different polymers as well as additives, which reflects the diversity of sources and emissions to the environment (Rochman et al. 2019). The dominant microplastic polymers which are detected in the marine environment include polyethylene (PE), polypropylene (PP), polystyrene (PS), polyethylene terephthalate (PET), polyamide, and polyvinylchloride (PVC) (Hantoro et al. 2019). Primary microplastics are manufactured intentionally for a range of commercial uses (e.g., microbeads) whereas secondary microplastics originate from parent material such as textiles and discarded plastic items and are either generated through the use of plastic products or fragmentation following their loss to different environmental compartments. Plastic debris can enter the ocean from ships and fishing gear, as well as from atmospheric deposition, river transport, stormwater, sewage effluents, etc. (Browne et al. 2011; Napper and Thompson 2016; Lebreton et al. 2017; Allen et al. 2019). Plastics and microplastics have been identified in the oceans, from coastal zones to offshore areas, such as oceanic gyres (Eriksen et al. 2014; Jiang et al. 2020;), as well as in remote areas including the Arctic (e.g., Cózar et al. 2017). The ubiquitous nature of plastics in the ocean is an obvious concern for marine ecosystems and their inhabitants. In particular, the progression from macro- to microplastics at sea is a result of physical erosion and UV action and increases the bioavailability of smaller-sized particles to a wide array of marine organisms (Browne et al. 2008; Wright et al. 2013). Plastics have long been reported associated with marine organisms, from the first study of plastic ingestion by fish (Carpenter et al. 1972) to mariculture sites where boring worms facilitate the generation of microplastics from polystyrene buoys (Jang et al. 2018). Many investigations have been conducted to further understand the interaction between marine organisms and microplastics with several studies focusing on microplastic uptake, ingestion, exposure, and metabolic dynamics (Roch et al. 2020). Fibers are routinely identified as the most common microplastic type reported in fish, accounting for 58–87% of the plastic morphologies observed (Walkinshaw et al. 2020). Fragments, films, and fibers are also frequently found in fish and shellfish while microplastics in the forms of spheres are less common. The physical impacts of microplastic ingestion on marine organisms can include oxidative stress, inflammation, and potentially starvation, while less is known regarding the chemical effects of ingestion. The bioavailability and potential toxicity of microplastics are size dependent, with smaller particles able to penetrate further into an organism (Browne et al. 2008), with the potential of the release of associated co-contaminants (Bakir et al. 2016; Batel et al. 2016). It is widely accepted that ingestion is the main route of microplastics uptake in marine biota; however, it has recently also been demonstrated that surface scavenging appears to be an alternative route, as demonstrated in mussels (Kolandhasamy et al. 2018). Microplastics can also be taken up through respiration via the gills (Watts et al. 2016; Franzellitti et al. 2019) and have additionally been demonstrated to be maternally transferred to eggs in zebra fish (Pitt et al. 2018).

Plastics in aquatic environments have been shown to affect an organism's health (such as behavioral changes and reduced growth rates); however, there is limited information on the effects of microplastics in seafood on human health. Lusher et al. (2017a) reported that more than 220 species of marine organisms including zooplankton, bivalves, crustaceans, fish, marine mammals, sea turtles, and seabirds had been shown to have ingested plastics, and more recently this number of species has increased to 690 (Carbery et al. 2018). Here we focus on those species which are commonly consumed by humans, such as bivalves, echinoderms, gastropods, cephalopods, crustaceans, and finfish. The specific objectives of this chapter are to (1) outline the major sources, fate, and transport dynamics of microplastics in marine ecosystems, (2) provide a critical assessment and synthesis of microplastics in seafood taxa commonly consumed by humans, (3) discuss the implications of microplastics with regard to human health risk assessments, and (4) suggest future research priorities and recommendations for assessing microplastics in marine ecosystems in the context of global food security and ocean and human health.

5.2 Fate and Transport of Microplastics in Marine Ecosystems.

The fate and transport of microplastics in the context of physical and biological oceanography has recently been reviewed by van Sebille et al. (2020), as well as by Thushari and Senevirathna (2020) with older reviews and critical papers developed by Andrady (2011), Wright et al. (2013), Galloway et al. (2017), and Wieczorek et al. (2019). The ocean can be both a source and sink for microplastics (Allen et al. 2020), and important themes within the cycling and degradation of microplastic particles (Weinstein et al. 2016) include the importance of transport from land via rivers (Lebreton et al. 2017; Hurley et al. 2018), the role of seafloor ocean circulation patterns as a driver of microplastic hotspots (Kane et al. 2020), and the concept of marine snow which has been identified as an important mechanism for transporting microplastic particles from the water column to the sediment (Porter et al. 2018). Moreover, fishing gear and other sources of macroplastics can degrade into microplastics via biological, chemical, and physical processes (Davidson 2012). Although settling of microplastic particles to the ocean floor is well-documented, recent research has shown that episodic events such as flooding (Hurley et al. 2018) and typhoons (Wang et al. 2019) are important drivers regarding the distribution and abundance of microplastics in coastal marine ecosystems.

5.3 Microplastic in Bivalves

Bivalves are by far the most investigated seafood species (Smith et al. 2018; Walkinshaw et al. 2020). Much of the investigations were performed for the purpose of uptake of microplastics from the environment, as filtering puts bivalves at an increased risk of microplastic intake from the water column (Li et al. 2019). Early investigations focused on blue mussels (*Mytilus* spp.), with wild and market brought samples presenting contamination levels of up to 7.2 microplastics per gram (Abidli et al. 2019; Bråte et al. 2018; Cho et al. 2019; De Witte et al. 2014; Li et al. 2015, 2016; Renzi et al. 2018a; van Cauwenberghe and Janssen 2014; van Cauwenberghe et al. 2015; Vandermeersch et al. 2015). Other bivalve species which have been investigated for microplastic uptake include clams (*Venerupis philippinarum*), oysters (*Crassostrea gigas*), and scallops (*Patinopecten yessoensis*) (Abidli et al. 2019; Cho et al. 2019, 2020; Davidson and Dudas 2016; Li et al. 2015; Rochman et al. 2015; van Cauwenberghe and Janssen 2014).

Microplastic fibers are often the most dominant morphology reported in bivalves. For example, fibers accounted for 80% of microplastics in mussels (*Mytilus edulis, Perna viridis*) from China (Qu et al. 2018), 90% of microplastics in Manila clams (*V. philippinarum*) from British Columbia (Davidson and Dudas 2016), and 99% in razor clams (*Siliqua patula*) and Pacific oyster (*Crassostrea gigas*) from Oregon, USA (Baechler et al. 2020a). One of the hypotheses behind the observed high abundance of this type of microplastic is that fibers are likely harder to remove from digestive tracts. Ward et al. (2019) reported that larger spheres are rejected at higher numbers (98%) than smaller spheres (10–30%). Fragments were most common in blue mussels and Pacific oyster from the French Atlantic coast (Phuong et al. 2018) as well as those from Korea, where EPS fragments likely originated from the high abundance of aquaculture facilities in the region (Cho et al. 2020). De Witte et al. (2014) reported that there was a high prevalence of fibers in blue mussels collected from quaysides related to fishing activities.

Microplastics in bivalves are likely dependent on several factors including, but not limited to, culture conditions and contamination levels in the environment, depuration procedures, filtration capabilities, as well as the tissues targeted for investigation. Some investigations, in distinct parts of the world, have found that bivalves sampled from highly contaminated areas or within the vicinity of urban sources of microplastics contained higher numbers of microplastics (Bråte et al. 2018; Qu et al. 2018; Cho et al. 2020). However, conversely, some investigations have reported no difference in microplastic exposure in bivalves related to sources (Covernton et al. 2019; Phuong et al. 2018).

There have been some reported differences between the occurrence of microplastics in market purchased (80%) and wild-caught individual bivalves (40%)

(Ding et al. 2018). Similarly, farmed mussels displayed higher concentrations of microplastics than wild mussels (75 items and 34 items per mussel, respectively) (Mathalon and Hill 2014), although no difference was observed for wild and cultured Manila clams (*Venerupis philippinarum*) in British Columbia (Davidson and Dudas 2016). The use of depurations procedures appears to reduce the number of microplastics identified in bivalve species (van Cauwenberghe and Janssen 2014 – *Mytilus edulis* and *Crassostrea gigas*; Birnstiel et al. 2019 – *Perna perna*), which would hold significance in the preparation of mussels for consumption. The seasonality of sampling could also play a role in observed microplastic concentrations in marine biota. A significant seasonal variation was observed during summer for oyster samples which contained more microplastics; however, this trend was not detected for razor clams (Baechler et al. 2020c).

Particle selection by bivalves, related to size and morphology, will influence which particles are internalized both pre- and post-ingestion (Ward et al. 2018). Gut retention times, which are known to vary between bivalve species and the age of individuals, have shown, in general, that as particle size decreases, accumulation increases (Browne et al. 2008; Ward and Kach 2009; Ward et al. 2019). Much of the work performed on bivalves is based on the sampling and processing of whole organisms, with no differentiation between and among tissue types; this makes it impossible to determine whether microplastics were internalized by individuals, had migrated from gills and guts to visceral tissue, or were in the process of being egested (e.g., as pseudofeces). Kolandhasamy et al. (2018) reported that microplastic fibers can accumulate on the foot and mantle of blue mussels.

Consequences of microplastic intake/uptake by bivalves indicate that microplastics can directly affect bivalve physiology but also indirectly change the structure of their habitats, impairing food resources and facilitate the efficient transfer of organic pollutants (Zhang et al. 2019a). Other observed implications include negative effects on filtration activity (Green et al. 2019; Xu et al. 2017), feeding behavior (Wegner et al. 2012), and reproduction (Sussarellu et al. 2016; Gardon et al. 2018). It is important to highlight that effects are mostly studied using uniform particles, mostly spheres so these may not be truly representative of environmentally relevant microplastic exposure regimes (see Gomes et al. 2021, Chap. 7, this volume).

5.4 Microplastics in Echinoderms

Sea urchins and sea cucumbers are the main echinoderms consumed as food item, and few studies have been conducted on the abundance of microplastics in these marine organisms. Of the heart urchins (*Brissopsis lyrifera*) analyzed, 40% were found to contain microplastics in their soft tissue, primarily in the form of flakes (90%, the remaining 10% as fibers). In most cases the number of particles present was 1/individual (Bour et al. 2018). It is noteworthy that this study was conducted for an ecological assessment of the influence of habitat, feeding mode, and trophic level on microplastic abundance in benthic and epibenthic organism and that this species is not commonly consumed. Feng et al. (2020) reported a higher prevalence of microplastics (in 90% of the individuals) in four species of sea urchins (*Strongylocentrotus intermedius*, *Temnopleurus hardwickii*, *Temnopleurus reevesii*, and *Hemicentrotus pulcherrimus*) harvested from 12 sites along the northern China coast. The average abundance of microplastics (predominantly as fibers) in soft tissue from sea urchins from all sites was 5 particles/individual (1.1 particles/g), considerably higher than reported in heart urchins from the Oslofjord, Norway (Bour et al. 2018). Higher detection rates and abundances were found in sea urchins from Dalian, China (Feng et al. 2020). The tissue of relevance in urchins with regards to seafood safety is the gonads, and while whole soft tissue of heart urchins was analyzed for microplastics (Bour et al. 2018), the abundance in urchins from the Yellow Sea was assessed in gonads, coelomic fluid, and the gut. Gonads and coelomic fluid contained significantly lower number of particles/individual than the gut in all four species of urchin; however, this difference was not evident when normalized to wet weight in three of the species, and it only remained significantly lower in *S. intermedius* (Feng et al. 2020).

Microplastic ingestion has been reported in several species of sea cucumber including *Holothuria grisea*, *Cucumaria frondosa*, *Holothuria floridana*, *Thyonella gemmata* (Graham and Thompson 2009), *Holothuria tubulosa* (Renzi et al. 2018b), *Holothuria mexicana*, *Actinopyga agassizi* (Plee and Pomory 2020), and *Apostichopus japonicus* (Mohsen et al. 2019). Sea cucumbers are commonly eaten in Asia, and farming is widespread to meet consumer demand. The body wall of sea cucumbers is typically eaten raw in Japan and boiled, pickled, or salted in China, and the internal organs (gonads, respiratory trees, and intestines) are also edible (Kiew and Don 2011). Iwalaye et al. (2020) reported microplastic particles in the intestines, coelomic fluid, and respiratory trees of the *Holothuria cinerascens* and that uptake was both via the feeding tentacles and the respiratory trees. The most abundant microplastics found in farmed sea cucumbers (*Apostichopus japonicus*) from eight locations in the Bohai Sea and Yellow Sea in China were cellophane microfibers (Mohsen et al. 2019).

5.5 Microplastics in Gastropods

Limited research has been carried out on microplastics in marine gastropods. Xu et al. (2020) analyzed nine species of gastropods from shores in Hong Kong for microplastics, with the highest abundance found in sea snails (*Batillaria multiformis*, 5.4 ± 1.2 particles/g wet weight) and the lowest observed in Chameleon nerite snails (*Nerita chamaeleon*, 1.50 ± 0.2 particles/g wet weight). The common periwinkle (*Littorina littorea*) sampled from four different locations in Galway Bay, Ireland, contained between 0.6 and 2.8 microplastics/g wet weight of soft tissue, and commercial common periwinkles, intended for human consumption, contained on average 2.2 microplastic s/g wet weight soft tissue (Doyle et al. 2019). Most of the microplastics (97%) recorded in periwinkles were fibers. Similarly, fibers accounted for more than half of the total microplastics present in the girdled horn shell (*Cerithidea cingulata*), whereas film was the most abundant microplastic (approximately 44%) in *Thais mutabilis* from the Persian Gulf region. The mean number of total microplastics was 13 and 20 particles/g wet soft tissue weight for *C. cingulata* and *T. mutabilis*, respectively (Naji et al. 2018). Lower levels of microplastic contamination were reported in periwinkles (*Littorina* spp.) from two sites on the eastern coast of Thailand with an average of 0.17 particles/g wet weight and 0.23 particles/g wet weight, with no contamination observed in periwinkles from Bangasaen, the third site investigated (Thushari et al. 2017).

5.6 Microplastics in Cephalopods

Cephalopods are the seafood phylum which have received the least focus with regard to microplastic contamination. Oliveira et al. (2020) investigated the levels of microplastics in the stomach, caecum/intestine, and digestive gland of cuttlefish (*Sepia officinalis*); however, tissue relevant for consumer exposure was not included in this study. Microplastic contamination in Indian squid (*Uroteuthis duvaucelii*) was found in 18% of the individuals examined, with an average of 0.2 microplastic particles/individual and 0.008 microplastic particles/g wet weight of edible tissue (Peng et al. 2020).

5.7 Microplastics in Crustaceans

Most biota-based studies have examined microplastics in the organisms' gut, which is not generally an organ consumed directly by humans. However, shellfish including crustaceans and mollusks are an exception since these are frequently eaten either whole or with their gut removed. The risk of ingesting microplastics from other tissues, such as muscle, depends on the ability to cross the intestinal barrier and subsequent accumulation (Zeytin et al. 2020).

To date, most literature on crustaceans which are commonly harvested for human consumption has focused on wild individuals, rather than those that are farmed. Investigations generally have not focused on crustaceans in the context of seafood safety but rather from an environmental contaminant perspective. For example, there have been numerous investigations into langoustine, *Nephrops norvegicus*, which are also commercially exploited. *N. norvegicus*, sampled from the Clyde Sea area, were shown to contain more microplastic fibers in their gut than individuals from the North Sea and Minch where only a small percentage of individuals contained microplastic, predominantly as single-strand fibers (Welden and Cowie 2016). Other commercially relevant species, such as spinous spider crabs (*Maja squinado*), shrimps, and prawns, have been observed to contain microplastics (Welden et al. 2018; Devriese et al. 2015; Zhang et al. 2019b; Cau et al. 2019).

Many crustaceans are harvested from coastal environments, which may be close to sources of microplastic contamination, including the influence of terrestrial plastic sources. As an example, shrimp (*A. antennatus*) from the Mediterranean had an average occurrence of microplastics equal to 39.2%; however, those in close vicinity to urban areas had 100% presence of microplastics (Carreras-Colom et al. 2018). The same % occurrence trend was observed between remote populations (<40%) of *N. norvegicus* compared to those sampled near Glasgow (84%) in the Clyde Sea (Welden and Cowie 2016).

Additionally, no spatial pattern was observed in a similar study of *N. norvegicus* in Irish waters (Hara et al. 2020). Both *N. norvegicus* and *Aristeus antennatus* were investigated in the Mediterranean Sea from depths between 270 and 660 meters (Cau et al. 2019). The authors reported a significant difference in the size and composition of microplastics identified between the two species and suggested that the nonselective feeding strategy of *N. norvegicus* likely led to a higher degree of exposure to microplastics and hence a higher measured abundance. Nonselective feeding is an example of direct uptake of microplastics from the environment. Organisms can also ingest microplastics which have been internalized by prey species, a concept commonly referred to as trophic transfer. Laboratory studies on this topic performed with shore crabs (*Carcinus maenas*) fed mussels which had been exposed to microplastics showed that polystyrene microspheres could accumulate in the foregut of the crabs (Watts et al. 2015).

Fibers and fragments are the most often reported particle type in crustaceans sampled from the wild, with fiber bundles reported across many species (Welden and Cowie 2016; Cau et al. 2019; McGoran et al. 2020). In most studies, stomachs were often the target organ of microplastics investigations, but other tissues are starting to be considered further, as these may have relevance for human exposure, especially when stomachs are removed prior to cooking and consumption. As an example, microplastics have been found in different tissues of wild-caught *Portunus gracilimanus* and *P. trituberculatus* (Zhang et al. 2019b).

5.8 Microplastics in Finfish

Evaluating microplastic occurrence and abundance in finfish is fundamental to understanding how plastics and their associated chemical compounds affect and potentially impact wild fisheries that are relied upon by humans as an important source of food and nutrition (Rochman et al. 2015; Barboza et al. 2018; FAO 2020; Lusher and Welden 2020). The topic of microplastics in the marine environment, including information on finfish, has been reviewed by several authors (Andrady 2011; Cole et al. 2011; Hidalgo-Ruz et al. 2012; Wright et al. 2013; Gall and Thompson 2015; Galloway et al. 2017; Baechler et al. 2020a, b; Thushari and Senevirathna 2020; Wang et al. 2020; Walkinshaw et al. 2020). Microplastics exposure in finfish is largely a result of plastics being mistaken for natural prey items, via ingestion of contaminated prey items or by passive uptake through gills (Lusher et al. 2016; Watts et al. 2015; Nelms et al. 2018; Roch et al. 2020). Trophic transfer of microplastics may also expose predaceous fish to microplastics (Farrell and Nelson 2013; Setälä et al. 2014; Lusher et al. 2016; Baechler et al. 2020a), and microplastics have frequently been detected in finfish gastrointestinal tracts (e.g., Lusher et al. 2017b,). The methodological challenges with identifying particles within fillet muscle tissue have limited the number of published studies thus far, although they have been identified albeit at extremely low concentrations (Zeytin et al. 2020).

Many species of edible demersal, pelagic, and reef fish, sampled from across the globe, have been found to contain microplastics (e.g., Bellas et al. 2016; Rummel et al. 2016; Bråte et al. 2016; Lusher et al. 2013; Tanaka and Takada 2016; Rochman et al. 2015; Neves et al. 2015; Critchell and Hoogenboom 2018; Abbasi et al. 2018; Su et al. 2018). The percentages of different fish species which have been found to contain microplastics in their gut vary greatly: 0.9% Peruvian anchovy, 2.8% Atlantic cod, 8.8% Atlantic herring, 9.4% Skipjack tuna, 24.5% Jack and Horse mackerel, 23.3% Pacific chub mackerel, 23.4% Yellowfin tuna, and 76.6% Japanese anchovy (Neves et al. 2015; Lusher et al. 2013; Güven et al. 2017; Ogonowski et al. 2017; Rummel et al. 2016; Hermsen et al. 2018; Rochman et al. 2015; Choy and Drazen 2013; Markic et al. 2018; Bråte et al. 2016; Liboiron et al. 2016; Tanaka and Takada 2016). Several studies have examined the microplastic particle prevalence in fish with different feeding ecology (Foekema et al. 2013; Lusher et al. 2013). Lusher et al. (2013) did not find any significant difference between the abundance of plastic ingested by pelagic and demersal fish. Of the 24 fish species examined from the Beibu Gulf, one of the world's largest fishing grounds, in the South China Sea, 12 species contained microplastics (Koongolla et al. 2020). The abundance of microplastics varied from 0.027 to 1 item per individual, and most was present in fish stomach (57.7%) and less in intestines and gills (34.6% and 7.7%, respectively). Nine of the 11 fish species sampled from Zhoushan fishing grounds in the East China Sea were found to contain microplastics, with 23 different polymer types identified, and the highest number of items was 8 in a single individual (Zhang et al. 2019a). It is challenging to compare all the studies listed above, as many different methods have been utilized by researchers to determine the presence or absence of microplastics across these species. Some trends in the methods used have previously been described, with visually searching the most common method (Lusher et al. 2017b); however, the lack of standards and incomplete reporting of data, and quality control procedures have also been highlighted (Hermsen et al. 2018). Differences in sampling and analytical methods may lead to different values being observed and are important to consider when evaluating trends across regions, ecosystem types, and species.

The microplastic content of wild fish has been more widely studied than aquaculture species. A recent review of microplastics in seafood found that data were lacking for four of the ten most cultured aquatic food species, namely, grass carp, whiteleg shrimp, bighead carp, and catla (Walkinshaw et al. 2020).

5.9 Co-contaminants Associated with Microplastics in Seafood

The role of marine microplastics as vectors for major ocean pollutants was recently reviewed by Ziccardi et al. (2016), Santillo et al. (2017), and Amelia et al. (2021). Plastics are inherently complex in size, morphology, and polymer composition and may contain a range of additives, including plasticizers, stabilizers, pigments, fillers, and flame retardants which may leach out into the environment including air, water, and food, and in general, microplastics are now considered to represent a suite of co-contaminants (Rochman et al. 2019). More than 50% of plastics are associated with hazardous monomers, additives, and chemical byproducts (Lithner et al. 2011). Plastics have been shown to accumulate various organic and inorganic co-contaminants from the surrounding water column (Rochman et al. 2013, 2015). The high surface area to volume ratio of small particles and hydrophobic nature facilitate the sorption of chemicals on the plastic surface, forming a complex mixture of contaminants available to marine organisms (Rochman et al. 2013). Laboratory studies have demonstrated that continuous exposure to contaminated plastics can lead to the accumulation of plastic-associated co-contaminants in fish (Rochman et al. 2013; Wardrop et al. 2016).

Both field and modeling studies suggest that transfer of environmental pollutants through microplastics are negligible compared to other routes of uptake (Gouin et al. 2011; Bakir et al. 2016; Espinosa et al. 2018; Koelmans et al. 2016; Ziccardi et al. 2016; Lohmann 2017; Smith et al. 2018). Nonetheless, caution is warranted as many of the chemicals sorbed onto microplastics are known to be potent toxicants to humans and marine biota, triggering adverse effects such as endocrine disruption, neurological disorders, and reduced reproductive success (GESAMP 2016). An example of this is the investigation by Barboza et al. (2020a) who reported significantly higher concentrations of bisphenols in fish with microplastics compared to individuals with no microplastics. However, none of the fish species investigated (European seabass *Dicentrarchus labrax*, Atlantic horse mackerel *Trachurus trachurus*, and Atlantic chub mackerel *Scomber colias*) contained bisphenol A levels which would lead to an exceedance of the Tolerable daily Intake established by the European Food Safety Authority (EFSA) (Barboza et al. 2020a).

5.10 Microplastic Uptake and Toxicity in Humans

The uptake of microplastics is dependent on size, morphology, solubility, and surface charge and chemistry. Microplastics <130 μm in diameter can potentially translocate into human tissue (EFSA 2016), and particles sized 1.5 μm and below can penetrate capillaries (Yoo et al. 2011). Proposed mechanisms for uptake of microplastics include endocytotic and paracellular transfer across epithelial tissues (Wright and Kelly 2017). It is estimated that 90% of ingested microplastics are excreted from the body (EFSA 2016); however, the remaining microplastics may be detrimental to human health, and further research is required to develop a more comprehensive understanding regarding public health aspects of microplastic pollution.

Oxidative stress and subsequent inflammation are both thought to be the main mechanisms of particle toxicity (Feng et al. 2016). Other potential biological responses to microplastic exposure include genotoxicity, apoptosis, and necrosis, which could ultimately lead to tissue damage, fibrosis, and carcinogenesis (Wright and Kelly 2017). The extent of potential adverse effects is dependent on particle size, and nanoparticles have been found to generate more reactive oxygen species (ROS) than larger particles and are more likely to be translocated (Stone et al. 2007). Consequently, potential health effects of microplastics largely depend on particle characteristics, and it is envisaged that nanoplastics are likely more deleterious than microplastics (Feng et al. 2016).

5.11 Consequences of Microplastics in Marine Animals

More than 690 marine species from different trophic levels have been reported to contain plastic debris; however, the transfer of microplastics and associated co-contaminants, from seafood to humans, and the implications for seafood safety have received limited attention to date (Carbery et al. 2018; Lusher et al. 2017a; Walkinshaw et al. 2020). Most studies conducted have considered the environmental rather than the potential human health impacts of micro- and nanoplastics. Effects of micro- and nanoplastic exposure reported in marine organisms include reduced growth, impacted energy metabolism, feeding behavior, and locomotion, effects on the immune system, and hormonal regulation, physiological stress, oxidative stress, inflammation, aberrant development, cell death, general toxicity, and altered lipid metabolism (Kögel et al. 2019). In humans, it is evidenced that consumers may be exposed to microplastics from seafood consumption; however, the risks remain unclear (Smith et al. 2018; VKM 2019).

Shellfish and small fish that are consumed whole are the seafoods which are likely to give the highest exposure risk since the gastrointestinal tract, which generally contains the highest microplastic concentrations, are consumed (van Raamsdonk et al. 2020). In contrast most fish species are filleted, and most crustaceans have their digestive tracts removed before consumption, thereby reducing microplastic exposure. Similarly, bivalves, shellfish, and other lower trophically positioned marine organisms are probably the most important seafood source of dietary exposure to microplastics (Walkinshaw et al. 2020). It has been estimated that the average European shellfish consumer may ingest up to 11,000 microplastics per year based on levels in mussels and oysters (van Cauwenberghe and Janssen 2014). A systematic review and meta-analysis of microplastic contamination in seafood reported a maximum annual consumption of 55,000 microplastic particles, with mollusks from Asia being the most heavily contaminated (Danopoulos et al. 2020).

The presence of several types of microplastics in human stool from different countries has been reported, with PP and PET being the most abundant types (Schwabl et al. 2019), indicating human exposure. However, there is currently no indisputable evidence on the effects of microplastics on human health (Toussaint et al. 2019). Potential health impacts can result directly such as tissue damage but also indirectly from environmental contaminants associated with microplastics or associated microorganisms (Oberbeckmann et al. 2015).

While the focus of the scientific literature has primarily been on human exposure to microplastics from seafood consumption, much less data is available on the occurrence of microplastics in other food groups, so their relative contribution is unknown which is important from a risk assessment perspective (Wright and Kelly 2017). Data on microplastics in foods (Kwon et al. 2020) other than seafood include sugar, salt, honey, and drinking water and beer (Karbalaei et al. 2018), whereas there are significant data gaps for plant- and terrestrial animal-derived foods (van Raamsdonk et al. 2020). To date there are also limited data on microplastic levels in freshwater fish (Collard et al. 2019) and terrestrial foods (e.g., vegetables, poultry). In addition to food and drinking water, inhalation is a potential route of exposure, and atmospheric fallout is thus also an important source of microplastic exposure (Dris et al. 2016; Allen et al. 2019). Catarino et al. (2018) concluded that the potential for microplastic ingestion from shellfish consumption was minimal, especially when compared to general air exposure from household dust (123–4620 particles/year/capita and 13,731–68,415 particles/year/capita from food versus dust, respectively). Similarly, Rist et al. (2018) highlighted that food and beverages likely only constitute a minor exposure pathway to human microplastic exposure. Based on the current knowledge on microplastics in seafood, there is no evidence that food safety is compromised (Gamarro et al. 2020).

The extent to which microplastics present in foods contribute to human exposure is not well understood, especially as studies evaluating microplastics and associated chemical exposure to humans are not consistent (Rist et al. 2018; Barboza et al. 2020a, b). Human exposure estimates in the USA to microplastics in food (seafood, sugars, salts, honey), drinking water, alcohol, and air found that inhalation was the main route of exposure for adults whereas drinking water was the main source for children (Cox et al. 2019). However, this study did not include major food groups such as meats, grains, and vegetables due to a lack of empirical data.

5.12 Challenges and Priorities in Marine Microplastic Research

Risk characterization including information on the particle size-dependent toxicokinetics and dynamics of microplastics is needed to calculate evidence-based guidance or tolerable weekly intakes to support realistic human health and exposure risk assessments. Discrepancies exist in the sampling, extraction, identification, and quantification of microplastics (Collard et al. 2019), and there is a need for harmonization of current procedures (Hartmann et al. 2019; Cowger et al. 2020). An effective risk assessment of the human health effects of microplastics requires reliable human exposure data which is currently limited (Toussaint et al. 2019). Knowledge gaps regarding the uptake and potential human health effects of microplastic exposure have been highlighted (EFSA 2016; Wright and Kelly 2017; van Raamsdonk et al. 2020).

Importantly, there is currently still a lack of harmonized and proven methods for microplastics which can compromise the level to which microplastic contamination in seafood species (and other foods) can be compared. Some recommendations have been presented which focus on the methods that – thus far – have proven efficient at isolating microplastics from biota tissues (e.g., Dehaut et al. 2019; Lusher et al. 2020; Ribeiro et al. 2020). The field of microplastic research has been moving very rapidly, with several advancements in methods emerging in parallel. It is therefore of great urgency to coordinate an effort to compare the field and laboratory-based methods to one another to determine the level of comparability and overall effectiveness. This is easier said than done. Currently, laboratory comparisons have been limited to scientific approaches conducted by individual research groups (e.g., Catarino et al. 2017; Karlsson et al. 2017; Thiele et al. 2019; Yu et al. 2019; Jaafar et al. 2020; Ribeiro et al. 2020), rather than between different institutions. Some interlaboratory efforts have been made, but these have generally focused on clean water samples, rather than complex matrices such as seafood material and biological tissues (e.g., ongoing EU-JRC and SCCWRP intercalibration exercises). Similarly, there are different reporting criteria that have been applied to the study of microplastics in biota, and the quantification of the microplastics is not standardized which presents some important challenges to this subdiscipline of environmental chemistry. Different measurement units are often used (e.g., numbers per weight or per individual) highlighting the need for harmonization and standardization.

Moving forward, methods will need to be adopted that are truly reproducible and that can be validated and compared using standard reference materials. This requires that validation and feasibility assessments are undertaken while also supporting initiatives that promote scientific discovery and method development. There are several methods that are promising, and utilizing these novel approaches will allow for the development of more robust and comparable methods across different sectors/regions within the sphere of microplastic research. Unfortunately, several methods are focused on the larger fraction of microplastics (e.g., >100 μm), and method development is still required for accurately detecting smaller microplastics (<20 μm) and nanoplastics (<1 μm). Methods that focus on smaller fractions are needed to support risk characterization and exposure assessments in marine biota and humans.

5.13 Future Recommendations and Conclusions

Microplastics are ubiquitously found in seafood, and the importance of standardized and harmonized methods for the effective biomonitoring of farmed and wild seafood species including bivalves and finfish has become evident (Lusher et al. 2017b; Hartmann et al. 2019; Ribeiro et al. 2020). Lower trophically positioned organisms may be at the highest risk of contamination from microplastics, and currently there is insufficient evidence to conduct realistic and meaningful human health risk assessments. Moreover, several seafood species from wild fisheries and aquaculture are not well studied in the context of global food security including commonly consumed taxa (Lusher et al. 2017a; Walkinshaw et al. 2020). Microplastic pollution and exposure to plastics and their associated co-contaminants via seafood consumption will likely serve as effective themes to help link the IOC-UNESCO's Decade of Ocean Science for Sustainable Development (2021–2030) with the UN Decade of Action on Nutrition (2016–2025) and to gather critical stakeholders and develop important sustainable development strategies to support ocean and human health. In conclusion, the effects of microplastics on food security are still largely unknown, and further research and robust biomonitoring efforts on seafood are required to elucidate potential impacts.

References

Abbasi S, Soltani N, Keshavarzi B, Moore F, Turner A, Hassanaghaei M (2018) Microplastics in different tissues of fish and prawn from the Musa Estuary, Persian Gulf. Chemosphere 205:80–87

Abidli S, Lahbib Y, El Menif NT (2019) Microplastics in commercial molluscs from the lagoon of Bizerte (Northern Tunisia). Mar Pollut Bull 142:243–252

Allen S, Allen D, Phoenix VR (2019) Atmospheric transport and deposition of microplastics in a remote mountain catchment. Nat Geosci 12:339–344

Allen S, Allen D, Moss K, Le Roux G, Phoenix VR, Sonke JE (2020) Examination of the ocean as a source for atmospheric microplastics. PLoS One 15(5):e0232746. https://doi.org/10.1371/journal.pone.0232746

Amelia TSM, Khalik WMAWM, Ong MC, Shao YT, Pan H-J, Bhubalan K (2021) Marine microplastics as vectors of major ocean pollutants and its hazards to the marine ecosystem and humans. Prog Earth Planet Sci 8:12

Andrady AL (2011) Microplastics in the marine environment. Mar Pollut Bull 62:1596–1605

Baechler BR, Granek EF, Hunter MV, Conn KE (2020a) Microplastic concentrations in two Oregon bivalve species: spatial, temporal, and species variability. Limnol Oceanogr Lett 5:54–65

Baechler BR, Stienbarger CD, Horn DA, Joseph J, Taylor AR, Granek EF, Brander SM (2020b) Microplastic occurrence and effects in commercially harvested North American finfish and shellfish: current knowledge and future directions. Limnol Oceanogr Lett 5:113–136

Baechler BR, Granek EF, Mazzone SJ, Nielsen-Pincus M, Brander SM (2020c) Microplastic exposure by razor clam in recreational harvester-consumers along a sparsely populated coastline. Front Mar Sci 7:980

Bakir A, O'Connor IA, Rowland SJ, Hendriks AJ, Thompson RC (2016) Relative importance of microplastics as a pathway for the transfer of hydrophobic organic chemicals to marine life. Environ Pollut 219:56–65. https://doi.org/10.1016/j.envpol.2016.09.046

Barboza LGA, Vethaak AD, Lavorante BRBO, Lundebye A-K, Guilhermino L (2018) Marine microplastic debris: an emerging issue for food security, food safety and human health. Mar Pollut Bull 133:336–348

Barboza LGA, Cunha SC, Monteiro C, Fernandes JO, Guilhermino L (2020a) Bisphenol A and its analogs in muscle and liver of fish from the North East Atlantic Ocean in relation to microplastic contamination. Exposure and risk to human consumers. J Hazard Mater 393:122419

Barboza L, Lopes C, Oliveira P, Bessa F, Otero V, Henriques B, Raimundo J, Caetano M, Vale C, Guilhermino L (2020b) Microplastics in wild fish from North East Atlantic Ocean and its potential for causing neurotoxic effects, lipid oxidative damage, and human health risks associated with ingestion exposure. Sci Tot Environ 717:134625

Batel A, Linti F, Scherer M, Erdinger L, Braunbeck T (2016) Transfer of benzo[a]pyrene from microplastics to Artemia nauplii and further to zebrafish via a trophic food web experiment: CYP1A induction and visual tracking of persistent organic pollutants. Environ Toxicol Chem 35(7):1656–1666

Bellas J, Martínez-Armental J, Martínez-Cámara A, Besada V, Martínez-Gómez C (2016) Ingestion of microplastics by demersal fish from the Spanish Atlantic and Mediterranean coasts. Mar Pollut Bull 109:55–60. https://doi.org/10.1016/j.marpolbul.2016.06.026

Birnstiel S, Soares-Gomes A, da Gama BA (2019) Depuration reduces microplastic content in wild and farmed mussels. Mar Pollut Bull 140:241–247

Bour A, Avio CG, Gorbi S, Regoli F, Hylland K (2018) Presence of microplastics in benthic and epibenthic organisms: influence of habitat, feeding mode and trophic level. Environ Pollut 243:1217–1225. https://doi.org/10.1016/j.envpol.2018.09.115

Bråte ILN, Eidsvoll DP, Steindal CC, Thomas KV (2016) Plastic ingestion by Atlantic cod (Gadus morhua) from the Norwegian coast. Mar Pollut Bull 112:105–110. https://doi.org/10.1016/j.marpolbul.2016.08.034

Bråte ILN, Hurley R, Iversen K, Beyer J, Thomas KV, Steindal CC, Green NW, Olsen M, Lusher A (2018) Mytilus spp. as sentinels for monitoring microplastic pollution in Norwegian coastal waters: a qualitative and quantitative study. Environ Pollut 243:383–393

Browne MA, Dissanayake A, Galloway TS, Lowe DM, Thompson RC (2008) Ingested microscopic plastic translocates to the circulatory system of the mussel, Mytilus edulis (L). Environ Sci Technol 42:5026–5031

Browne M, Crump P, Niven SJ, Teuten E, Tonkin A, Galloway T, Thompson R (2011) Accumulation of microplastic on shorelines worldwide: sources and sinks. Environ Sci Technol 45. https://doi.org/10.1021/es201811s

Carbery M, O'Connor W, Palanisami T (2018) Trophic transfer of microplastics and mixed contaminants in the marine food web and implications for human health. Environ Int 115:4. https://doi.org/10.1016/j.envint.2018.03.007

Carpenter EJ, Anderson SJ, Harvey GR, Miklas HP, Peck BB (1972) Polystyrene spherules in coastal waters. Science 178(4062):749–750. https://doi.org/10.1126/science.178.4062.749

Carreras-Colom E, Constenla M, Soler-Membrives A, Cartes JE, Baeza M, Padrós F, Carrassón M (2018) Spatial occurrence and effects of microplastic ingestion on the deep-water shrimp Aristeus antennatus. Mar Pollut Bull 133:44–52

Catarino AI, Thompson R, Sanderson W, Henry TB (2017) Development and optimization of a standard method for extraction of microplastics in mussels by enzyme digestion of soft tissues. Environ Toxicol Chem 36(4):947–951

Catarino AI, Macchia V, Sanderson WG, Thompson RC, Henry TB (2018) Low levels of microplastics (MP) in wild mussels indicate that MP ingestion by humans is minimal compared to exposure via household fibres fallout during a meal. Environ Polllut 237:675–684

Cau A, Avio CG, Dessì C, Follesa MC, Moccia D, Regoli F, Pusceddu A (2019) Microplastics in the crustaceans Nephrops norvegicus and Aristeus antennatus: flagship species for deep-sea environments? Environ Polllut 255:113107

Cho Y, Shim WJ, Jang M, Han GM, Hong SH (2019) Abundance and characteristics of microplastics in market bivalves from South Korea. Environ Pollut 245:1107–1116

Cho Y, Shim WJ, Jang M, Han GM, Hong SH (2020) Nationwide monitoring of microplastics in bivalves from the coastal environment of Korea. Environ Pollut 270:116175

Choy CA, Drazen JC (2013) Plastic for dinner? Observations of frequent debris ingestion by pelagic predatory fishes from the central North Pacific. Mar Ecol Prog Ser 485:155–163

Cole M, Lindeque P, Halsband C, Galloway SC (2011) Microplastics as contaminants in the marine environment: a review. Mar Pollut Bull 62:2588–2597

Collard F, Gasperi J, Gabrielsen GW, Tassin B (2019) Particle ingestion by wild freshwater fish: a critical review. Environ Sci Technol 53:12974–12988. https://doi.org/10.1021/acs.est.9b03083

Covernton GA, Collicutt B, Gurney-Smith HJ, Pearce CM, Dower JF, Ross PS, Dudas SE (2019) Microplastics in bivalves and their habitat in relation to shellfish aquaculture proximity in coastal British Columbia, Canada. Aquac Environ Interact 11:357–374. https://doi.org/10.3354/aei00316

Cowger W, Booth AM, Hamilton BM, Thaysen C, Primpke S, Munno K, Lusher AL, Dehaut A, Vaz VP, Liboiron M, Devriese LI, Hermabessiere L, Rochman C, Athey SN, Lynch JM, De Frond H, Gray A, Jones OAH, Brander B, Steele C, Moore S, Sanchez A, Nel H (2020) Reporting guidelines to increase the reproducibility and comparability of research on microplastics. Appl Spectrosc 74:1066–1077. https://doi.org/10.1177/0003702820930292

Cox KD, Covernton GA, Davies HL, Dower JF, Juanes F, Dudas SE (2019) Human consumption of microplastics. Environ Sci Technol 53:7068. https://doi.org/10.1021/acs.est.9b01517

Cózar A, Martí E, Duarte CM, García-de-Lomas J, van Sebille E, Ballatore TJ, Eguíluz VM, González-Gordillo JI, Pedrotti ML, Echevarría F, Troublè R, Irigoien X (2017) The Arctic Ocean as a dead end for floating plastics in the North Atlantic branch of the thermohaline circulation. Sci Adv 3(4):e1600582

Critchell K, Hoogenboom MO (2018) Effects of microplastic exposure on the body condition and behaviour of planktivorous reef fish (Acanthochromis polyacanthus). PLoS One 13(3):e0193308. https://doi.org/10.1371/journal.pone.0193308

Danopoulos E, Jenner LC, Twiddy M, Rotchell JM (2020) Microplastic contamination of seafood intended for human consumption: a systematic review and meta-analysis. Environ H Persp 126002-1. https://doi.org/10.1289/EHP7171

Davidson TM (2012) Boring crustaceans damage polystyrene floats under docks polluting marine waters with microplastic. Mar Pollut Bull 64(9):1821–1828

Davidson K, Dudas SE (2016) Microplastic ingestion by wild and cultured Manila clams (Venerupis philippinarum) from Baynes Sound, British Columbia. Arch Environ Contam Toxicol 71(2):147–156

De Witte B, Devriese L, Bekaert K, Hoffman S, Vandermeersch G, Cooreman K, Robbens J (2014) Quality assessment of the blue mussel (Mytilus edulis): comparison between commercial and wild types. Mar Pollut Bull 85(1):146–155

Dehaut A, Hermabessiere L, Duflos G (2019) Current frontiers and recommendations for the study of microplastics in seafood. Trends Anal Chem 116:346–359

Devriese LI, Van der Meulen MD, Maes T, Bekaert K, Paul-Pont I, Frère L, Vethaak AD (2015) Microplastic contamination in brown shrimp (Crangon crangon, Linnaeus 1758) from coastal waters of the southern North Sea and channel area. Mar Pollut Bull 98(1–2):179–187

Ding JF, Li JX, Sun CJ, He CF, Jiang FH, Gao FL, Zheng L (2018) Separation and identification of microplastics in digestive system of bivalves. Chin J Anal Chem 46(5):690–697

Doyle D, Gammell M, Frias J, Griffin G, Nash R (2019) Low levels of microplastics recorded from the common periwinkle, Littorina littorea on the west coast of Ireland. Mar Pollut Bull 149:110645. https://doi.org/10.1016/j.marpolbul.2019.110645

Dris R, Gasperi J, Saad M, Mirande C, Tassin B (2016) Synthetic fibres in atmospheric fallout: a source of microplastics in the environment? Mar Pollut Bull 104:290–293. https://doi.org/10.1016/j.marpolbul.2016.01.006

EFSA (2016) Presence of microplastics and nanoplastics in food, with particular focus on seafood. Panel on contaminants in the food chain. EFSA J 14:e04501. https://doi.org/10.2903/j.efsa.2016.4501

Eriksen M, Lebreton LC, Carson HS, Thiel M, Moore CJ, Borerro JC, Galgani F, Ryan PG, Reisser J (2014) Plastic pollution in the world's oceans: more than 5 trillion plastic pieces weighing over 250,000 tons afloat at sea. PLoS One 9(12):e111913

Espinosa C, García Beltrán JM, Esteban MA, Cuesta A (2018) In vitro effects of virgin microplastics on fish head-kidney leucocyte activities. Environ Pollut 235:30–38. https://doi.org/10.1016/j.envpol.2017.12.054

FAO (2017) The state of food security and nutrition in the world 2017. Building resilience for peace and food security. FAO, Rome. https://doi.org/10.1080/15226514.2012.751351

FAO (2018) The state of world fisheries and aquaculture meeting the sustainable development goals. Food and Agriculture Organization of the United Nations. https://doi.org/10.1364/OE.17.003331

FAO (2020) The state of world fisheries and aquaculture 2020. Sustainability in action. Rome. doi: https://doi.org/10.4060/ca9229en

Farrell P, Nelson K (2013) Trophic level transfer of microplastic: Mytilus edulis (L.) to Carcinus maenas (L.). Environ Pollut 177:1–3

Feng S, Gao D, Liao F, Zhou F, Wang X (2016) The health effects of ambient PM2.5 and potential mechanisms. Ecotoxicol Environ Saf 128:67–74. https://doi.org/10.1016/j.ecoenv.2016.01.030

Feng Z, Wang R, Zhang T, Wang J, Huang W, Li J, Xu J, Gao G (2020) Microplastics in specific tissues of wild sea urchins along the coastal areas of northern China. Sci Total Environ 728:138660. https://doi.org/10.1016/j.scitotenv.2020.138660

Foekema EM, De Gruijter C, Mergia MT, van Franeker JA, Murk AJ, Koelmans AA (2013) Plastic in north sea fish. Environ Sci Technol 47(15):8818–8824. https://doi.org/10.1021/es400931b

Franzellitti S, Canesi L, Auguste M, Wathsala RHGR, Fabbri E (2019) Microplastic exposure and effects in aquatic organisms: a physiological perspective. Environ Toxicol Pharmacol 68:37–51

Gall SC, Thompson RC (2015) The impact of debris on marine life. Mar Pollut Bull 92:170–179

Galloway TS (2015) Micro- and nano-plastics and human health. In: Marine Anthropogenic Litter. https://doi.org/10.1007/978-3-319-16510-3_13

Galloway TS, Cole M, Lewis C (2017) Interactions of microplastic debris throughout the marine ecosystem. Nat Ecol Evol 1:1–8

Gamarro EG, Ryder J, Elvevoll EO, Olsen RL (2020) Microplastics in fish and shellfish – a threat to seafood safety? J Aquat Food Product Technol 29:417–425. https://doi.org/10.1080/10498850.2020.1739793

Gardon T, Reisser C, Soyez C, Quillien V, Le Moullac G (2018) Microplastics affect energy balance and gametogenesis in the pearl oyster Pinctada margaritifera. Environ Sci Technol 52(9):5277–5286

GESAMP (2016) Sources, fate and effects of microplastics in the marine environment: part two of a global assessment. In: Kershaw PJ, Rochman CM (eds) Joint Group of Experts on the Scientific Aspects of Marine Environmental Protection, Rep. Stud. GESAMP No. 93. 220 p

GESAMP (2019) Guidelines or the monitoring and assessment of plastic litter and microplastics in the ocean. In: Kershaw PJ, Turra A, Galgani F (eds) IMO/FAO/UNESCO-IOC/UNIDO/WMO/IAEA/UN/UNEP/UNDP/ISA Joint Group of Experts on the Scientific Aspects of Marine Environmental Protection, Rep. Stud. GESAMP No. 99. 130p

Gouin T, Roche N, Lohmann R, Hodges G (2011) A thermodynamic approach for assessing the environmental exposure of chemicals absorbed to microplastic. Environ Sci Technol 45:1466–1472

Graham ER, Thompson JT (2009) Deposit- and suspension-feeding sea cucumbers (Echinodermata) ingest plastic fragments. J Exp Mar Biol Ecol 368:22–29. https://doi.org/10.1016/j.jembe.2008.09.007

Green DS, Colgan TJ, Thompson RC, Carolan JC (2019) Exposure to microplastics reduces attachment strength and alters the haemolymph proteome of blue mussels (Mytilus edulis). Environ Pollut 246:423–434

Güven O, Gökdağ K, Jovanović B, Kıdeyş AE (2017) Microplastic litter composition of the Turkish territorial waters of the Mediterranean Sea, and its occurrence in the gastrointestinal tract of fish. Environ Pollut 223:286–294. https://doi.org/10.1016/j.envpol.2017.01.025

Hantoro I, Löhr AJ, Van Belleghem F, Widianarko B, Ragas A (2019) Microplastics in coastal areas and seafood: implications for food safety. Food Addit Contam 36(5):674–711

Hara J, Frias J, Nash R (2020) Quantification of microplastic ingestion by the decapod crustacean Nephrops norvegicus from Irish waters. Mar Pollut Bull 152:110905

Hartmann NB, Hüffer T, Thompson RC, Hassellöv M, Verschoor A, Daugaard AE, Rist S, Karlsson T, Brennholt N, Cole M, Herrling MP, Hess MC, Ivleva NP, Lusher AL, Wagner M (2019) Are we speaking the same language? Recommendations for a definition and categorization framework for plastic debris. Environ Sci Technol 53(3):1039–1047

Hermsen E, Mintenig SM, Besseling E, Koelmans AA (2018) Quality criteria for the analysis of microplastic in biota samples: a critical review. Environ Sci Technol 52(18):10230–10240

Hidalgo-Ruz V, Gutow L, Thompson RC, Thiel M (2012) Microplastics in the marine environment: a review of the methods used for identification and quantification. Environ Sci Technol 46:3060–3075

Hurley R, Woodward J, Rothwell JJ (2018) Microplastic contamination of river beds significantly reduced by catchment-wide flooding. Nat Geosci 11:251–257. https://doi.org/10.1038/s41561-018-0080-1

Iwalaye OA, Moodley GK, Robertson-Andersson DV (2020) The possible routes of microplastics uptake in sea cucumber Holothuria cinerascens (Brandt, 1835). Environ Pollut 264:114644. https://doi.org/10.1016/j.envpol.2020.114644

Jaafar N, Musa SM, Azfaralariff A, Mohamed M, Yusoff AH, Lazim AM (2020) Improving the efficiency of post-digestion method in extracting microplastics from gastrointestinal tract and gills of fish. Chemosphere 260:127649

Jang M, Shim WJ, Han GM, Song YK, Hong SH (2018) Formation of microplastics by polychaetes (Marphysa sanguinea) inhabiting expanded polystyrene marine debris. Mar Pollut Bull 131:365–369

Jiang Y, Yang F, Zhao Y, Wang J (2020) Greenland Sea Gyre increases microplastic pollution in the surface waters of the Nordic Seas. Sci Total Environ 712:136484

Kane IA, Clare MA, Miramontes E, Wogelius R, Rothwell JJ, Garreau P, Pohl F (2020) Seafloor microplastic hotspots controlled by deep-sea circulation. Science 368:1140–1145

Karbalaei S, Hanachi P, Walker TR, Cole M (2018) Occurrence, sources, human health impacts and mitigation of microplastic pollution. Environ Sci Pollut Res Int 25:36046–36063. https://doi.org/10.1007/s11356-018-3508-7

Karlsson TM, Vethaak AD, Almroth BC, Ariese F, van Velzen M, Hassellöv M, Leslie HA (2017) Screening for microplastics in sediment, water, marine invertebrates and fish: method development and microplastic accumulation. Mar Pollut Bull 122(1–2):403–408

Kiew PL, Don MM (2011) Jewel of the seabed: sea cucumbers as nutritional and drug candidates. Int J Food Sci Nutr 63(5):616–636. https://doi.org/10.3109/09637486.2011.641944

Koelmans AA, Bakir A, Burton GA, Janssen CR (2016) Microplastic as a vector for chemicals in the aquatic environment: critical review and model-supported reinterpretation of empirical studies. Environ Sci Technol 50(7):3315–3326. https://doi.org/10.1021/acs.est.5b06069

Kögel T, Bjorøy Ø, Toto B, Bienfait AM, Sanden M (2019) Micro- and nanoplastic toxicity on aquatic life: determining factors. Sci Total Environ 709:136050. https://doi.org/10.1016/j.scitotenv.2019.136050

Kolandhasamy P, Su L, Li J, Qu X, Jabeen K, Shi H (2018) Adherence of microplastics to soft tissue of mussels: a novel way to uptake microplastics beyond ingestion. Sci Total Environ 610-611:635–640. https://doi.org/10.1016/j.scitotenv.2017.08.053

Koongolla JB, Lin L, Pan Y-F, Yang C-P, Sun D-R, Liu S, Xu X-R, Maharana D, Huang J-S, Li H-X (2020) Occurrence of microplastics in gastrointestinal tracts and gills of fish from Beibu Gulf, South China Sea. Environ Pollut 258:113734. https://doi.org/10.1016/j.envpol.2019.113734

Kwon J-H, Kim J-W, Pham TD, Tarafdar A, Hong S, Chun S-H, Lee S-H, Kang D-Y, Kim J-Y, Kim S-B, Jung J (2020) Microplastics in food: a review on analytical methods and challenges. Int J Environ Res Public Health 17(18):6710. https://doi.org/10.3390/ijerph17186710

Lebreton L, van der Zwet J, Damsteeg J, Slat B, Andrady A, Reisser J (2017) River plastic emissions to the world's oceans. Nat Commun 8:15611. https://doi.org/10.1038/ncomms15611

Li J, Yang D, Li L, Jabeen K, Shi H (2015) Microplastics in commercial bivalves from China. Environ Pollut 207:190–195

Li J, Qu X, Su L, Zhang W, Yang D, Kolandhasamy P, Shi H (2016) Microplastics in mussels along the coastal waters of China. Environ Pollut 214:177–184

Li J, Lusher AL, Rotchell JM, Deudero S, Turra A, Bråte ILN, Sun C, Hossain MS, Li Q, Kolandhasamy P, Shi H (2019) Using mussel as a global bioindicator of coastal microplastic pollution. Environ Pollut 244:522–533

Liboiron M, Liboiron F, Wells E, Richárd N, Zahara A, Mather C, Bradshaw H, Murichi J (2016) Low plastic ingestion rate in Atlantic cod (Gadus morhua) from Newfoundland destined for human consumption collected through citizen science methods. Mar Pollut Bull 113(1–2):428–437. https://doi.org/10.1016/j.marpolbul.2016.10.043

Lithner D, Larsson A, Dave G (2011) Environmental and health hazard ranking and assessement of plastic polymers based on chemical composition. Sci Total Environ 409:3309–3324

Lohmann R (2017) Microplastics are not important for the cycling and bioaccumulation of organic pollutants in the oceans-but should microplastics be considered POPs themselves? Integr Environ Assess Manag 13(3):460–465. https://doi.org/10.1002/ieam.1914

Lusher A, Welden NAC (2020) Microplastic impact in fisheries and aquaculture. In: Rocha-Santos T et al (eds) Handbook of microplastics in the environment. Springer, Cham

Lusher AL, McHugh M, Thompson RC (2013) Occurrence of microplastics in the gastrointestinal tract of pelagic and demersal fish from the English Channel. Mar Pollut Bull 67:94–99. https://doi.org/10.1016/j.marpolbul.2012.11.028

Lusher AL, O'Donnell C, Officer R, O'Connor I (2016) Microplastic interactions with North Atlantic mesopelagic fish. ICES J Mar Sci 73:1214–1225

Lusher A, Hollman P, Mendoza-Hill J (2017a) Microplastics in fisheries and aquaculture – status of knowledge on their occurrence and implications for aquatic organisms and food safety. FAO Fisheries and Aquaculture Technical Paper, 615

Lusher AL, Welden NA, Sobral P, Cole M (2017b) Sampling, isolating and identifying microplastics ingested by fish and invertebrates. Anal Methods 9(9):1346–1360

Lusher AL, Munno K, Hermabessiere L, Carr S (2020) Isolation and extraction of microplastics from environmental samples: an evaluation of practical approaches and recommendations for further harmonization. Appl Spectrosc 74(9):1049–1065

Markic A, Niemand C, Bridson JH, Mazouni-Gaertner N, Gaertner JC, Eriksen M, Bowen M (2018) Double trouble in the South Pacific subtropical gyre: increased plastic ingestion by fish in the oceanic accumulation zone. Mar Pollut Bull 136:547–564. https://doi.org/10.1016/j.marpolbul.2018.09.031

Mathalon A, Hill P (2014) Microplastic fibers in the intertidal ecosystem surrounding Halifax Harbor, Nova Scotia. Mar Pollut Bull 81(1):69–79

McGoran AR, Clark PF, Smith BD, Morritt D (2020) High prevalence of plastic ingestion by Eriocheir sinensis and Carcinus maenas (Crustacea: Decapoda: Brachyura) in the Thames estuary. Environ Pollut 265:114972

Mohsen M, Wang Q, Zhang L, Sun L, Lin C, Yang H (2019) Microplastic ingestion by the farmed sea cucumber Apostichopus japonicus in China. Environ Pollut 245:1071–1078. https://doi.org/10.1016/j.envpol.2018.11.083

Naji A, Nuri M, Vethaak AD (2018) Microplastics contamination in molluscs from the northern part of the Persian Gulf. Environ Pollut 235:113–120. https://doi.org/10.1016/j.envpol.2017.12.046

Napper IE, Thompson RC (2016) Release of synthetic microplastic plasticfibres fromdomestic washing machines: effects of fabric type and washing conditions. Mar Pollut Bull 112:39–45. https://doi.org/10.1016/j.marpolbul.2016.09.025

Nelms SE, Galloway TS, Godley BJ, Jarvis DS, Lindeque PK (2018) Investigating microplastic trophic transfer in marine top predators. Environ Pollut 238:999–1007. https://doi.org/10.1016/j.envpol.2018.02.016

Neves D, Sobral P, Ferreira JL, Pereira T (2015) Ingestion of microplastics by commercial fish off the Portuguese coast. Mar Pollut Bull 101(1):119–126. https://doi.org/10.1016/j.marpolbul.2015.11.008

Oberbeckmann SL, Martin GJ, Labrenz M (2015) Marine microplastic-associated biofilms – a review. Environ Chem 12(5):551–562. https://doi.org/10.1071/EN15069

Ogonowski M, Wenman V, Danielsson S Gorokhova E (2017) Ingested microplastic is not correlated to HOC concentrations in Baltic Sea herring. In: Proceedings of the 15th international conference on environmental science and technology, Rhodes, Greece, 31 August–2 September

Oliveira AR, Sardinha-Silva A, Andrews PLR, Green D, Cooke GM, Hall S, Blackburn K, Sykes AV (2020) Microplastics presence in cultured and wild-caught cuttlefish, Sepia officinalis. Mar Pollut Bull 160:111553. https://doi.org/10.1016/j.marpolbul.2020.111553

Peng L, Fu D, Qi H, Lan CQ, Yu H, Ge C (2020) Micro- and nano-plastics in marine environment: source, distribution and threats—a review. Sci Total Environ 698:134254. https://doi.org/10.1016/j.scitotenv.2019.134254

Phuong NN, Poirier L, Pham QT, Lagarde F, Zalouk-Vergnoux A (2018) Factors influencing the microplastic contamination of bivalves from the French Atlantic coast: location, season and/or mode of life? Mar Pollut Bull 129(2):664–674

Pitt JA, Trevisan R, Massarsky A, Kozal JS, Levin ED, Di Giulio RT (2018) Maternal transfer of nanoplastics to offspring in zebrafish (*Danio rerio*): a case study with nanopolystyrene. Sci Total Environ 643:324–334. https://doi.org/10.1016/j.scitotenv.2018.06.186

Plee TA, Pomory CM (2020) Microplastics in sandy environments in the Florida keys and the panhandle of Florida, and the ingestion by sea cucumbers (Echinodermata: Holothuroidea) and sand dollars (Echinodermata: Echinoidea). Mar Pollut Bull 158:111437. https://doi.org/10.1016/j.marpolbul.2020.111437

Porter A, Lyons BP, Galloway TS, Lewis C (2018) Role of marine snows in microplastic fate and bioavailability. Environ Sci Technol 52:7111–7119

Qu X, Su L, Li H, Liang M, Shi H (2018) Assessing the relationship between the abundance and properties of microplastics in water and in mussels. Sci Total Environ 621:679–686

Renzi M, Guerranti C, Blašković A (2018a) Microplastic contents from maricultured and natural mussels. Mar Pollut Bull 131:248–251

Renzi M, Blašković A, Bernardi G, Russo GF (2018b) Plastic litter transfer from sediments towards marine trophic webs: a case study on holothurians. Mar Pollut Bull 135:376–385. https://doi.org/10.1016/j.marpolbul.2018.07.038

Ribeiro F, Okoffo ED, O'Brien JW, Fraissinet-Tachet S, O'Brien S, Gallen M, Samanipour S, Kaserzon S, Mueller JF, Galloway T, Thomas KV (2020) Quantitative analysis of selected plastics in high-commercial-value Australian seafood by pyrolysis gas chromatography mass spectrometry. Environ Sci Technol 54:9408–9417

Rist S, Almroth BC, Hartmann NB, Karlsson TM (2018) A critical perspective on early communications concerning human health aspects of microplastics. Sci Total Environ 626:720–726

Roch S, Friedrich C, Brinker A (2020) Uptake routes of microplastics in fishes: practical and theoretical approaches to test existing theories. Sci Rep 10:3896

Rochman CM, Hoh E, Kurobe T, The SJ (2013) Ingested plastic transfers hazardous chemicals to fish and induces hepatic stress. Sci Rep 3:3263. https://doi.org/10.1038/srep03263

Rochman CM, Tahir A, Williams SL, Baxa DV, Lam R, Miller JT, The F-C, Werorilangi S, Teh SJ (2015) Anthropogenic debris in seafood: plastic debris and fibers from textiles in fish and bivalves sold for human consumption. Sci Rep 5:14340

Rochman CM, Brookson C, Bikker J, Djuric N, Earn A, Bucci K, Athey S, Huntington A, McIlwraith H, Munno K, De Frond H, Kolomijeca A, Erdle L, Grbic J, Bayoumi M, Borrelle SB, Wu T, Santoro S, Werbowski LM, Zhu X, Giles RK, Hamilton BM, Thaysen C, Kaura A, Klasios N, Ead L, Kim J, Sherlock C, Ho A, Hung C (2019) Rethinking microplastics as a diverse contaminant suite. Environ Toxicol Chem 38:703–711

Rummel CD, Löder MG, Fricke NF, Lang T, Griebeler EM, Janke M, Gerdts G (2016) Plastic ingestion by pelagic and demersal fish from the North Sea and Baltic Sea. Mar Pollut Bull 102(1):134–141. https://doi.org/10.1016/j.marpolbul.2015.11.043

Santillo D, Miller K, Johnston P (2017) Microplastics as contaminants in commercially important seafood species. Integr Environ Assess Manag 13:516–521

Schwabl P, Koppel S, Konigshofer P, Bucsics T, Trauner M, Reiberger T, Liebmann B (2019) Detection of various microplastics in human stool: a prospective case series. Ann Intern Med. https://doi.org/10.7326/M19-0618

Setälä O, Fleming-Lehtinen V, Lehtiniemi M (2014) Ingestion and transfer of microplastics in the planktonic food web. Environ Pollut 185:77–83

Smith M, Love DC, Rochman CM, Neff RA (2018) Microplastics in seafood and the implications for human health. Curr Environ Health Rep 5:375. https://doi.org/10.1007/s40572-018-0206-z

Stone V, Johnston H, Clift MJD (2007) Air pollution, ultrafine and nanoparticle toxicology: cellular and molecular interactions. IEEE Trans Nanobioscience 6(4):331–340. https://doi.org/10.1109/TNB.2007.909005

Su L, Deng H, Li B, Chen Q, Pettigrove V, Wu C, Shi H (2018) The occurrence of microplastic in specific organs in commercially caught fishes from coast and estuary area of East China. J Hazard Mater. https://doi.org/10.1016/j.jhazmat.2018.11.024

Sussarellu R, Suquet M, Thomas Y, Lambert C, Fabioux C, Pernet ME, Le Goïc N, Quillien V, Mingant C, Epelboin Y, Corporeau C, Guyomarch J, Robbens J, Paul-Pont I, Soudant P, Huvet A (2016) Oyster reproduction is affected by exposure to polystyrene microplastics. Proc Natl Acad Sci U S A 113(9):2430–2435. https://doi.org/10.1073/pnas.1519019113

Tanaka K, Takada H (2016) Microplastic fragments and microbeads in digestive tracts of planktivorous fish from urban coastal waters. Sci Rep 6:34351. https://doi.org/10.1038/srep34351

Thiele CJ, Hudson MD, Russell AE (2019) Evaluation of existing methods to extract microplastics from bivalve tissue: adapted KOH digestion protocol improves filtration at single-digit pore size. Mar Pollut Bull 142:384–393

Thushari G, Senevirathna J (2020) Plastic pollution in the marine environment. Heliyon 6(8):e04709. https://doi.org/10.1016/j.heliyon.2020.e04709

Thushari GGN, Senevirathna JDM, Yakupitiyage A, Chavanich S (2017) Effects of microplastics on sessile invertebrates in the eastern coast of Thailand: an approach to coastal zone conservation. Mar Pollut Bull 124:349–355. https://doi.org/10.1016/j.marpolbul.2017.06.010

Toussaint B, Raffael B, Angers-Loustau A, Gilliland D, Kestens V, Petrillo M, Rio-Echevarria IM, Van den Eede G (2019) Review of micro- and nanoplastic contamination in the food chain. Food Addit Contam A 36:639. https://doi.org/10.1080/19440049.2019.1583381

van Cauwenberghe L, Janssen CR (2014) Microplastics in bivalves cultured for human consumption. Environ Pollut 193:65–70

van Cauwenberghe L, Claessens M, Vandegehuchte MB, Janssen CR (2015) Microplastics are taken up by mussels (Mytilus edulis) and lugworms (Arenicola marina) living in natural habitats. Environ Pollut 199:10–17

van Raamsdonk LWD, van der Zande M, Koelmans AA, Hoogenboom RLAP, Peters RJB, Groot MJ, Peijnenburg ACM, Weesepoel YJA (2020) Current insights into monitoring, bioaccumulation, and potential health effects of microplastics present in the food chain. Foods. https://doi.org/10.3390/foods9010072

van Sebille E, Aliani S, Law KL, Maximenko N, Alsina JM, Bagaev A, Bergmann M, Chapron B, Chubarenko I, Cózar A, Delandmeter P, Egger M, Fox-Kemper B, Garaba SP, Goddijn-Murphy L, Hardesty BD, Hoffman MJ, Isobe A, Jongedijk CE, Kaandorp MLA, Khatmullina L, Koelmans AA, Kukulka T, Laufkötter C, Lebreton L, Lobelle D, Maes C, Martinez-

Vicente V, Morales Maqueda MA, Poulain-Zarcos M, Rodríguez E, Ryan PG, Shanks AL, Shim WJ, Suaria G, Thiel M, van den Bremer TS, Wichmann D (2020) The physical oceanography of the transport of floating marine debris. Environ Res Lett 15:023003. https://doi.org/10.1088/1748-9326/ab6d7d

Vandermeersch G, Van Cauwenberghe L, Janssen CR, Marques A, Granby K, Fait G, Kotterman MJ, Diogène J, Bekaert K, Robbens J, Devriese L (2015) A critical view on microplastic quantification in aquatic organisms. Environ Res 143:46–55. https://doi.org/10.1016/j.envres.2015.07.016

VKM (2019) Microplastics; occurrence, levels and implications for environment and human health related to food. Scientific opinion of the Scientific Steering Committee of the Norwegian Scientific Committee for Food and Environment (No. 2019:16). Norwegian Scientific Committee for Food and Environment (VKM), Oslo, Norway

Walkinshaw C, Lindeque PK, Thompson R, Tolhurst T, Cole M (2020) Microplastics and seafood: lower trophic organisms at highest risk of contamination. Ecotoxicol. Environ Saf 190. https://doi.org/10.1016/j.ecoenv.2019.110066

Wang J, Lu L, Wang M, Jiang T, Liu X, Ru S (2019) Typhoons increase the abundance of microplastics in the marine environment and cultured organisms: a case study in Sanggou Bay, China. Sci Total Environ 667:1–8. https://doi.org/10.1016/j.scitotenv.2019.02.367

Wang W, Ge J, Yu X (2020) Bioavailability and toxicity of microplastics to fish species: a review. Ecotoxicol Environ Saf 189:109913. https://doi.org/10.1016/j.ecoenv.2019.109913

Ward JE, Kach DJ (2009) Marine aggregates facilitate ingestion of nanoparticles by suspension-feeding bivalves. Mar Environ Res 68(3):137–142

Ward JE, Rosa M, Shumway SE (2018) Capture, ingestion, and egestion of microplastics by suspension-feeding bivalves: a 40-year history. Mar Microplast Pollut Contr 2(1):39–49. https://doi.org/10.1139/anc-2018-0027@anc-mmpc.issue1

Ward JE, Zhao S, Holohan BA, Mladinich KM, Griffin TW, Wozniak J, Shumway SE (2019) Selective ingestion and egestion of plastic particles by the blue mussel (Mytilus edulis) and eastern oyster (Crassostrea virginica): implications for using bivalves as bioindicators of microplastic pollution. Environ Sci Technol 53(15):8776–8784

Wardrop P, Shimeta J, Nugegoda D, Morrison PD, Miranda A, Tang M, Clarke BO (2016) Chemical pollutants sorbed to ingested microbeads from personal care products accumulate in fish. Environ Sci Technol 50:4037–4044. https://doi.org/10.1021/acs.est.5b06280

Watts AJR, Urbina MA, Corr S, Lewis C, Galloway T (2015) Ingestion of plastic microfibers by the crab Carcinus maenas and its effect on food consumption and energy balance. Environ Sci Technol 49(24):14597–14604. https://doi.org/10.1021/acs.est.5b04026

Watts AJR, Urbina MA, Goodhead R, Moger J, Lewis C, Galloway TS (2016) Effect of microplastic on the gills of the shore crab Carcinus maenas. Environ Sci Technol 50:5364–5369. https://doi.org/10.1021/acs.est.6b01187

Wegner A, Besseling E, Foekema EM, Kamermans P, Koelmans AA (2012) Effects of nanopolystyrene on the feeding behavior of the blue mussel (Mytilus edulis L.). Environ Toxicol Chem 31(11):2490–2497

Weinstein JE, Crocker BK, Gray AD (2016) From macroplastic to microplastic: degradation of high-density polyethylene, polypropylene, and polystyrene in a salt marsh habitat. Environ Toxicol Chem 35:1632–1640. https://doi.org/10.1002/etc.3432

Welden NA, Cowie PR (2016) Environment and gut morphology influence microplastic retention in langoustine, Nephrops norvegicus. Environ Pollut 214:859–865

Welden NA, Abylkhani B, Howarth LM (2018) The effects of trophic transfer and environmental factors on microplastic uptake by plaice, Pleuronectes plastessa, and spider crab, Maja squinado. Environ Pollut 239:351–358

Wieczorek AM, Croot PL, Lombard F, Sheahan JN, Doyle TK (2019) Microplastic ingestion by gelatinous zooplankton may lower efficiency of the biological pump. Environ Sci Technol 53(9):5387–53955. https://doi.org/10.1021/acs.est.8b07174

Wright SL, Kelly FJ (2017) Plastic and human health: a micro issue? Environ Sci Technol 51(12):6634–6647. https://doi.org/10.1021/acs.est.7b00423

Wright SL, Thompson RC, Galloway TS (2013) The physical impacts of microplastics on marine organisms: a review. Environ Pollut 178:483–492

Xu XY, Lee WT, Chan AKY, Lo HS, Shin PKS, Cheung SG (2017) Microplastic ingestion reduces energy intake in the clam Atactodea striata. Mar Pollut Bull 124(2):798–802

Xu X, Wong CY, Tam NFY, Lo H-S, Cheung S-G (2020) Microplastics in invertebrates on soft shores in Hong Kong: influence of habitat, taxa and feeding mode. Sci Tot Environ 715:13699. https doi.org/10.1016/j.scitotenv.2020.136999

Yoo JW, Doshi N, Mitragotri S (2011) Adaptive micro and nanoparticles: temporal control over carrier properties to facilitate drug delivery. Adv Drug Deliv Rev 63(14–15):1247–1256

Yu Z, Peng B, Liu LY, Wong CS, Zeng EY (2019) Development and validation of an efficient method for processing microplastics in biota samples. Environ Toxicol Chem 38(7):1400–1408

Zeytin S, Wagner G, Mackay-Roberts N, Gerdts G, Schuirmann E, Klockmann S, Slater M (2020) Quantifying microplastic translocation from feed to the fillet in European sea bass Dicentrarchus labrax. Mar Pollut Bull 156:111210

Zhang F, Man YB, Mo WY, Man KY, Wong MH (2019a) Direct and indirect effects of microplastics on bivalves, with a focus on edible species: A mini-review. Crit Rev Environ Sci Technol 1–35. https://doi.org/10.1080/10643389.2019.1700752

Zhang F, Wang X, Xu J, Zhu L, Peng G, Xu P, Li D (2019b) Food-web transfer of microplastics between wild caught fish and crustaceans in East China Sea. Mar Pollut Bull 146:173–182. https://doi.org/10.1016/j.marpolbul.2019.05.061

Ziccardi LM, Edgington A, Hentz K, Kulacki KJ, Driscoll SK (2016) Microplastics as vectors for bioaccumulation of hydrophobic organic chemicals in the marine environment: a state-of-the-science review. Environ Toxicol Chem 35:1667–1676

6

Chemical Risk Assessment and MP Vector Effect

Albert A. Koelmans, Noël J. Diepens and Nur Hazimah Mohamed Nor

Abstract The concern that in nature, ingestion of microplastic (MP) increases exposure of organisms to plastic-associated chemicals (the 'MP vector effect') plays an important role in the current picture of the risks of microplastic for the environment and human health. An increasing number of studies on this topic have been conducted using a wide variety of approaches and techniques. At present, the MP vector effect is usually framed as 'complex', 'under debate' or 'controversial'. Studies that critically discuss the approaches and techniques used to study the MP vector effect, and that provide suggestions for the harmonization needed to advance this debate, are scarce. Furthermore, only a few studies have strived at interpreting study outcomes in the light of environmentally relevant conditions. This constitutes a major research gap, because these are the conditions that are most relevant when informing risk assessment and management decisions. Based on a review of 61 publications, we propose evaluation criteria and guidance for MP vector studies and discuss current study designs using these criteria. The criteria are designed such that studies, which fulfil them, will be relevant to inform risk assessment. By critically reviewing the existing literature in the light of these criteria, a weight of evidence assessment is provided. We demonstrate that several studies did not meet the standards for their conclusions on the MP vector effect to stand, whereas others provided overwhelming evidence that the vector effect is unlikely to affect chemical risks under present natural conditions.

A. A. Koelmans (✉) · N. J. Diepens · N. H. Mohamed Nor
Aquatic Ecology and Water Quality Management Group, Wageningen University & Research, Wageningen, The Netherlands
e-mail: bart.koelmans@wur.nl; noel.diepens@wur.nl; Hazimah.mohamednor@wur.nl

6.1 Introduction

Plastics in the environment contain mixtures of chemicals. These chemicals stem from deliberate additions during manufacture of the plastic or from the ambient water via absorption or both. Plastic particles are known to be ingested by organisms, including humans (World Health Organization 2019), but not all particles can be ingested by all organisms. Exposure, bioavailability, feeding behaviour and size of the plastic item in relation to size of the mouth opening determine whether a plastic item can be taken up (Jâms et al. 2020; Koelmans et al. 2020). Plastic is persistent, and ingestion of sufficiently high particle concentrations has been demonstrated to cause effects on small organisms like some zooplankton and invertebrates, most likely due to the reduced caloric value of the ingested material (Gerdes et al. 2019a; de Ruijter et al. 2020). Besides this mechanism of reduced caloric value due to dilution of ingested material, concerns have been raised due to the fact that chemicals on the plastic are ingested with the plastic. This has led to an increasing number of studies that investigate the potential of microplastic to increase exposure of organisms to plastic-associated chemicals. Henceforth, we will refer to these studies as MP vector studies. MP vector studies have used a wide variety of approaches and techniques and have been summarized in several reviews (Rochman 2015, 2019; Koelmans et al. 2016; Ziccardi et al. 2016; Hartmann et al. 2017; Burns and Boxall 2018). However, none of these have focused on quality assurance and study design criteria.

In the recent literature, MP vector studies are often framed as 'controversial', 'complex' or 'under debate'(Gassel and Rochman 2019). This suggests that apparently there is no consensus in the scientific community on the nature or the relevance of the MP vector effect. We argue that there are three main reasons why this could be the case, reasons that may explain why it takes so long before consensus is reached.

The first reason relates to confusion about when a study is to be considered relevant in this context, a question which in the literature on chemical risks often is referred to as the 'so what' question. To date, studies have addressed detailed mechanisms, specific exposure scenarios in the lab, have either measured effects on uptake or on biological endpoints in the lab and have evaluated uptake under natural conditions either by field studies or by model scenario analysis. Studies mostly addressed whether chemicals *can* (*potentially*) be taken up from plastic under some specific conditions, but not if they *will* be taken up under natural conditions. Few studies have strived at interpreting results in the context of environmentally relevant conditions, and none of them addressed to what extent ingestion of MP would actually increase the chemical risks for organisms. An increase in chemical risk would exist if, due to the MP vector effect, exposure to chemicals (e.g., predicted exposure concentrations (PEC)) would exceed the toxicity thresholds that are known for these chemicals (predicted no effect concentration (PNEC)). Putting study results in the context of such actual risks is relevant, because if a detected vector effect would not increase risks of the chemicals, then it might be less important to address it and there would be less reason for concern. With no studies actually demonstrating the occurrence of an MP vector effect on chemical risks under natural conditions, the evidence base is remarkably thin. This constitutes a research gap, because these are the conditions that would be most relevant when informing risk assessment and management decisions. We thus propose relevance for chemical risk assessment as an overarching umbrella criterion to evaluate the setup of MP vector studies.

The second reason is that the discussion of the topic may have lacked a common understanding of the processes at play, especially in the earlier studies. Most of them address a part of the exposure conditions that are relevant to the occurrence of MP vector effects and their possible implications for effects on risks in the environment (e.g. Rochman et al. 2013; Chua et al. 2014; Wardrop et al. 2016; Granby et al. 2018; Kühn et al. 2020). Only few studies use theoretical frameworks to add greater depth to the interpretation of study outcomes. Such frameworks allow for inter- and extrapolation across chemical properties, microplastic characteristics, biological traits and environmental conditions (Koelmans et al. 2016; Bakir et al. 2016; Rochman et al. 2017; Lee et al. 2019; Mohamed Nor and Koelmans 2019; Wang et al. 2020a, b). As these frameworks often come in the form of mathematical models, they are not automatically adopted by researchers that use laboratory or field observational studies as their primary research tools.

The third reason relates to the general quality of some of the microplastic research, which has been framed as limiting in recent literature (Lenz et al. 2016; Hermsen et al. 2018; Koelmans et al. 2019; Markic et al. 2019; de Ruijter et al. 2020; Provencher et al. 2020). Microplastic research is a young and fast-growing field in the environmental sciences, and this is reflected in the wide variety of approaches used when sampling and analysing microplastic and in effect testing methods. The same diversity and therefore incomparability of approaches apply to MP vector studies.

In this review, we discuss the potential of studies and study designs to inform risk assessment of plastic-associated chemicals, emphasizing organic chemicals. Heavy metals bind to microplastic particles as well, but their sorption affinity is limited compared to that for sorption to natural particles like sediment (e.g. Besson et al. 2020). First, we propose evaluation criteria for MP vector studies. The criteria are designed such that studies that fulfil them can be considered relevant to inform risk assessment. Second, we critically review the existing literature in the light of these criteria, thereby providing a weight of evidence assessment. Finally, we discuss several key references from the literature as examples of how they can be used to inform plastic-associated chemical risk assessment given the criteria and recommendations.

Literature was selected through reference as well as cited reference searches for the aforementioned six existing MP vector effect reviews. This method assumes that all relevant MP vector studies either cite or are cited (in) at least one of these six reviews, published between 2015 and 2019.

We emphasize that we reviewed the existing literature only for the aim set for the present review: to assess the weight of evidence for the chemical vector effect to occur under environmentally relevant conditions and with respect to implications for chemical risks. Part of the reviewed papers may not have had this specific aim. They often were performed to verify the validity of underlying mechanisms, or they were meant to study *potential* effects rather than to demonstrate actual chemical risks under field conditions. Therefore, the present retrospective assessment does not necessarily disqualify studies as such, as they could have had other aims than mimicking the environmental realism with respect to the chemical risks stemming from the MP vector effect that we aim to pursue here.

6.2 Guidance for Microplastic Vector Studies in the Context of Chemical Risk Assessment

Several criteria can be used when interpreting results from studies investigating the MP vector effect with respect to their relevance for conditions occurring in nature (Table 6.1). The criteria can be placed into two broad categories: (a) characteristics of a study in providing evidence for the occurrence of an MP vector effect in nature and (b) relevance of the setup and outcomes for risk assessment of chemically contaminated MP. The difference is that the latter category would need reflection on whether effect thresholds for chemical toxicity are exceeded. For instance, if an MP vector effect is detected, concentrations of environmentally relevant chemical mixtures still may be too low to exceed such toxicity threshold concentrations. Furthermore, even if ingestion of plastic would increase actual risks because the plastic acts as a source for some plastic-associated chemicals, the same ingested plastic could at the same time act as a sink ('cleaning agent') for chemicals other than these plastic-associated ones, i.e. chemicals which are present in the gastrointestinal tract from other sources (Koelmans 2015). An increasing number of studies has demonstrated that this is possible, from empirical evidence as well as through model scenario analysis (Koelmans et al. 2013b, 2016; Devriese et al. 2017; Scopetani et al. 2018; Mohamed Nor and Koelmans 2019; Heinrich and Braunbeck 2020; Thaysen et al. 2020). In terms of risks of the overall chemical mixture that animals are exposed to, the cleaning phenomenon could compensate for the increased exposure of plastic-associated chemicals, possibly leading to net zero or even less effects of the chemical mixture. Below, we further detail these criteria for laboratory studies, field studies, in vitro studies and modelling studies.

6.2.1 Criteria for In Vivo Laboratory Studies

An often used setup of laboratory MP vector studies is that MP is first contaminated with chemicals, after which test organisms are exposed to these contaminated MPs and to controls without MP for comparison. An MP vector effect is then indicated if body burdens are higher in the MP treatments than in the control, whereas a risk is

Table 6.1 Quality criteria for studies investigating the microplastic vector effect in the context of chemical risk assessment

Criterion		Guidance	Type of study for which the criterion is relevant [a]
1	Preparation of microplastic with associated chemicals	Use long-term field-contaminated microplastic, or laboratory spiking with equilibration time of at least a month, and/or corrections for non-equilibrium by means of kinetic sorption modelling	L
2	Resemblance of natural exposure pathways	Chemical exposure via all pathways that are relevant to the organism under consideration is assessed, either through measurement or modelling, preferably through both	L, F, M
3	Verification of chemical exposure	Chemical exposure should be assessed for all exposure pathways, either through measurement or modelling	L, F
4	Concentration gradient	No or limited concentration gradient; if the study aims to mimic natural conditions Maximum gradient; if the study aims to assess sorption kinetic parameters	L, M
5	Ingestion	Ingestion should be demonstrated	L, F
6	Evidence from correlations	Rule out multiple causation Assess correlations on the level of individual chemicals Account for measurement error	F
7	Reversibility of chemical transfer	Parameter fitting and data interpretation should account for bidirectional chemical transfer between ingested plastic and biota tissue	L, F, M
8	Model validity	The model needs to be consistent with empirical data, with current knowledge, and with design criteria	M
9	Threshold effect concentration	If a microplastic vector effect is detected, it should be assessed whether it leads to exceedance of a chemical threshold effect concentration	L, F, M
10	Mixture toxicity	If a microplastic vector effect is detected, it should be assessed whether the increased ('vector effect') and decreased ('cleaning effect') chemical exposures due to microplastic ingestion still lead to a net exceedance of a chemical threshold effect concentration for the chemical mixture	L, F, M

[a]*L* laboratory study, *F* field study, *M* model study

indicated if the increased exposure is the reason for exceeding an effect threshold. Several factors define the weight of evidence from such studies (Table 6.1).

Preparation of MP with Associated Chemicals Several studies have spiked chemicals, e.g. persistent organic pollutants (POPs) to MP, in order to mimic environmentally relevant MP. This should be done for long enough time, because in the environment the far majority of microplastic particles has an age in time scale of more than months. For instance, Rochman et al. (2013) deployed MP pellets in San Diego Bay for 3 months, which yields environmentally relevant concentrations that are in (near-)equilibrium with the ambient water. Equilibrium has been argued to be the most relevant state for POPs on MP (Koelmans et al. 2016; Lohmann 2017; De Frond et al. 2019; Seidensticker et al. 2019) and implies that the chemicals are relatively well diffused inside the polymer phase, rendering desorption to be slower compared to non-equilibrium situations. After all, when POPs are pre-sorbed for short times only, they desorb from outer sorption domains, which is faster. This means that any detected vector effect would be overestimated. Sorption times as short as 72 h have been applied in some studies (e.g. Kleinteich et al. (2018), Chua et al. (2014)) which thus would lead to higher desorption rates than in nature. In principle, such non-equilibrium artefacts can be corrected via modelling, where the parameters obtained from a non-equilibrium setup can be used in a scenario analysis for equilibrium conditions (Mohamed Nor and Koelmans 2019; Seidensticker et al. 2019).

Sometimes it is argued that additives can have higher than equilibrium concentrations, suggesting that the aforementioned long sorption times would not be needed for such studies. However, microplastic is, by definition, a mixture of the smaller plastic in the environment, largely originating from slow fragmentation, with time scales longer than chemical desorption time scales. The default state for additives thus also is chemical equilibrium with ambient water (Koelmans et al. 2016; Lohmann 2017; De Frond et al. 2019; Seidensticker et al. 2019), a condition which is not always met in laboratory studies mimicking scenarios with additives. For a credible study, we advise an equilibration time of a month or longer, or corrections by means of kinetic sorption modelling when shorter times are used.

Resemblance of Natural Exposure Pathways In the environment, exposure of organisms to chemicals occurs via multiple pathways, simultaneously (Fig. 6.1). For aquatic organisms, the most important pathways are dermal uptake via skin and/or gills, and water and food ingestion, whereas for air-breathing organisms, inhalation can be relevant. If MP is present, ingestion of MP might contribute to total uptake. If this occurs, it does not necessarily imply that it is a significant vector. It has been demonstrated that transfer of chemicals via ingestion of MP often is negligible compared to the sum of the uptake via the other pathways (Koelmans et al. 2013b; Bakir et al. 2016). Recent experimental studies have confirmed this experimentally, e.g. for lugworms (Besseling et al. 2017), seabirds (Herzke et al. 2016), daphnids (Horton et al. 2018), marine phytoplankton and zooplankton (Beiras et al. 2018, 2019; Beiras and Tato 2019; Sørensen et al. 2020). These observations comply with theory as was shown in several studies where experimental and natural conditions were simulated using numerical modelling (Bakir et al. 2016; Lee et al. 2019). In contrast, some of the key studies in the literature did not address other exposure pathways (e.g. Browne et al. (2013); Rochman et al. (2013); Chua et al. (2014); Wardrop et al. (2016)). These studies have only one possible outcome, namely, that the ingestion of microplastic is important for chemical uptake because other exposure mechanisms that would occur in nature have been disabled. These

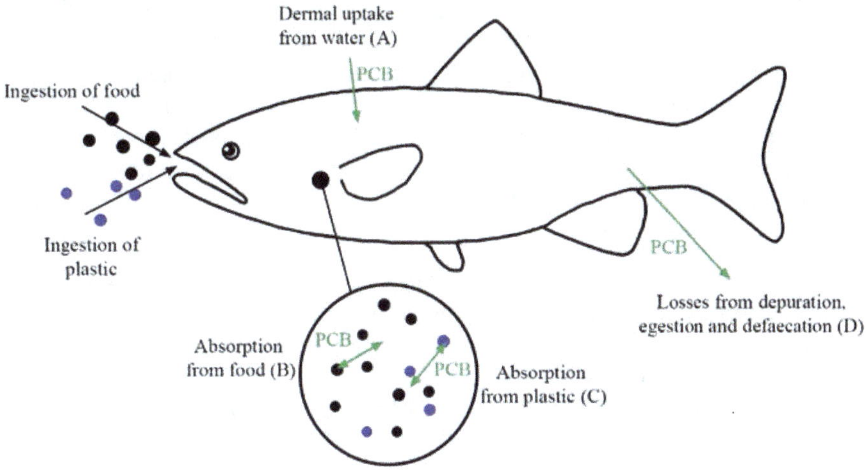

Fig. 6.1 Diagram illustrating the uptake and elimination pathways for chemicals by and from fish. PCB is used to represent chemicals in general. Processes A, B, C and D represent the concomitant terms in Eq. (6.1)

studies have provided support for the MP vector effect paradigm, even though they did not necessarily seek relevance with respect to natural conditions. While the data from these studies are relevant under the conditions used, their deviation from natural conditions makes them less useful for risk assessment related to the MP vector chemical effect. In summary, experimental designs should ideally include, quantify and discuss all pathways relevant in nature, either based on measurement or on measurement in combination with modelling.

Control, Measurement and Assessment of Exposure Multiple pathways thus should be included in MP vector study designs, and these should be quantified in order to assess the relative importance of the MP vector effect. One approach is to assess this via measurement. This requires assessment of chemical concentrations in plastic, water and ingested food throughout the experiment (e.g., Besseling et al. 2017; Rehse et al. 2018; Gerdes et al. 2019b). Many studies have interpreted data based on nominal concentrations of chemicals on plastic, which however is not fully reliable. After all, when chemicals are spiked on the plastic as described in the previous section, or when they also are present in added food, they can (partly) desorb or re-equilibrate before MP ingestion has taken place. This implies that the original, nominal concentrations on plastic do not apply anymore, and actual exposure is unknown. This hampers data interpretation and has led to uncertainty with respect to the applicability of the results of some studies. This especially occurs for studies that used gradient between MP sorbed with chemicals and clean water (Rochman et al. 2013; Chua et al. 2014; Wardrop et al. 2016; Beckingham and Ghosh 2017). In a recent review, Burns and Boxall (2018) re-analysed exposure in a study where *Oryzias latipes* were exposed to MP associated with organic contaminants sorbed from San Diego Bay (Rochman et al. 2013) and assessed that there was insufficient chemical mass on the MP to explain mass measured in fish for most chemicals, suggesting that the cod oil in the diet also was a source. MP vector studies also have suggested that chemicals evaporated from their experimental systems, further limiting data interpretation (Rochman et al. 2017). There are established approaches to measure aqueous phase concentrations, such as with passive samplers, and these have already been applied in some MP vector studies (Besseling et al. 2017; Horton et al. 2018; Wang et al. 2019).

The second approach to assess the relative magnitude of the MP vector effect compared to competing uptake pathways is by calculation. Several studies have demonstrated how results from experiments that did not as such fully comply to natural conditions can still be used in model scenario analysis that were environmentally relevant (Koelmans et al. 2016; Rochman et al. 2017). An example of such an assessment is provided as a case study in Section 4 of this review.

Resemblance of Chemical Concentration Gradients as They Would Occur in Nature Several studies have mentioned non-equilibrium exposure as a limitation in the interpretation of their data (e.g. Beckingham and Ghosh 2017; Sleight et al. 2017; Rochman et al. 2017). Furthermore, it has been argued that chemical equilibrium is the most likely state for chemicals on small microplastic (Koelmans et al. 2016; Diepens and Koelmans 2018; De Frond et al. 2019; Seidensticker et al. 2019). Microplastic is small by definition and 'old' for the far majority of the particles, because its main process of formation is the slow process of embrittlement, erosion and fragmentation. Smaller and older particles are closer to equilibrium, due to shorter intrapolymer diffusion path lengths and longer sorption time scales (Seidensticker et al. 2019). For microplastic, sorption kinetic time scales range from weeks to months. Residence times of the particles in the environment are long (90% being older than 2 years in the ocean (reviewed in Koelmans et al. (2016)), which is one of the main reasons for the concerns surrounding microplastic. In combination, these factors cause equilibrium to be the default state for plastic-associated chemicals. The same applies to other small particles in the ocean, like detritus, black carbon and sediment or suspended solid organic matter, which all would have sorption half-lives of days to weeks at most (Schwarzenbach et al. 2005). This however also applies to organisms at the base of the food chain, as well as, for instance, fish eggs and larvae. In later stages of their life cycle, fish consume contaminated prey keeping the organism body burden close to equilibrium or even higher due to biomagnification. This implies that for chemicals in ingested microplastics, there is no concentration gradient that would drive transfer to the organism, to begin with. For MP vector studies to be environmentally relevant, this means that plastic, organisms, water and other compartments (food, sediment) should all include a scenario where chemicals are contaminated to the same extent (equal fugacity), ideally by pre-equilibration of the chemicals prior to exposure to microplastic. Although second best, studies could still use a large enough gradient, in order to be able detect an effect in the first place. However, in such a case, results should be calculated back to a natural conditions scenario. Alternatively, the bias caused by testing chemical uptake from contaminated MP by clean organisms can be balanced by also testing the opposite treatment, i.e. by using (relatively) clean MP ingested by contaminated test organisms (Koelmans et al. 2013b; Rummel et al. 2016; Scopetani et al. 2018; Wang et al. 2020b). Studies that make an effort to use pre-equilibrated exposure media and organisms (e.g. Besseling et al. (2013); Besseling et al. (2017)) or that otherwise account for vector as well as cleaning mechanisms combined, score higher, whereas evidence from studies that, for instance, force chemical transfer by using clean test animals would receive lower weight.

Assessment of Ingestion Laboratory MP vector studies have been performed with a wide variety of test organisms, covering functional groups such as phytoplankton, zooplankton, (benthic) invertebrates and fish. The essence of the MP vector hypothesis relates to chemical transfer upon ingestion of the contaminated particles. Therefore, we argue that MP vector studies should ideally provide evidence that the test animals actually ingested the particles and preferably also at which rate.

6.2.2 Criteria for Field Studies

Studies have attempted to use observed co-occurrences or correlations between field data on plastic densities or of chemical concentrations in plastic, with chemical concentrations in organisms, as evidence supporting the MP vector hypothesis. The main challenge here is that, whereas in laboratory experiments typically all conditions are kept the same except for the one being researched, in nature, (a) multiple mechanisms are at play simultaneously, (b) there is no control over them, and (c) they may remain partly unknown. Given this, several factors define the weight of evidence from such studies (Table 6.1).

Resemblance of Natural Exposure Pathways Various researchers and expert groups have defined multiple processes and uptake pathways that can explain the occurrence of chemicals (e.g. POPs, PBTs, additives) in organisms, one of which is by ingestion of MP and subsequent chemical desorption inside the organism (GESAMP 2015; Rochman 2015; Koelmans et al. 2016). This means that studies should motivate why the uptake flux via these combined other parallel processes is small compared to the chemical uptake flux via microplastic ingestion. Pathways should be assessed such that it is relevant for the organism studied. However, studies have often neglected to consider such parallel pathways. For instance, the effect of microplastic ingestion on bioaccumulation of phthalates by the fin whale was speculated from detection of plastic in plankton samples and from phthalates being detected in the same plankton samples and fin whale (Fossi et al. 2012). However, the plankton to microplastic number concentration ratios were higher than 1000. Given these high ratios and residence times of microplastic in the oceans, causing chemical desorption from microplastic and subsequent uptake by plankton, it is most likely that dermal absorption and ingestion of phthalate-contaminated plankton were the dominant pathways, rather than from direct ingestion of the microplastic. A follow-up study analysed the case using theoretical modelling and indeed found that 99.2 % of phthalate uptake was by plankton ingestion and only 0.8 % by microplastic ingestion (Panti et al. 2016). Another example is the study by Tanaka et al. (2013), who found some chemicals in seabirds that were absent in prey and thus concluded that the ingested plastic must have been the source. However, the prey organisms in which the chemicals were not detected were collected 7 years after the year of collection of the seabirds, rendering the comparison inconclusive (Burns and Boxall 2018; Tanaka et al. 2013). There are few field studies that quantify the fluxes of the parallel pathways based on field data (Koelmans et al. 2014; Herzke et al. 2016; Panti et al. 2017), which all conclude a negligible relevance of the MP vector effect. As an opposite example, for a 'hot spot' location in the North Pacific accumulation zone, Chen et al. (2017) calculated MP mass to biomass ratios and determined that MPs outweigh prey in the same size range (0.5–5 mm) by 40 times (and by 180 times if all buoyant plastic and biota > 0.5 mm are considered). They thereby suggested that MP at the surface may make up a significant dietary contribution. A further example of such an analysis is provided as a case study in Section 4. In summary, like empirical studies, field studies should account for all exposure pathways in order to assess the relative importance of the MP ingestion vector effect, either based on measurement or on measurement in combination with modelling.

Evidence from Correlations Some studies have found evidence for the MP vector effect in observed correlations between field data on plastic densities or of chemical concentrations in plastic, with chemical concentrations in organisms (Rochman et al. 2014; Panti et al. 2017; Gassel and Rochman 2019). There are a few pitfalls with the use of such correlations. First, correlations show that variables are related, but they do not reveal the causality of the relationship. For example, in the oceans, high plastic concentrations with high chemical concentrations will inevitably lead to high chemical concentrations in the seawater and thus in biota, via phase partitioning. These concentrations thus will always be correlated. However, this also occurs without ingestion, thus minimizing the weight of evidence from such correlations. Given that (a) MP ingestion is not often verified (Koelmans et al. 2016), (b) studies that assessed ingested MP reveal that MP levels are very low (Foekema et al. 2013; Hermsen et al. 2017; Markic et al. 2019), and (c) if they would be ingested, a gradient for transfer would be lacking (see above), chemical accumulation via the water or ingestion of zooplankton is a far more plausible explanation. Furthermore, bioaccumulation is driven by many factors and not by plastic density alone. For instance, phytoplankton blooms or dissolved organic carbon (DOC) concentrations are highly dynamic and have been demonstrated to affect chemical concentrations in the oceans (Dachs et al. 2002; Jurado et al. 2004). Bioaccumulation thus is multifactorial and without considering all factors at play, i.e. by using univariate correlations, no conclusions on implications of MP can be drawn. Co-occurrence or correlations of high chemical concentrations in fish with higher plastic density are difficult to interpret if plastic densities apply to the surface, whereas the fish species reside (for considerable time) in another habitat, for instance, below the surface, like for myctophids (e.g. Gassel and Rochman 2019). Another pitfall relates to the use of summed concentrations of the investigated chemicals (e.g. ΣPCBs; Rochman et al. 2014; Gassel and Rochman 2019). The sums all have a different proportion of each of the individual chemicals in the mixture. Therefore, each data point in the correlation resembles something different, another variable, rendering the correlation to be unclear. Instead, we argue that for an unambiguous measurement of chemical trends, such correlations should only use concentrations of the same one chemical. A final evaluation criterion relates to accounting for error and uncertainty in the data used in the correlation. As mentioned, the dependent variable, which is often the summed chemical concentration, suffers from uncertainty related to the variability in congener composition which is not accounted for. However, regression analysis, as used in these types of studies, considers the independent variable (e.g. MP density) as if it contained no error itself. This is problematic when these data points actually are prone to error, which is often the case. For instance, Gassel and Rochman (2019) correlated sum chemical concentrations with plastic concentrations interpolated by modelling, which can be assumed to have considerable uncertainty. In these cases, it is recommended that weighted regression procedures are applied. In studies that use such correlations, such pitfalls are typically not discussed. In summary, we recommend that studies that provide evidence based on correlations, to rule out parallel causal explanations for the observations, use analyses of individual chemicals and take measurement error into account.

Assessment of Ingestion Similar to laboratory MP vector studies, field studies that conclude that ingestion of MP increase chemical concentration in biota should provide evidence that the test animals actually ingested the particles and preferably also at which rate.

6.2.3 Criteria for In Vitro Studies

Several scientists have studied chemical transfer kinetics between plastic particles and water or plastic particles and gut fluids or stomach oil (e.g. Teuten et al. (2007); Bakir et al. (2014); Beckingham and Ghosh (2017); Lee et al. (2019); Martin and Turner (2019); Mohamed Nor and Koelmans et al. (2019); Kühn et al. (2020)). Besides the common quality assurance (QA) criteria that apply to any study that needs the analysis of chemical concentrations on plastic, there are several criteria that need to be fulfilled in order to make a study relevant with respect to risks caused by the MP vector effect (Table 6.1). Chemical desorption kinetics in artificial gut fluids have been assessed in order to show the MP vector effect. Such studies should either use long-term adsorption prior to desorption in order to not overestimate the desorption rates, as mentioned in Section 2.1, or the level of non-equilibrium should be accounted for in the desorption kinetic parameter estimation (Mohamed Nor and Koelmans 2019). Because sorption involves a dynamic equilibrium, desorption studies should also account for backward sorption in their setup and data analysis. Kinetic modelling to acquire the rate constants may need to take intrapolymer diffusive mass transfer and/or biphasic behaviour into account (Seidensticker et al. 2017, 2019; Town et al. 2018; Lee et al. 2019). Kinetic parameters should be provided with significance level and/or error estimate, preferably using multiple time points in the desorption curve allowing for a rigorous estimation of parameter values (Mohamed Nor and Koelmans 2019). Furthermore, besides exchange fluxes between MP and gut fluid, the fluxes from food should be assessed in order to assure that the role of ingested MP is not negligible as compared to the parallel exposure pathways. For instance, Kühn et al. (2020) observed leaching of five chemicals from MP into stomach oil sampled from seabirds. However, the uptake of chemicals from normal food and the concentrations of background chemicals in the oil were not measured or modelled, nor were kinetic parameters estimated, precluding conclusions regarding the relevance of the vector effect under environmentally realistic conditions. Furthermore, the in vitro transfer data cannot be assumed relevant for natural conditions in vivo where removal processes such as metabolization and elimination would reduce the gradient for transfer and where bioaccumulation by seabirds is likely to be at steady state. The kinetic data thus have to be put in the context of all relevant processes that govern the uptake and elimination of the chemicals of interest. Models have been developed that allow for such a context-dependent evaluation using the kinetic parameters obtained from in vitro desorption studies (Bakir et al. 2016; Koelmans et al. 2013b, 2016). Studies that account for all these aspects would be most relevant for assessing the relevance of the MP vector effect in the context of chemical risk assessment.

6.2.4 Criteria for Model Scenario Studies

Several studies have applied models to calibrate parameters relevant for the MP vector effect (Koelmans et al. 2016; Mohamed Nor and Koelmans 2019) or to evaluate the importance of the MP vector effect based on empirical data (Gouin et al. 2011; Koelmans et al. 2013b, 2014, 2016; Bakir et al. 2016; Herzke et al. 2016; Rochman et al. 2017; Wang et al. 2019; Wang et al. 2020a, b). As a starting point for the evaluation of studies that used models, we adopt the criteria for a valid model suggested by Rykiel Jr (1996): 'Validation is establishing the truth of a model in the sense of (a) consistency with data, (b) accordance with current knowledge, (c) conformance with design criteria'. Consistency with data means that MP vector models should be demonstrated to concur with observations. This implies that calibrated parameters comply with theory or that parameter sets purely obtained from theory comply with observations. Model frameworks that lack such consistency would score lower. Accordance with current knowledge means that the model application or parts of it should not be in conflict with established and widely accepted evidence-based theoretical concepts or that no such concept is overlooked in case it would be relevant. For instance, model scenario studies informing chemical risk assessments about the relevance of the MP vector effect should account for reversible exchange in the gut, should take into account what is the gradient occurring under natural conditions and should account for all pathways that are relevant under natural conditions. Conformance with design criteria means that a model's complexity, approach and output comply with what a model is meant to do in a certain context. For instance, a model should not be over-parameterized or should it be too simple with respect to the dominant processes that drive the behaviour of the system under consideration.

6.3 Weight of Evidence Supporting the Microplastic Vector Hypothesis in the Context of Chemical Risk

The literature provides varying types of evidence for the MP vector effect and its implications on chemical risks. In 2016, we provided a first evaluation of 13 key laboratory, field and model studies that were available at that time (Koelmans et al. 2016). The evaluation was based on whether studies used realistic MP concentrations or MP fractions in the diet, whether MP ingestion was confirmed and whether all environmentally relevant uptake processes were accounted for. In 2018, a similar analysis was published by Burns and Boxall (2018), who evaluated 18 studies and ranked them into 3 evidence categories: 'demonstrated', 'inconclusive' and 'not supported'.

The review in this section can be seen as an extended and detailed version of these earlier evaluations. Here, we evaluate the weight of evidence for the MP vector effect from 61 studies, based on 2 criteria: (a) is the evidence conclusive enough, given the recommendations discussed in the previous section, and if so, (b) does the study provide evidence for an MP vector effect that would affect chemical risk in nature. Following Burns and Boxall (2018), the same three weight of evidence categories were distinguished.

6.3.1 Weight of Evidence from In Vivo Laboratory Studies

We reviewed 30 studies providing evidence with respect to the MP vector effect (Table 6.2). None of them explicitly discussed the results with respect to implications for chemical risks as defined by the PEC/PNEC approach (Koelmans et al. 2017). Twenty-one out of thirty studies were evaluated as inconclusive with respect to detection of a chemical vector effect caused by microplastic ingestion. However, there appears to be a trend among studies over time. Earlier studies more often neglect environmentally relevant exposure conditions by focusing on chemical uptake from microplastic alone, resulting in a low (i.e. 'inconclusive') weight of evidence score. Later studies more often include co-exposure from water or also from food or other background particles empirically and/or via modelling, which makes them more relevant for environmental conditions, resulting in a higher weight of evidence (conclusive) score. All of the latter studies, however, conclude that a evidence, also because many of these simulations were done with validated models or backed up with experimental data provided in the same study. Most of the studies concluded that a vector effect is not supported, due to lack of gradient or due to other pathways being more important. Some studies might have overlooked some

Table 6.2 Overview of laboratory in vivo studies addressing the role of microplastic ingestion by organisms on bioaccumulation of plastic-associated chemicals

Study	Demonstrated	Inconclusive	Not supported	Comments
Besseling et al. (2013)		X		Measured environmentally relevant exposure of PCBs to *A. marina* in the lab, including all pathways. Treatments with MP showed higher bioaccumulation. However, aqueous exposure and organism lipids were not measured and could have differed among treatments, which limits interpretation with regard to the MP vector effect
Browne et al. (2013)		X		Exposed initially clean *A. marina* to MP or sand spiked with phenanthrene, nonylphenol, triclosan or PBDE-47. Co-exposure from water and food was not considered. Considering the experimental design, uptake from water could have occurred as well. Found more uptake from sand than from ingested plastic. This may be explained from the plastic being a more efficient cause for elimination from the test organisms, via particle egestion/defaecation, compared to sand
Rochman et al. (2013)		X		Test organisms *Oryzias latipes* were exposed to MPs that were enriched with environmental contaminants sorbed from San Diego Bay. Uptake from water was not accounted for and not measured. Test organisms were not at equilibrium at the start of the experiment. Not clear which part of uptake was from contaminated food or from water
Chua et al. (2014)		X		Amphipods *Allorchestes compressa* were exposed to PBDEs in the presence or absence of MP. Co-exposure from water and food was not considered, and initially clean organisms were used. Considering the experimental design, uptake from water could have occurred as well. Used unrealistically high plastic concentrations

(continued)

Table 6.2 (continued)

Study	Demonstrated	Inconclusive	Not supported	Comments
Avio et al. (2015)		X		Mussels (*Mytilus galloprovincialis*) were exposed to MP with and without sorbed pyrene. Uptake from water and natural food was not considered, and initially clean organisms were used. Considering the experimental design, uptake from water could have occurred as well. Used very high plastic concentrations
Wardrop et al. (2016)		X		Rainbow fish (*Melanotaenia fluviatilis*) were exposed to MP with and without sorbed PBDEs. Co-exposure from water and food was not considered, and initially clean organisms were used. Considering the experimental design, uptake from water could have occurred as well. Used very high plastic concentrations
Rummel et al. (2016)		X		Studied 'cleaning' of *Oncorhynchus mykiss* by microplastic ingestion (reversed vector effect), by feeding PCB-contaminated fish a diet with uncontaminated PE. No observable effect of plastic ingestion on the rate of depuration of PCBs was reported. However, due to the wide confidence intervals in the depuration rate constants, the study had a low sensitivity to detect treatment effects. Hence, the observations may not have been sufficient to make the assumption that the cleaning effect was not present
Besseling et al. (2017)			X	Mimicked environmentally relevant exposures of *A. marina* to PCBs in sediment, MP and water after 6-week equilibration, including all pathways, and assessed ingestion. Aqueous exposure was measured using passive samplers, and organism lipids were measured. Data interpretation was aided by biodynamic modelling based on Koelmans et al. (2013b) (Table 6.4). Bioaccumulation factors did not differ between MP and no-MP treatments, providing evidence for the absence of a vector effect, which was supported by model simulations

(continued)

Table 6.2 (continued)

Study	Demonstrated	Inconclusive	Not supported	Comments
Beckingham and Ghosh (2017)		X		In vivo study on bioaccumulation of PCBs with *Lumbriculus variegatus*. Did not consider leaching of PCBs from polypropylene into sediment and water. Did not quantify the relevance of the polypropylene as vector compared to the indirect exposure through desorption to the surrounding media. Did not assess ingestion of microplastic in the worms
Devriese et al. (2017)		X		Norway lobster *Nephrops norvegicus* fed with (short-term spiked) PCB-loaded MPs showed limited additional PCB uptake after 3-week ingestion of PE MP, but not of PS MP. PCBs were MP surface spiked, gelatine food contained PCBs as well, and uptake from water could have occurred as well, impeding differentiation between accumulation from food, water and plastic
Sleight et al. (2017)		X		Tested the hypothesis that MP can act as a vector of chemicals from pelagic to benthic habitats. MP bound chemical bioavailability for zebrafish larvae *Danio rerio* was assessed via gene expression. Aqueous exposure was not assessed. Sorption equilibrium was assumed in data interpretation, yet not experimentally verified and unlikely given the short (5d) exposure time. MP and chemical concentrations were considerably higher than environmentally relevant concentrations
Rochman et al. (2017)		X		Tested how the interaction between microplastics and PCBs could affect a prey species *Corbicula fluminea* and its predator *Acipenser transmontanus*. Exposure was to initially clean organisms, via plastic ingestion alone, but not quantified. PCBs could have desorbed to the clean water followed by volatilization of PCBs from microplastics from aeration in the tanks. PCBs were not detected in test animals, so treatment effects on bioaccumulation could not be assessed

(continued)

Table 6.2 (continued)

Study	Demonstrated	Inconclusive	Not supported	Comments
Horton et al. (2018)		X		Environmentally relevant co-exposure of *Daphnia magna* to chemicals in MP and water. However, the systems were not at equilibrium partly due to addition of the test organisms. MP concentrations were much higher than in the environment. No effects of MP on chemical toxicity were found
Guven et al. (2018)		X		Tested acute effect of pyrene and MP on swimming and predatory performance of a tropical fish (*Lates calcarifer*). Not clear if test organisms, MP and pyrene were in equilibrium as effects were recorded between 1 and 24 h. fish were not fed, eliminating the realism of this exposure pathway. Not clear how pyrene was measured. The authors confirm that short test duration may have restricted pyrene absorption to the MP
Batel et al. (2018)		X		Tested and confirmed transfer of BaP from MP by simple attachment to clean adult zebrafish (*Danio rerio*) gills and zebrafish embryos. BaP equilibration with MP was for only 24 h. BaP may have desorbed to the water during exposure. Role of waterborne exposure not assessed. No environmentally realistic concentration gradient due to use of clean organisms. Very high MP and BaP levels
Beiras et al. (2018)			X	Tested ingestion and contact with PE MP; did not find acute toxicity on marine zooplankton (*Brachionus plicatilis*, *Tigriopus fulvus*, *Acartia clausi*, *Mytilus galloprovincialis*, *Paracentrotus lividus*, *Oryzias melastigma*). Spiking of MP was for 2 d only. Contaminated MP was dosed to clean organisms, and no other parallel chemical exposure pathways were taken into account. Despite these conditions of artificially favouring an effect of MP, no effect was found. This means the results can be considered as not supporting the occurrence of an MP vector effect for the species tested

(continued)

Table 6.2 (continued)

Study	Demonstrated	Inconclusive	Not supported	Comments
Scopetani et al. (2018)			X	Clean *Talitrus saltator* were fed with uncontaminated fish food mixed with (short term) contaminated MPs, or contaminated *Talitrus saltator* was fed with contaminated food in combination with clean MP. The first treatment showed a vector effect, whereas the second treatment showed that MP cleaned the organism. Spiking of MP was short term and desorption in the water could have occurred. Given the detection of opposite direction chemical transport, the conclusion of a limited relevance for a vector effect in the environment is considered correct
Barboza et al. (2018)		X		Effects of *Dicentrarchus labrax* juveniles upon exposure to MP and $HgCl_2$. MP polymer type and associated chemicals not known. Clean fish were used and no food was added, causing the hg speciation and exposure to be different from natural conditions, leaving uncertainty with respect to how MP would make a difference in reality
Rehse et al. (2018)			X	Analysed how the presence of polyamide MP modifies effects of bisphenol A (BPA) on *Daphnia magna*. BPA exposure was via water only or via water plus ingested MP. EC_{50} values were the same, which demonstrates that an MP vector effect did not occur
Beiras and Tato (2019)			X	MP and nonylphenol effects on urchin larvae (*Paracentrotus lividus*). Conditions aimed at maximizing the relevance of the particulate phase for chemical uptake in the test species: Natural particles were not present, and MP loads tested were far above the environmental levels ever found in marine waters. With MP and food present, there were no differences in chemical effect thresholds as compared to when no MPs were present, indicating that no vector effect occurred

(continued)

Table 6.2 (continued)

Study	Demonstrated	Inconclusive	Not supported	Comments
Wang et al. (2019)			X	Earthworm *Eisenia fetida* exposed to PE and PS particles in agricultural soil; PCB-, PAH- and MP-contaminated soil; and PCB/PAH-contaminated soil. Data interpretation was aided by biodynamic modelling based on Koelmans et al. (2013b) (Table 6.4). No MP vector effect was found
Beiras et al. (2019)			X	Sea urchin pluteus and copepod nauplius larvae actively ingest MP particles. MP did not increase accumulation of organic chemicals in sea urchin larvae. MP did not increase the toxicity or 4-n-NP or 4-MBC to zooplankton. PE microplastics did not act as vectors of hydrophobic chemicals to zooplankton
Sørensen et al. (2020)			X	*Acartia tonsa* and *Calanus finmarchicus* were exposed to ingestible and non-ingestible PE microbeads, spiked with fluoranthene and phenanthrene, which also were present in the water phase. Under this co-exposure scenario, bioaccumulation factors were the same for systems with versus without MP, and for systems with ingestible versus non-ingestible MP, indicating that no MP vector effect occurred
Thaysen et al. (2020)		X		Studied the direction of transfer between ingested plastic and biota lipids of brominated flame retardants in *Larus delawarensis*. It was found that the concentration gradients were opposite for different chemicals, and thus would lead to bidirectional transfer dependent on the chemical, yet with transfer from bird to plastic ('cleaning') occurring for a higher number of compounds

(continued)

Table 6.2 (continued)

Study	Demonstrated	Inconclusive	Not supported	Comments
Bartonitz et al. (2020)		X		*Gammarus roeseli* were exposed to phenanthrene in water, as well as in the presence of MP or sediment. Due to their high concentrations, MP (and sediment) particles reduced phenanthrene toxicity due to sorption to the particles, and absence of a vector effect was concluded. In nature, however, MP would not reduce aqueous phase concentrations at realistic MP concentrations, rendering the results inconclusive
Xia et al. (2020)		X		Exposed *Chlamys farreri* to BDE209 with versus without presence of PS MP. Bioconcentration factors (BCF) were the same; however, depuration was faster in the presence of MP, demonstrating the 'cleaning' effect. The conditions of MP and BDE209 pre-equilibration were not fully clear; co-exposure from the water was not quantified; water was renewed and scallops removed every day, which was not accounted for in the modelling; clean test organisms were used; and no food was present during exposure
Coffin et al. (2020)			X	Exposed *Atractoscion nobilis* to environmentally relevant MP (0.32 particles/L) and benzo(a)pyrene-sorbed MP concentrations. Co-exposure via food or water was not included. No effects of the presence of MP were observed in 5-d exposure, indicating that no MP vector effect occurred
Tanaka et al. (2020)		X		Feeding experiment under environmentally relevant conditions, in which PE pellets contaminated with 5 additives were fed to *Calonectris leucomelas* chicks. The MP vector effect was demonstrated under these experimental conditions. However test animals were clean at start, and the chemicals were not present in the diet, whereas co-exposure via the diet is plausible in nature where most chemicals are ubiquitous

(continued)

Table 6.2 (continued)

Study	Demonstrated	Inconclusive	Not supported	Comments
Wang et al. (2020a)		X		Earthworm *Eisenia fetida* exposed to five MP polymer types, spiked with PCB and PAH, over 28 d. Chemical transfer to the worms was observed with dermal exposure generally dominating overexposure via ingested plastic ('vector'). Addition of clean MP reduced transfer ('cleaning'). Data interpretation was aided by biodynamic modelling based on Koelmans et al (2013b) (Table 6.4). It was concluded that MP can act as a source and a sink but that predictions based on short-term non-equilibrium conditions may both be representative of natural conditions
Wang et al. (2020b)		X		Earthworm *Eisenia fetida* exposed to PCB-contaminated soil and clean MP (PE of three sizes) or to clean soil and PCB-contaminated MP, for 28 d. Both treatments demonstrated less uptake, i.e. 'cleaning' compared to non-MP controls. Data interpretation was aided by biodynamic modelling based on Koelmans et al (2013b) (Table 6.4). Chemical uptake via MP ingestion was smaller than via the parallel pathways. It was concluded that MP can act as a source and a sink but that any effect in nature would be small due to low MP concentration

General approach, type of evidence for an effect on chemical risk and environmental realism are summarized

vector effect is of little relevance under natural conditions (Table 6.2). Four studies explicitly aimed for finding empirical evidence for the 'cleaning' effect by microplastic ingestion (Rummel et al. 2016; Scopetani et al. 2018; Thaysen et al. 2020; Heinrich and Braunbeck 2020), as predicted based on first principles (Gouin et al. 2011; Koelmans et al. 2013b). A cleaning effect was, however, also suggested to occur in the study by Devriese et al. (2017). Browne et al. (2013) found lower chemical body burdens in *A. marina* after chemical uptake via ingested microplastic as compared to uptake via sand, suggesting more efficient 'cleaning' via the plastic as compared to sand. Recently, Xia et al. (2020) reported a statistically significant 30% to 45% faster depuration of BDE-209 from marine scallops in treatments with MP. Bioconcentration factors (BCF) were 6 to 14% higher in the presence of MP, differences which, however, were not statistically significant. An MP 'cleaning' effect was thus demonstrated, in contrast to an MP vector effect.

6.3.2 *Weight of Evidence from Field Studies*

We evaluated six field studies, of which only one was considered to provide conclusive evidence (Herzke et al. 2016), evidence which in this case was not supportive of a chemical vector effect (Table 6.3). The other studies all suffered from the pitfalls of causality, i.e. did not address alternative yet possibly occurring mechanisms that could also explain the observations, such as uptake via water or food.

Table 6.3 Overview of field studies addressing the role of microplastic ingestion by organisms on bioaccumulation of plastic-associated chemicals

Study	Demonstrated	Inconclusive	Not supported	Comments
Fossi et al. (2012)		X		Effect of microplastic ingestion on bioaccumulation of phthalates was speculated from detection of plastic in plankton samples, and phthalates detected in the same plankton samples and in fin whale (*Balaenoptera physalus*). The plankton to microplastic number concentration ratio was 1600 (Ligurian Sea) to 18000 (Sardinian Sea). Given these ratios and ageing of plastic in the oceans causing chemical desorption and uptake by plankton, uptake via plankton is more likely to occur than via plastic ingestion
Gassel et al. (2013)		X		Ingestion of plastic was speculated to best explain the detection of nonylphenol in 6 out of 19 fish individuals (*Seriola lalandi*), given the detection of two plastic particles in 2 out of the 19 fish individuals. Given the data provided, the study shows that plastic may have been the source of the nonylphenol. However, fish (n=19) and plastic (n=2) sample sizes were very low, and parallel pathways may have contributed to the uptake
Tanaka et al. (2013)		X		PBDE found in seabirds (*Puffinus tenuirostris*) were also present in plastic in the stomach, but not in prey, suggesting that the ingested plastic was the source of the PBDEs. However, the prey samples were taken 7 yrs later and >1000km away. The authors acknowledged these caveats

(continued)

Table 6.3 (continued)

Study	Demonstrated	Inconclusive	Not supported	Comments
Rochman et al. (2014)		X		Myctophid sampled at stations with greater plastic densities had larger concentrations of BDEs # 183–209 in their tissues suggesting that these chemicals are indicative of plastic contamination in the marine environment. Plastic was not measured in the fish, and the BDEs might as well have accumulated from water or the plankton diet. No strong conclusion on the role of ingestion was drawn
Herzke et al. (2016)			X	Combined three lines of evidence: (a) correlations among POP concentrations, differences in tissue concentrations of POPs between plastic ingestion subgroups, (b) fugacity calculations and (c) bioaccumulation modelling, to show that MP did not act as a vector of POPs for *Fulmarus glacialis*
Gassel and Rochman (2019)		X		Examined relationships among chemical contaminants and MP in lanternfish (Myctophidae). Lower chlorinated PCBs, higher in gyre fish, correlated with higher modelled plastic density. Data normality and uncertainty in modelled MP densities data potentially affecting significance levels were not taken into account in regression analysis. Exposure pathways other than MP that could also explain the observed differences were not considered

General approach, type of evidence for an effect on chemical risk and environmental realism are summarized

6.3.3 *Weight of Evidence from In Vitro Studies*

Thirteen in vitro studies reported on the chemical release kinetics from microplastic in the context of the chemical vector effect (Table 6.4). In general, most in vitro studies only reported the amount bioavailable or leached from plastic after exposure in the simulated gut fluid or stomach oil. These bioavailability percentages which are reported generally are not relevant in nature as they also depend on external factors such as fugacities in other compartments in the gut, which would be different in nature. Exchange fluxes with other components in the gut, like food organic matter or micelles, are also usually neglected. Generally, analysis of experiment results with models that describe the data are lacking. This limits the applicability of the findings as they cannot be extrapolated to the actual environment. All studies except two (Mohamed Nor and Koelmans 2019; Kühn et al. 2020) had used a clean gut

Table 6.4 Overview of laboratory in vitro studies aiming to assess the role of microplastic ingestion by organisms on bioaccumulation of plastic-associated chemicals

Study	Demonstrated	Inconclusive	Not supported	Comments
Teuten et al. (2007)		X		Compared desorption rates of sorbed phenanthrene from plastic and sediment in seawater and sodium taurocholate that simulated the gut fluid of *Arenicola marina*. Model analysis was limited to pseudo first-order rate analysis to determine initial desorption. Considered theoretical calculations using equilibrium partitioning method with different compartments to understand their respective relative contributions. This analysis overestimates the amount bioaccessible to benthic organisms as the environment is more dynamic and it also neglected other pathways
Bakir et al. (2014)		X		Investigated the desorption rates of chemicals from PVC and PE under simulated gut conditions. Model analysis was limited as only first-order rate kinetics was considered, neglecting backward sorption and bimodal behaviour. Chemicals were sorbed for a short time, but non-equilibrium desorption kinetics were not considered. Assessment with maximum concentration gradient is not relevant in nature. Other pathways in the gut are neglected
Turner and Lau (2016)		X		Evaluated bioaccessibility of elements indicative of halogenated flame retardants in a simulated digestion fluid of seabirds. Model analysis was limited and obtained rate constants using a diffusion-controlled and parabolic model. Assessment with maximum concentration gradient is not relevant in nature. Evaluation of the results in the context of the actual environment is based on total percentage leached from plastic which is not adequate as the gut of the seabirds are already contaminated from other sources

(continued)

Table 6.4 (continued)

Study	Demonstrated	Inconclusive	Not supported	Comments
Beckingham and Ghosh (2017)		X		Compared gut fluid solubilization between polypropylene and other natural and anthropogenic organic particles. The model remained limited and data was analysed assuming that equilibrium was reached for all congeners in the system after 4h. Maximum concentration gradient between contaminated particles and gut fluid was not environmentally relevant
Massos and Turner (2017)		X		Evaluated bioaccessibility of Cd, Pb and Br from beached microplastics in a physiologically based extraction test. Considered only percentage bioaccessible (over total available in the plastic) which does not reflect what happens in nature as the results are only applicable for the case when the organism do not have any accumulated element. Compared bioaccessibility of the elements through plastic against dietary concentrations and did not consider that the percentages may change in the presence of food
Turner (2018)		X		Evaluated bioaccessibility of hazardous elements in simulated digestion fluid of northern fulmar. Model analysis was limited and did not consider accounting for individual exchange fluxes as it uses the pseudo first-order diffusion model. Did not consider the scenario in which the elements have bioaccumulated in the fulmar. Maximum concentration gradients and evaluation of bioaccessibility percentage based on maximum leached (equilibrium concentrations) as in the experiment are not relevant in nature
Lee et al. (2019)		X		Evaluated relevance of ingested microplastics to overall transfer of chemicals into fish using simulated intestinal fluid and model analysis. Model analysis did not include error estimates and did not reflect experimental results as it was not calibrated. Did not evaluate the results in the context of the actual environment

(continued)

Table 6.4 (continued)

Study	Demonstrated	Inconclusive	Not supported	Comments
Coffin et al. (2019)		X		Evaluated leaching of plastic additives from commonly ingested plastic items in gut mimic models for fish and seabirds. Used virgin plastic items, maximizing the concentration gradient between plastic and gut fluid, which is not relevant in nature. Single time point statistical analysis of the concentration leaching out from plastic and its effects on estrogenicity only reflects the vector effect under such experimental conditions. Results were not evaluated in the context of other relevant processes
Guo et al. (2019)		X		Investigated leaching of flame retardants from acrylonitrile-butadiene-styrene (ABS) in simulated avian digestive systems and the effect of co-ingesting sediment. Did not consider further evaluation of results with models or calculations with chemical characteristics (K_{OW}). Results indicated chemicals transferred from plastic to sediment. However, the study did not consider sorption capacities of the components in the system and its exchange fluxes
Martin and Turner (2019)		X		Evaluated the mobilization of cadmium under simulated digestive conditions over 6h. Model was limited and did not consider the other phases and exchange fluxes. Applicability of the rate constants obtained by fitting a second-order diffusion model to the experimental data was not discussed. Evaluation of environmental implication is not relevant in nature as there is no clean sediment in nature. Sediment used in experiment was collected from a protected and unpolluted location which is not representative of other areas

(continued)

Table 6.4 (continued)

Study	Demonstrated	Inconclusive	Not supported	Comments
Mohamed Nor and Koelmans (2019)			X	Evaluated chemicals exchange of PCBs in gut fluid mimic systems considering different scenarios representing different fugacity levels between plastic and organism. Considered all exchange fluxes in the system and bimodal behaviour in model analysis, showing error estimates for parameters. Considered exchange fluxes with other components in the gut such as food. Discussed bioavailability from ingested plastic to organisms under different scenarios
Heinrich and Braunbeck (2020)		X		Studied the effect of PE MP addition to rainbow trout RTL-W1 cells dosed with 7-ethoxyresorufin-O-deethylase (EROD) inducers by addition. The addition of MP reduced EROD activity. The authors concluded that the presence of MP can reduce the amount of bioavailable pollutants in situ ('cleaning' effect) but that it remains unclear to what extent this mechanism occurs under natural conditions
Kühn et al. (2020)		X		Observed leaching of 5 (out of 15) chemicals to stomach oil sampled from fulmars. Results of different experiments were said to be inconsistent. Concentrations in the oil could have decreased due to biodegradation, increasing the gradient for transfer. Biomagnification from regular food was not addressed. Chemical transfer from the oil to the plastic was not considered. It was acknowledged that the experimental data should not be compared with models simulating natural conditions

General approach, type of evidence and environmental realism are summarized

system, which is unrealistic. This may result in an overestimation of the amount leached from plastic. Kühn et al. (2020) used natural contaminated stomach oil and indeed only found limited transfer for five and no transfer for the other ten chemicals studied. Only few studies relate experimental observations with the chemical and sorbent characteristics. Most studies did not provide error estimates or significance levels for their parameter estimations. Due to these limitations, 12 out of 13 studies were evaluated as inconclusive. The one study providing conclusive evidence demonstrated that the occurrence of a vector effect was context-dependent

(Mohamed Nor and Koelmans 2019). They showed that the boundary conditions, i.e. where either the MP or the food was initially contaminated, or both, determined whether transfer occurred and in which direction.

6.3.4 *Weight of Evidence from Modelling Studies*

Twelve studies used models to investigate the microplastic vector effect (Table 6.5). Most of them were explicitly applied to simulate exposure to plastic-associated chemicals under natural conditions, taking multiple exposure pathways into account, or to simulate realistic experiments that also had multiple exposure pathways covered in their experimental design. These studies generally provided conclusive

Table 6.5 Overview of studies that used simulation models to address the role of microplastic ingestion by organisms on bioaccumulation of plastic-associated chemicals

Study	Demonstrated	Inconclusive	Not supported	Comments
Teuten et al. (2007)		X[a]		An equilibrium model scenario predicted a cleaning effect of plastics in reducing contaminant concentrations in benthic organisms. Another scenario predicted enrichment of chemical concentrations on plastic in the surface microlayer (SML) leading to higher exposure to benthic organisms. The acclaimed SML enrichment, however, has been argued to be based on a misinterpretation, as discussed in Koelmans (2015). Desorption during microplastic settling and burial was not accounted for. Uptake by sediment or food ingestion was not considered
Gouin et al. (2011)			X[a]	Considered all known accumulation pathways in order to quantitatively assess the relative importance of plastic ingestion to total bioaccumulation. A worst case was considered by assuming concentrations in plastic and tissue to be at steady state. Provided mechanistic evidence based on first principles
Koelmans et al. (2013b)			X	Considered all known accumulation pathways in order to quantitatively assess the relative importance of plastic ingestion to total bioaccumulation for a previously published dataset. Provided mechanistic evidence based on first principles, as well as causal evidence on the treatment level by validation with empirical data

(continued)

Table 6.5 (continued)

Study	Demonstrated	Inconclusive	Not supported	Comments
Koelmans et al. (2014)			X[a]	Compared bioaccumulation due to plastic ingestion only, with total observed bioaccumulation in the field. Provided mechanistic evidence based on first principles. The model was validated elsewhere (Koelmans et al. 2013b, 2016)
Bakir et al. (2016)			X	Considered all known accumulation pathways in order to quantitatively assess the relative importance of plastic ingestion to total bioaccumulation. Neglected backward sorption. The 50% scenario can be considered unrealistic or worst case. Provided mechanistic evidence based on first principles
Koelmans et al. (2016)			X[a]	Simulated a series of published experiments using spiked plastic and clean organisms. Provided mechanistic model validation based on three lines of evidence: (a) intrapolymer diffusion, (b) in vitro desorption kinetic data to artificial gut fluids and (c) evaluation against experimental data from three published datasets. Simulations representing natural exposure conditions demonstrated the vector effect to be negligible
Herzke et al. (2016)			X	Modelled chemical fluxes of microplastic ingested by seabirds, accounting for all uptake pathways. Revealed that plastic was more likely to act as a passive sampler than as a vector for chemicals
Rochman et al. (2017)		X		Used the model developed by Koelmans et al. (2013b) to simulate chemical concentrations from a dietary uptake experiment, which however could not be measured due to detection limit problems. Only uptake via plastic was addressed; uptake from food or water was not accounted for
Besseling et al. (2017)			X	Modelled experimental data on chemical uptake by lugworms accounting for all pathways (plastic water, food). Aqueous exposure was assessed using passive samplers. The ingestion vector effect was demonstrated to be irrelevant

(continued)

Table 6.5 (continued)

Study	Demonstrated	Inconclusive	Not supported	Comments
Diepens and Koelmans (2018)		X (PAHs)	X (POP)	Reported a food web model (MICROWEB) based on the model by Koelmans et al. (2013b), which has been validated for POPs. For POPs no vector effect was predicted across all trophic levels of the food web (evidence for 'non-supported'). For degrading compounds, a vector effect was predicted, which, however, has not yet been experimentally validated and therefore was rated 'inconclusive'
Lee et al. (2019)			X	Considered all known pathways to assess the contribution of plastic ingestion to chemical bioaccumulation with an uncertainty analysis using Monte-Carlo simulation. Did not use results from in vitro experiments in model to obtain a more refined model
Wang et al. (2019)			X	Modelled experimental data on chemical uptake by earthworms (*Eisenia fetida*) accounting for all pathways (plastic, water, food), using the model by Koelmans et al. (2013b). Concentrations in soil (food) and water were measured. Modelled and empirical data agreed well and no vector effect was found
Wang et al. (2020a)		X		Modelled experimental data on chemical uptake by earthworms (*Eisenia fetida*) accounting for all pathways (plastic, water, food), using the model by Koelmans et al. (2013b) for contaminated plastic – clean worm scenarios ('vector effect'). Chemical transfer to the worms was observed with dermal exposure generally dominating overexposure through MP ingestion. A minor vector effect thus was found, but it remained unclear to what extent this would apply under natural conditions

(continued)

Table 6.5 (continued)

Study	Demonstrated	Inconclusive	Not supported	Comments
Wang et al. (2020b)		X		Modelled experimental data on chemical uptake by earthworms (*Eisenia fetida*) accounting for all pathways (plastic, water, food), using the model by Koelmans et al. (2013b) for a contaminated plastic – clean worm scenario ('vector effect'), as well as the opposite ('cleaning effect'). Both effects were found implying that the vector effect is context-dependent

[a]Similarly assessed earlier by Burns and Boxall (2018)
General approach, type of evidence for an effect on chemical risk and environmental realism are summarized

aspects, such as release kinetics, gut retention times or reversible (backward) sorption, aspects that however would lead to an overestimation of the importance of the vector effect. Studies that considered such 'worst-case' conditions but still concluded that a vector effect was minor were evaluated as conclusive. Diepens and Koelmans (2018) provided theory suggesting that chemicals present at higher than equilibrium concentrations, like additives, in some cases can (over-)compensate for the simultaneously occurring cleaning effect, leading to a net MP vector effect if plastic mass concentrations are high. Furthermore, they calculated that degradable compounds such as PAH may also be prone to a vector effect due to the chemicals being less bioavailable for metabolization during gut passage. It must be noted that the credibility of their model applies to all scenarios modelled (equilibrium and 'under'-equilibrium) and that the scenarios that represent equilibrium conditions for persistent (i.e. non-metabolizable) chemicals are more likely to be relevant (Koelmans et al. 2016; De Frond et al. 2019).

6.4 Risk Assessment of Plastic-Associated Chemicals: A Case Study Illustrating the Relevance of the MP Vector Effect for Risks of Plastic-Associated Chemicals in San Diego Bay

The previous sections have shown that studies addressing the MP vector effect often did not put their results in an environmentally realistic context or did not address results in the context of chemical risks. Here, the notion 'risk' is not meant as probability, hazard, harm or threat but as the extent to which exposure chemical concentrations exceed a known effect threshold for the species for which the risk is then assessed. We use this definition because it is objective and quantitative and because it is the metric that is used for risk assessment of chemicals in regulatory frameworks (Koelmans et al. 2017). As mentioned, this context of actual risks is relevant, because if a detected MP vector effect would not affect risks of the chemicals, then it might be less urgent to study it and there might be less reason for concern.

Description of the Case For the case study, we use the pioneering laboratory experiment by Rochman et al. (2013) where *Oryzias latipes* (Japanese medaka) were exposed to MP loaded with HOCs sorbed from San Diego Bay seawater. Note that this Asian species was used as test organisms and was not claimed to live in San Diego Bay. For the case study, we use the PCB data from their study and calculate back what medaka would have been exposed to, when they would actually reside in San Diego Bay. In the laboratory study, the observed increase in PCB bioaccumulation by medaka was ascribed to the ingestion of MP. The MP was equilibrated in San Diego Bay, which yielded field relevant PCB concentrations on the plastic and which also contributed to the environmental relevance of the study. The fact that the MPs acquired PCBs from the seawater implies that there were PCBs in the water. Therefore, in the bay, medaka would have been exposed to plastic-associated PCBs as well as to the same PCBs dissolved in the water. Additionally, fish would also be exposed to PCBs in food (prey), which, similar to the MPs, would have absorbed or bioconcentrated PCBs from the ambient water. In the laboratory experiment, however, aqueous phase concentrations were kept at zero. Furthermore, in the experiment, fish was fed PCB contaminated cod liver oil. PCB concentrations were known, but these were not equal to concentrations that would be present in natural food at equilibrium in the bay. Furthermore, in the laboratory experiment, fish were already contaminated with PCBs, but at lower than equilibrium concentrations, because uptake was observed, in the controls as well as in the plastic treatments. In contrast, in the bay, chemical concentrations in small fish like medaka would have been at steady state.

Calculation of the Contribution of Plastic in the Laboratory Experiment Koelmans et al. (2016) modelled the experiment published by Rochman et al. (2013) with aqueous concentration set to zero and ingestion rate and concentrations in food set at the values reported in the study. This model framework has been developed, applied and validated (Table 6.5) in a series of studies (Koelmans et al. 2013a, b, 2014, 2016; Koelmans 2015; Rochman et al. 2017; Wang et al. 2019, 2020a, b). A similar model has been applied by Bakir et al. (2016) to assess the relative contribution of chemical uptake pathways for marine worms, fish and seabirds. This modelling shows that in the experiment (Rochman et al. 2013), the contribution of the ingested MP to bioaccumulation by medaka in the experiment can be estimated to range from 3% to 100%, depending on the PCB congener (Table 6.6).

Table 6.6 Percentage uptake from water, food and plastic for PCBs in the lab experiment by Rochman et al. (2013), modelled according to data provided

	PCB18	PCB28	PCB52	PCB44	PCB101	PCB123	PCB118	PCB153	PCB138	PCB187
Water	0.0	0.0	0.0	0.0	0.0	0.0	0.0	0.0	0.0	0.0
Food	62.7	0.0	28.6	0.0	22.4	96.8	38.3	40.1	40.1	58.2
MP	37.3	100	71.4	100	77.6	3.22	61.7	59.9	59.9	41.8

Calculation of the Contribution of Plastic Occurring Under Natural Conditions in San Diego Bay Using the model with the same parameter values, it can be calculated what the contribution of MP ingestion would be, if (a) PCB exposure from water is included as it would occur in San Diego Bay, if (b) exposure from natural food is included as it would occur in the bay and if (c) the background concentration in fish from the bay would be taken into account. Altogether, this calculation thus demonstrates what the contribution of MP ingestion would have been under natural, non-laboratory conditions. This calculation uses the simple mass balance principle as follows (Fig. 1):

$$\begin{aligned} &UPTAKE\ BY\ FISH(t) = \\ &(A)UPTAKE\ FROM\ WATER + \\ &(B)UPTAKE\ FROM\ FOOD + \\ &(C)UPTAKE\ FROM\ PLASTIC - \\ &(D)LOSSES\ FROM\ DEPURATION, EGESTION, DEFAECATION. \end{aligned} \tag{6.1}$$

In mathematical form with terms A, B, C and D in the same order (Koelmans et al. 2013a, b, 2014, 2016, Koelmans 2015):

$$\frac{dC_{B,t}}{dt} = k_{derm} C_W + IR \times S_{FOOD} a_{FOOD} C_{FOOD} + IR \times S_{PL} C_{PLR,t} - k_{loss} C_{B,t} \tag{6.2}$$

The fraction uptake from plastic is calculated as term C divided by the sum of the uptake terms A, B and C.

For uptake from water (term A), the absorption rate constant is needed, which is taken from Hendriks et al. (2001). The chemical concentration in the water (C_W) is obtained from the chemical concentrations on the plastic (C_{PL}) as reported by Rochman et al. (2013) and the equilibrium partition coefficient for PE, K_{PE} (Lohmann 2011), according to $C_W = C_{PL}/K_{PE}$.

Fish food (term B) can be considered as organic matter. The PCB concentration in the food at equilibrium as it would occur in the bay can be calculated from C_W calculated above and an organic matter (OM) partition coefficient K_{OM}, which can be calculated from the octanol-water partition coefficient K_{OW} ($\log K_{OM} = \log K_{OW} - 0.48$, (Seth et al. 1999)).

The contribution from ingested MP (term C) is modelled as described in Koelmans et al. (2016). This calculation accounts for partition coefficients to plastic, residence times for plastic in the gut and net absorption (leaching and readsorption) rates inside the organism. The fraction of plastic in the ingested material was kept at 10%, in accordance to the study (Rochman et al. 2013).

Loss from depuration, egestion and defaecation (term D) is also modelled as described in Koelmans et al. (2016). However, it does not play a role in the calculation of the percentage of chemical uptake from microplastic ingestion.

The background concentration in the hypothetical San Diego Bay medaka can be calculated from the same C_W (above) and the lipid normalized bioconcentration factor, which can be equated to K_{OW}. This assumes that these lipids are at equilibrium with water. In reality these lipids may have a higher than equilibrium concentration due to biomagnification. This implies that the present calculation would overestimate the relative contribution of plastic, because the actual fugacity gradient between the fish lipids and ingested plastics is not taken into account.

The calculation shows that in this environmentally realistic scenario, food intake is the major PCB uptake pathway, and ingestion of MP would make a negligible contribution to uptake by medaka (Table 6.7).

Validity of Assumptions and Calculations The concentrations in MP were taken from Rochman et al. (2017) and were measured after a 3-month exposure in San Diego Bay, which implies that the estimates of the aqueous concentrations in the bay (C_W) are quite accurate, given that PCB sorption to small MP reaches equilibrium in 3 months (De Frond et al. 2019; Mohamed Nor and Koelmans 2019). Sorption to small organic matter particles, phytoplankton or zooplankton is also within days (Koelmans et al. 1993, 1997), which implies that using a K_{OM} for estimating concentrations in food as they would occur in the bay is defensible. Given the small sizes of the early life stages of fish like medaka, they are subjected to fast equilibration kinetics while they grow from egg and larvae size to adult life stage, which implies PCB body burden are maintained at steady state.

Discussion The calculation suggests that the experiment by Rochman et al. (2013) occurred at non steady-state conditions due to zero concentration in the water and lower than equilibrium (compared to the plastic) concentration in the food. The fugacity was thus higher in the plastic than in food and especially water. Natural systems strive towards chemical equilibrium and steady state and reach such a state usually because of chemical transfer rates that are sufficiently fast compared to residence times of particles and animals in nature. The experimental conditions as in Rochman et al. (2013) were supplemented to align with the exposure scenario we

Table 6.7 Percentage uptake from water, food and plastic for PCBs in San Diego Bay

	PCB18	PCB28	PCB52	PCB44	PCB101	PCB123	PCB118	PCB153	PCB138	PCB187
Water	4.37	3.16	1.19	1.19	0.48	0.39	0.39	0.19	0.19	0.11
Food	95.5	96.7	98.7	98.7	99.4	99.5	99.5	99.6	99.6	99.7
MP	0.097	0.10	0.12	0.12	0.14	0.15	0.15	0.16	0.16	0.17

have chosen as relevant to the environment, by adding exposure from water, assuming sorption equilibrium between plastic and water and setting the fugacity in the food and fish lipids to the same value as the fugacity in the plastic exposed in the bay. This recalculation to such an equilibrium setting shows that the contribution to chemical transfer by ingestion of MP becomes very small. There is some uncertainty in the coefficients used, and newly added MP may, for some time, have a higher than equilibrium fugacity compared to the other media present. Still, even if concentrations in plastic would be one or two orders of magnitude higher and chemicals inside the microplastic would not re-partition to water and/or food, MP would still be the minor source. The present calculation is conservative because (a) biomagnification, which attenuates the gradient for uptake between ingested plastic and animal tissue (Diepens and Koelmans 2018), was not taken into account and (b) the fraction of plastic in the food was 10%, which is a rather high value for many habitats. Finally, the model-aided recalculation illustrates how information from experimental studies can inform questions related to other systems, for instance, environmentally relevant natural systems.

Implications for Chemical Risks Here we provide an example of the reasoning that would be needed to assess the implication of the MP vector effect on risk, again with San Diego Bay as an example. First of all, the results (Table 6.7) demonstrate that ingestion of MP contributes a negligible fraction to total PCB bioaccumulation. In other words, there is no MP vector effect in San Diego Bay, and thus there is no implication for risk. If however, ingestion of MP would have contributed for 50%, 90% or 99% of all uptake, exposure could be increased by, for instance, factors of 2, 10 and 100, respectively. Whether such MP vector effects would affect chemical risk then depends only on whether such increases would bring the risk characterization ratio (MEC/PNEC) to a value larger than 1. The sum ΣPCB concentration calculated for San Diego Bay based on the measured concentrations on MP is 1.8×10^{-5} μg/L, whereas the EPA water quality standard for ΣPCB is 0.03 μg/L, which is 1600 times higher than the ΣPCB aqueous concentration. This implies that even if MP ingestion would increase exposure by a factor of 100, MP ingestion would not have implications for the risks of the chemicals. Two disclaimers need to be mentioned. First, there are other HOCs in the water, such as PAHs and PBDEs, which could potentially change this chemical risk assessment. PAH do not readily bioaccumulate as they are metabolized by the fish but may still lead to higher exposure via an MP vector effect (Diepens and Koelmans 2018). PBDEs have a similar concentration and behaviour pattern as PCBs and thus would roughly double the ΣHOC risk profile, yet still rendering it negligible. Second, the specific outcome of this first case study is meant as an example of how model simulation can complement experimental data to render experimental studies more relevant for natural conditions. However, that does not imply that the conclusion for San Diego Bay can be generalized, as it is only the first analysis of this kind. Although the situation may be similar for many locations, it cannot be precluded that implications for risks are absent on all possible locations or in the future when MP concentrations increase.

6.5 Mitigation of Microplastic and Plastic-Associated Chemicals

The problem of global pollution with plastic debris has increased awareness within the public, the scientific community as well as policy makers. This has led to a wide debate on solutions for this problem. General consensus is that there is no single 'silver bullet' solution but that when we are to reduce the presence of plastic debris in the environment, this has to be achieved from a combination of measures (Alexy et al. 2019). These include reducing emissions, reducing littering, use of different and 'safe by design' types of polymers and products, close the waste cycle, recycling and cleaning (SAPEA 2019).

The present article addresses implications of chemicals associated with microplastic. This renders the question to what extent remediation approaches for MP-associated chemicals exist. Recently, De Frond et al. (2019) coined the idea of 'microplastic mitigation is chemical mitigation'. They argue that due to their sorptive properties, plastics have been suggested as a management tool to clean chemicals from the water column (Zhu et al. 2011; Tomei et al. 2015). Removing plastics that contain chemicals from the marine environment removes these chemicals, reducing exposure to wildlife and decreasing potential harm. The mitigation approach would be more efficient for locations with higher microplastic densities, i.e. more on beaches and in coastal zones than in oceanic gyres.

This approach is likely to be useful, especially for locations with high plastic pollution. There also may be limitations, related to how the chemicals are actually distributed in the system to be remediated. For efficient environmental remediation of chemicals, it needs to be assessed where the chemicals are. For instance, De Frond et al. (2019) expressed concern for chemicals in the marine environment, which thus renders the question where these chemicals actually reside under the specific conditions of marine systems. In their paper, they assume chemical equilibrium, thereby following the basic principles of environmental chemistry with respect to chemical kinetics and thermodynamics. When we apply these to the marine environment, we would consider microplastic particles as passive samplers, polymer particles being in equilibrium with the ambient water (Seidensticker et al. 2019). Consequently, aqueous phase concentrations can be calculated, because the plastic to water partition coefficients are known. Given the aqueous phase concentrations and the size of a marine compartment (i.e. coastal zone or oceanic gyre subsystem), the mass of chemical in that compartment can be assessed. Such calculations have been provided by Koelmans et al. (2016) who calculated that a negligible 2×10^{-4} % of chemical mass in the global ocean would be associated with plastic. De Frond et al. (2019) reported concentrations of PCBs on plastic of up to 757 ng/g, at a plastic density up to a maximum 249 g/m^2, albeit measured at different locations. Still, as a worst-case calculation, if we use 50% of these values as a proxy for the means of the actual concentrations and assume total suspended solids (TSS) concentrations of 10 mg/L and a mean PCB partition coefficient of 10^5 L/kg, then a water column of 60 m below such a square metre with 249 g of MP particles would hold the same mass of PCBs (i.e. the sum of the mass of PCBs sorbed on TSS and the mass dissolved in water) as all of these MP particles at the surface. Plastic removal would thus only remove a small fraction of the chemicals in marine systems. In coastal areas, there is more chemical mass on plastic than there is on plastic in the open ocean; however, nutrient loads are also higher because the anthropogenic activities leading to plastic pollution are the same ones that cause nutrient and organic pollution (Strokal et al. 2019). Therefore, the chemical fractions on plastic are even lower in coastal areas, which therefore limits the effectiveness of chemical mitigation by cleaning up (micro-)plastic.

6.6 General Discussion and Conclusion

The evaluation of 61 studies addressing the microplastic vector effect revealed that the evidence for the occurrence of the effect in nature is generally weak. This is because many studies remained inconclusive with respect to the hypothesized effect. Other studies were more conclusive; however, they generally provided evidence for the absence of a vector effect.

We suggested using a risk assessment perspective in order to provide context and meaning in case a vector effect would be detected. However, none of the reviewed studies provided such a risk assessment, that is, a quantification of to what extent the occurrence of a vector effect causes chemical exposure to exceed safety effect thresholds. As a proof of concept, we provided an analysis of existing data for San Diego Bay as a case study, where no implications for risks were predicted.

Finally, we emphasize that realizing that the weight of evidence for an MP vector effect to occur in the environment may be low is not the same as proving there is no risk from ingestion of plastic (e.g. Wardman et al. 2020). It has been demonstrated that ingestion of microplastic can lead to physical effects, effects that translate into ecological risks as soon as threshold effect concentrations are exceeded. Theoretically, plastic ingestion can also increase chemical risks. It is just a matter of plastic and chemical concentration levels that in the end determine to what extent chemical exposure is increased as compared to a chemically contaminated environment without plastic particles present. However, at present, the concentrations of microplastic generally seem too low and the alternative exposure pathways too important to cause ingestion of microplastic to make a difference with respect to the risks of chemicals.

References

Alexy P et al (2019) Managing the analytical challenges related to micro- and nanoplastics in the environment and food: filling the knowledge gaps. Food Addit Contam Part A 37:1–10. https://doi.org/10.1080/19440049.2019.1673905. Taylor & Francis

Avio CG et al (2015) Pollutants bioavailability and toxicological risk from microplastics to marine mussels. Environ Pollut 198:211–222. Available at: http://www.sciencedirect.com/science/article/pii/S0269749114005211

Bakir A, Rowland SJ, Thompson RC (2014) Enhanced desorption of persistent organic pollutants from microplastics under simulated physiological conditions. Environ Pollut 185:16–23

Bakir A et al (2016) Relative importance of microplastics as a pathway for the transfer of hydrophobic organic chemicals to marine life. Environ Pollut 219:56–65

Barboza LGA et al (2018) Microplastics increase mercury bioconcentration in gills and bioaccumulation in the liver, and cause oxidative stress and damage in Dicentrarchus labrax juveniles. Sci Rep 8(1):15655. Nature Publishing Group

Bartonitz A et al (2020) Modulation of PAH toxicity on the freshwater organism G. roeseli by microparticles. Environ Pollut. https://doi.org/10.1016/j.envpol.2020.113999

Batel A et al (2018) Microplastic accumulation patterns and transfer of benzo [a] pyrene to adult zebrafish (Danio rerio) gills and zebrafish embryos. Environ Pollut 235:918–930. Elsevier

Beckingham B, Ghosh U (2017) Differential bioavailability of polychlorinated biphenyls associated with environmental particles: microplastic in comparison to wood, coal and biochar. Environ Pollut 220:150–158

Beiras R, Tato T (2019) 'Microplastics do not increase toxicity of a hydrophobic organic chemical to marine plankton. Mar Pollut Bull 138:58–62. Elsevier

Beiras R et al (2018) Ingestion and contact with polyethylene microplastics does not cause acute toxicity on marine zooplankton. J Hazard Mater 360:452–460. Elsevier

Beiras R et al (2019) Polyethylene microplastics do not increase bioaccumulation or toxicity of nonylphenol and 4-MBC to marine zooplankton. Sci Total Environ 692:1–9. Elsevier

Besseling E et al (2013) Effects of microplastic on fitness and PCB bioaccumulation by the lugworm *Arenicola marina* (L.). 47(1):593–600. https://doi.org/10.1021/es302763x. American Chemical Society

Besseling E et al (2017) The effect of microplastic on the uptake of chemicals by the lugworm *Arenicola marina* (L.) under environmentally relevant exposure conditions. Environ Sci Technol 51(15):8795–8804. https://doi.org/10.1021/acs.est.7b02286. American Chemical Society

Besson M et al (2020) Preferential adsorption of Cd, Cs and Zn onto virgin polyethylene microplastic versus sediment particles. Mar Pollut Bull. https://doi.org/10.1016/j.marpolbul.2020.111223

Browne MA et al (2013) Microplastic moves pollutants and additives to worms, reducing functions linked to health and biodiversity. Curr Biol 23(23):2388–2392

Burns EE, Boxall ABA (2018) Microplastics in the aquatic environment: evidence for or against adverse impacts and major knowledge gaps. Environ Toxicol Chem 37(11):2776–2796. Wiley Online Library

Chen Q et al (2017) Pollutants in Plastics within the North Pacific Subtropical Gyre. Environ Sci Technol. ACS Publications

Chua EM et al (2014) Assimilation of polybrominated diphenyl ethers from microplastics by the marine amphipod, Allorchestes compressa. Environ Sci Technol 48(14):8127–8134

Coffin S et al (2019) Fish and seabird gut conditions enhance desorption of estrogenic chemicals from commonly-ingested plastic items. Environ Sci Technol 53(8):4588–4599. https://doi.org/10.1021/acs.est.8b07140

Coffin S et al (2020) Effects of short-term exposure to environmentally-relevant concentrations of benzo(a)pyrene-sorbed polystyrene to White seabass (Atractoscion nobilis)☆. Environ Pollut. https://doi.org/10.1016/j.envpol.2020.114617

Dachs J et al (2002) Oceanic biogeochemical controls on global dynamics of persistent organic pollutants. Environ Sci Technol 36(20):4229–4237. ACS Publications

De Frond HL et al (2019) Estimating the mass of chemicals associated with ocean plastic pollution to inform mitigation efforts. Integr Environ Assess Manag 15(4):596–606. Wiley Online Library de Ruijter VN et al (2020) Quality criteria for microplastic effect studies in the context of risk assessment: a critical review. Environ Sci Technol. https://pubs.acs.org/doi/10.1021/acs.est.0c03057

Devriese LI et al (2017) Bioaccumulation of PCBs from microplastics in Norway lobster (Nephrops norvegicus): an experimental study. Chemosphere 186:10–16

Diepens NJ, Koelmans AA (2018) Accumulation of plastic debris and associated contaminants in aquatic food webs. Environ Sci Technol. https://doi.org/10.1021/acs.est.8b02515

Foekema EM et al (2013) Plastic in north sea fish. Environ Sci Technol 47(15):8818–8824

Fossi MC et al (2012) Are baleen whales exposed to the threat of microplastics? A case study of the Mediterranean fin whale (Balaenoptera physalus). Mar Pollut Bull 64(11):2374–2379. Elsevier

Gassel M, Rochman CM (2019) The complex issue of chemicals and microplastic pollution: a case study in North Pacific lanternfish. Environ Pollut 248:1000–1009. Elsevier

Gassel M et al (2013) Detection of nonylphenol and persistent organic pollutants in fish from the North Pacific Central Gyre. Mar Pollut Bull 73(1):231–242. https://doi.org/10.1016/j.marpolbul.2013.05.014. Elsevier Ltd

Gerdes Z, Hermann M et al (2019a) A novel method for assessing microplastic effect in suspension through mixing test and reference materials. Sci Rep. https://doi.org/10.1038/s41598-019-47160-1

Gerdes Z, Ogonowski M et al (2019b) Microplastic-mediated transport of PCBs? A depuration study with Daphnia magna. PLoS One. https://doi.org/10.1371/journal.pone.0205378

GESAMP (2015) Sources, fate and effects of microplastics in the marine environment: a global assessment. Rep Stud GESAMP. https://doi.org/10.13140/RG.2.1.3803.7925

Gouin T et al (2011) A thermodynamic approach for assessing the environmental exposure of chemicals absorbed to microplastic. Environ Sci Technol 45(4):1466–1472

Granby K et al (2018) The influence of microplastics and halogenated contaminants in feed on toxicokinetics and gene expression in European seabass (Dicentrarchus labrax). Environ Res 164:430–443. Elsevier

Guo H et al (2019) The leaching of additive-derived flame retardants (FRs) from plastics in avian digestive fluids: the significant risk of highly lipophilic FRs. J Environ Sci (China) 85:200–207. https://doi.org/10.1016/j.jes.2019.06.013. Elsevier B.V.,

Guven O et al (2018) Microplastic does not magnify the acute effect of PAH pyrene on predatory performance of a tropical fish (Lates calcarifer). Aquat Toxicol 198:287–293. Elsevier

Hartmann NB et al (2017) Microplastics as vectors for environmental contaminants: exploring sorption, desorption, and transfer to biota. Integr Environ Assess Manag. https://doi.org/10.1002/ieam.1904

Heinrich P, Braunbeck T (2020) Microplastic particles reduce EROD-induction specifically by highly lipophilic compounds in RTL-W1 cells. Ecotoxicol Environ Saf. https://doi.org/10.1016/j.ecoenv.2019.110041

Hendriks AJ et al (2001) The power of size. 1. Rate constants and equilibrium ratios for accumulation of organic substances related to octanol-water partition ratio and species weight. Environ Toxicol Chem. https://doi.org/10.1002/etc.5620200703

Hermsen E et al (2017) Detection of low numbers of microplastics in North Sea fish using strict quality assurance criteria. Mar Pollut Bull 122(1–2):253–258. https://doi.org/10.1016/j.marpolbul.2017.06.051. Elsevier

Hermsen E et al (2018) Quality criteria for the analysis of microplastic in biota samples. Critical review. Environ Sci TechnolAmerican Chemical Society. https://doi.org/10.1021/acs.est.8b01611

Herzke D et al (2016) Negligible impact of ingested microplastics on tissue concentrations of persistent organic pollutants in northern fulmars off coastal Norway. Environ Sci Technol 50(4):1924–1933. https://doi.org/10.1021/acs.est.5b04663

Horton AA et al (2018) Acute toxicity of organic pesticides to Daphnia magna is unchanged by co-exposure to polystyrene microplastics. Ecotoxicol Environ Saf 166:26–34. Elsevier

Jâms IB et al (2020) Estimating the size distribution of plastics ingested by animals. Nat Commun. https://doi.org/10.1038/s41467-020-15406-6

Jurado, E. et al. (2004) Latitudinal and seasonal capacity of the surface oceans as a reservoir of polychlorinated biphenyls. Environ Pollut, 128(1–2), pp. 149–162. Elsevier

Kleinteich J et al (2018) Microplastics reduce short-term effects of environmental contaminants. Part II: polyethylene particles decrease the effect of polycyclic aromatic hydrocarbons on microorganisms. Int J Environ Res Public Health. Multidisciplinary Digital Publishing Institute 15(2):287

Koelmans AA (2015) Modeling the role of microplastics in bioaccumulation of organic chemicals to marine aquatic organisms. A critical review. In: Marine anthropogenic litter, pp 309–324. https://doi.org/10.1007/978-3-319-16510-3_11

Koelmans AA, De Lange HJ, Lijklema L (1993) Desorption of chlorobenzenes from natural suspended solids and sediments. Water Sci Technol:171–180

Koelmans AA et al (1997) Organic carbon normalisation of PCB, PAH and pesticide concentrations in suspended solids. Water Res. https://doi.org/10.1016/S0043-1354(96)00280-1

Koelmans AA et al (2013a) Erratum: plastic as a carrier of POPs to aquatic organisms: a model analysis (environmental science and technology (2013) 47 (7812-7820) DOI: 10.1021/es401169n). Environ Sci Technol. https://doi.org/10.1021/es403018h

Koelmans AA et al (2013b) Plastic as a carrier of POPs to aquatic organisms: a model analysis. Environ Sci Technol 47(14):7812–7820. https://doi.org/10.1021/es401169n

Koelmans AA, Besseling E, Foekema EM (2014) Leaching of plastic additives to marine organisms. Environ Pollut 187:49–54. https://doi.org/10.1016/j.envpol.2013.12.013

Koelmans AA et al (2016) Microplastic as a vector for chemicals in the aquatic environment: critical review and model-supported reinterpretation of empirical studies. Environ Sci Technol 50(7):3315–3326. https://doi.org/10.1021/acs.est.5b06069

Koelmans AA et al (2017) Risks of plastic debris: unravelling fact, opinion, perception, and belief. Environ Sci Technol. https://doi.org/10.1021/acs.est.7b02219

Koelmans AA et al (2019) Microplastics in freshwaters and drinking water: critical review and assessment of data quality. Water Res. https://doi.org/10.1016/j.watres.2019.02.054

Koelmans AA, Redondo-Hasselerharm PE, Mohamed Nur NH, Kooi M (2020) Solving the nonalignment of methods and approaches used in microplastic research in order to consistently characterize risk. Environ Sci Technol. https://doi.org/10.1021/acs.est.0c02982

Kühn S, Booth AM, Sørensen L, van Oyen A, van Franeker JA (2020) Transfer of additive chemicals from marine plastic debris to the stomach oil of northern fulmars. Front Environ Sci 8:138. https://www.frontiersin.org/article/10.3389/fenvs.2020.00138

Lee H, Lee HJ, Kwon JH (2019) Estimating microplastic-bound intake of hydrophobic organic chemicals by fish using measured desorption rates to artificial gut fluid. Sci Total Environ 651:162–170. https://doi.org/10.1016/j.scitotenv.2018.09.068

Lenz R, Enders K, Nielsen TG (2016) Microplastic exposure studies should be environmentally realistic. Proc Natl Acad Sci U S A. https://doi.org/10.1073/pnas.1606615113

Lohmann R (2011) Critical review of low-density polyethylene's partitioning and diffusion coefficients for trace organic contaminants and implications for its use as a passive sampler. Environ Sci Technol 46(2):606–618

Lohmann R (2017) Microplastics are not important for the cycling and bioaccumulation of organic pollutants in the oceans—but should microplastics be considered POPs themselves? Integr Environ Assess Manag 13(3):460–465. https://doi.org/10.1002/ieam.1914

Markic A et al (2019) Plastic ingestion by marine fish in the wild. Crit Rev Environ Sci Technol:1–41. Taylor & Francis

Martin K, Turner A (2019) Mobilization and bioaccessibility of cadmium in coastal sediment contaminated by microplastics. Mar Pollut Bull 146(July):940–944. https://doi.org/10.1016/j.marpolbul.2019.07.046

Massos A, Turner A (2017) Cadmium, lead and bromine in beached microplastics. Environ Pollut 227:139–145. https://doi.org/10.1016/j.envpol.2017.04.034. Elsevier Ltd

Mohamed Nor NH, Koelmans AA (2019) Transfer of PCBs from microplastics under simulated gut fluid conditions is biphasic and reversible. Environ Sci Technol 53:1874–1883. https://doi.org/10.1021/acs.est.8b05143. American Chemical Society

Panti C et al (2016) Microplastic as a vector of chemicals to baleen whales in the Mediterranean Sea: a model-supported analysis of available data.

Panti C et al (2017) Microplastic as a vector of chemicals to fin whale and basking shark in the Mediterranean Sea: a model-supported analysis of available data. In: Fate and impact of microplastics in marine ecosystems, pp 143–144. https://doi.org/10.1016/b978-0-12-812271-6.00140-x

Provencher JF et al (2020) Proceed with caution: the need to raise the publication bar for microplastics research. Sci Total Environ 748. https://doi.org/10.1016/j.scitotenv.2020.141426

Rehse S, Kloas W, Zarfl C (2018) Microplastics reduce short-term effects of environmental contaminants. Part I: effects of bisphenol a on freshwater zooplankton are lower in presence of polyamide particles. Int J Environ Res Public Health. https://doi.org/10.3390/ijerph15020280

Rochman CM (2015) The complex mixture, fate and toxicity of chemicals associated with plastic debris in the marine environment. In: Marine anthropogenic litter, pp 117–140. https://doi.org/10.1007/978-3-319-16510-3_5

Rochman CM (2019) The role of plastic debris as another source of hazardous chemicals in lower-trophic level organisms. In: Handbook of environmental chemistry. https://doi.org/10.1007/698_2016_17

Rochman CM et al (2013) Ingested plastic transfers hazardous chemicals to fish and induces hepatic stress. Sci Rep 3:838. https://doi.org/10.1038/srep03263. Nature Publishing Group

Rochman CM et al (2014) Polybrominated diphenyl ethers (PBDEs) in fish tissue may be an indicator of plastic contamination in marine habitats. Sci Total Environ 476:622–633. Elsevier

Rochman CM et al (2017) Direct and indirect effects of different types of microplastics on freshwater prey (Corbicula fluminea) and their predator (Acipenser transmontanus). PLoS OnePublic Library of Science 12(11):e0187664

Rummel CD et al (2016) No measurable "cleaning" of polychlorinated biphenyls from rainbow trout in a 9 week depuration study with dietary exposure to 40% polyethylene microspheres. Environ Sci: Processes Impacts 18(7):788–795

Rykiel EJ Jr (1996) Testing ecological models: the meaning of validation. Ecol Model 90(3):229–244. https://doi.org/10.1016/0304-3800(95)00152-2

SAPEA (2019) A scientific perspective on microplastics in nature and society | SAPEA. Evid Rev Rep. https://doi.org/10.26356/microplastics

Schwarzenbach RP, Gschwend PM, Imboden DM (2005) Environmental organic chemistry. Wiley

Scopetani C et al (2018) Ingested microplastic as a two-way transporter for PBDEs in Talitrus saltator. Environ Res 167(July):411–417. https://doi.org/10.1016/j.envres.2018.07.030

Seidensticker S et al (2017) Shift in mass transfer of wastewater contaminants from microplastics in the presence of dissolved substances. Environ Sci Technol 51(21):12254–12263. https://doi.org/10.1021/acs.est.7b02664

Seidensticker S et al (2019) Microplastic–contaminant interactions: influence of nonlinearity and coupled mass transfer. Environ Toxicol Chem. https://doi.org/10.1002/etc.4447

Seth R, Mackay D, Muncke J (1999) Estimating the organic carbon partition coefficient and its variability for hydrophobic chemicals. Environ Sci Technol. https://doi.org/10.1021/es980893j

Sleight VA et al (2017) Assessment of microplastic-sorbed contaminant bioavailability through analysis of biomarker gene expression in larval zebrafish. Mar Pollut Bull 116(1–2):291–297. Elsevier

Sørensen L et al (2020) Sorption of PAHs to microplastic and their bioavailability and toxicity to marine copepods under co-exposure conditions. Environ Pollut. https://doi.org/10.1016/j.envpol.2019.113844

Strokal M et al (2019) Global multi-pollutant modelling of water quality: scientific challenges and future directions. Curr Opin Environ Sustain. https://doi.org/10.1016/j.cosust.2018.11.004

Tanaka K et al (2013) Accumulation of plastic-derived chemicals in tissues of seabirds ingesting marine plastics. Mar Pollut Bull 69(1–2):219–222

Tanaka K et al (2020) In vivo accumulation of plastic-derived chemicals into seabird tissues. Curr Biol. https://doi.org/10.1016/j.cub.2019.12.037

Teuten EL et al (2007) Potential for plastics to transport hydrophobic contaminants. Environ Sci Technol 41(22):7759–7764

Thaysen C et al (2020) Bidirectional transfer of halogenated flame retardants between the gastrointestinal tract and ingested plastics in urban-adapted ring-billed gulls. Sci Total Environ. https://doi.org/10.1016/j.scitotenv.2020.138887

Tomei MC et al (2015) Rapid and effective decontamination of chlorophenol-contaminated soil by sorption into commercial polymers: concept demonstration and process modeling. J Environ Manag. https://doi.org/10.1016/j.jenvman.2014.11.014

Town RM, van Leeuwen HP, Blust R (2018) Biochemodynamic features of metal ions bound by micro- and nano-plastics in aquatic media. Front Chem. https://doi.org/10.3389/fchem.2018.00627

Turner A (2018) Mobilisation kinetics of hazardous elements in marine plastics subject to an avian physiologically-based extraction test. Environ Pollut 236:1020–1026. https://doi.org/10.1016/j.envpol.2018.01.023. Elsevier Ltd

Turner A, Lau KS (2016) Elemental concentrations and bioaccessibilities in beached plastic foam litter, with particular reference to lead in polyurethane. Mar Pollut Bull 112(1–2):265–270. https://doi.org/10.1016/j.marpolbul.2016.08.005. Elsevier Ltd

Wang J et al (2019) Negligible effects of microplastics on animal fitness and HOC bioaccumulation in earthworm Eisenia fetida in soil. Environ Pollut 249:776–784. Elsevier

Wang J et al (2020a) Accumulation of HOCs via precontaminated microplastics by earthworm Eisenia fetida in soil. Environ Sci Technol 54:11220–11229. https://doi.org/10.1021/acs.est.0c02922

Wang J et al (2020b) Microplastics as a vector for HOC bioaccumulation in earthworm Eisenia fetida in soil: importance of chemical diffusion and particle size. Environ Sci Technol 54(19):12154–12163. https://doi.org/10.1021/acs.est.0c03712

Wardman T et al (2020) Communicating the absence of evidence for microplastics risk: balancing sensation and reflection. Environ Int. https://doi.org/10.1016/j.envint.2020.106116

Wardrop P et al (2016) Chemical pollutants Sorbed to ingested microbeads from personal care products accumulate in fish. Environ Sci Technol 50(7):4037–4044. https://doi.org/10.1021/acs.est.5b06280

World Health Organization (2019) Microplastics in drinking-water. doi: ISBN: 978-92-4-151619-8

Xia B et al (2020) Polystyrene microplastics increase uptake, elimination and cytotoxicity of decabromodiphenyl ether (BDE-209) in the marine scallop Chlamys farreri. Environ Pollut. https://doi.org/10.1016/j.envpol.2019.113657

Zhu H et al (2011) Evaluation of electrospun polyvinyl chloride/polystyrene fibers as sorbent materials for oil spill cleanup. Environ Sci Technol. https://doi.org/10.1021/es2002343

Ziccardi LM et al (2016) Microplastics as vectors for bioaccumulation of hydrophobic organic chemicals in the marine environment: a state-of-the-science review. Environ Toxicol Chem 35(7):1667–1676. https://doi.org/10.1002/etc.3461

7

Micro- and Nanoplastics: Ecotoxicological Effects on Aquatic and Terrestrial Organisms

Tânia Gomes, Agathe Bour, Claire Coutris, Ana Catarina Almeida, Inger Lise Bråte, Raoul Wolf, Michael S. Bank and Amy L. Lusher

Abstract Plastic pollution is a widespread environmental problem that is currently one of the most discussed issues by scientists, policymakers and society at large. The potential ecotoxicological effects of plastic particles in a wide range of organisms have been investigated in a growing number of exposure studies over the past years. Nonetheless, many questions still remain regarding the overall effects of microplastics and nanoplastics on organisms from different ecosystem compartments, as well as the underlying mechanisms behind the observed toxicity. This chapter provides a comprehensive literature review on the ecotoxicological impacts of microplastics and nanoplastics in terrestrial and aquatic organisms in the context of particle characteristics, interactive toxicological effects, taxonomic gradients and with a focus on synergies with associated chemicals. Overall, a total of 220 references were reviewed for their fulfilment of specific quality criteria (e.g. experimental design, particle characteristics, ecotoxicological endpoints and findings), after which 175 were included in our assessment. The analysis of the reviewed studies

T. Gomes (✉) · A. C. Almeida · I. L. Bråte · R. Wolf · A. L. Lusher
Section of Ecotoxicology and Risk Assessment, Norwegian Institute for Water Research (NIVA), Oslo, Norway
e-mail: tania.gomes@niva.no; ana.catarina.almeida@niva.no; inger.lise.nerland@niva.no; raoul.wolf@niva.no; Amy.Lusher@niva.no

A. Bour
Department of Biological and Environmental Sciences, University of Gothenburg, Gothenburg, Sweden
e-mail: agathe.bour@bioenv.gu.se

C. Coutris
Division of Environment and Natural Resources, Norwegian Institute of Bioeconomy Research (NIBIO), Ås, Norway
e-mail: claire.coutris@nibio.no

M. S. Bank
Institute of Marine Research, Bergen, Norway

University of Massachusetts Amherst, Amherst, MA, USA
e-mail: Michael.Bank@hi.no; mbank@eco.umass.edu

revealed that organisms' responses were overall influenced by the physicochemical heterogeneity of the plastic particles used, for which distinct differences were attributed to polymer type, size, morphology and surface alterations. On the other hand, little attention has been paid to the role of additive chemicals in the overall toxicity. There is still little consistency regarding the biological impacts posed by plastic particles, with observed ecotoxicological effects being highly dependent on the environmental compartment assessed and specific morphological, physiological and behavioural traits of the species used. Nonetheless, evidence exists of impacts across successive levels of biological organization, covering effects from the subcellular level up to the ecosystem level. This review presents the important research gaps concerning the ecotoxicological impacts of plastic particles in different taxonomical groups, as well as recommendations on future research priorities needed to better understand the ecological risks of plastic particles in terrestrial and aquatic environments.

7.1 Introduction

Plastic particles are a widespread environmental problem and possibly an important human health issue that has recently garnered significant interest from scientists, policymakers, natural resource managers, media entities and the public (Prata et al. 2021; Thompson et al. 2004). The complexity of plastic pollution follows a dynamic environmental cycle (Bank and Hansson 2019, 2021), which involves bidirectional fluxes across different ecosystem compartments including the atmosphere, hydrosphere, biosphere as well as terrestrial environments (Vince and Hardesty 2017; Windsor et al. 2019). There has been an outburst of research into plastic pollution in recent years, with research focusing on sources, presence and transport in the environment (as presented in other chapters in this volume – e.g. Bank and Hansson 2021; Kallenbach et al. 2021; Lundebye et al. 2021). Despite this, many questions remain regarding the ecotoxicology of plastic particles and their overall effect on wild populations of biota from different ecosystem compartments (de Sá et al. 2018; Galloway et al. 2017; GESAMP 2020; Law and Thompson 2014; Prakash et al. 2020; VKM 2019).

Many of the challenges related to understanding the ecotoxicological consequences of plastic particles are inherently linked to their complex nature as environmental contaminants (Rochman et al. 2019). Microplastics are made up of different polymers and additives which can influence their impact on living organisms. Furthermore, microplastics can originate from many different sources. Some are specifically designed (primary microplastics), whereas others are formed through the breakdown of larger plastics (secondary microplastics) (Cole et al. 2011). The terminologies used to describe plastic particles can also hold significant weight in terms of how data is interpreted. Microplastics are most commonly defined by their size, being less than 5 mm (GESAMP 2019), although definitions used across different research fields introduce inconsistencies, especially with reference to their lower size limit (Hartmann et al. 2019). For the purpose of this chapter, we kept the definitions of microplastics as <5 mm in size (GESAMP 2019), even though much of the ecotoxicological data presented involved particles <1 mm in size. The lower size limit of microplastics is here defined as 1µm, following the definition set by Hartmann et al. (2019) in reference to nanoplastics (1–1000 nm).

A wide array of impacts and toxic effects have been reported for both microplastics and nanoplastics, and as a brief example, several studies have examined the direct and indirect effects of a broad range of size fractions on a range of different species. Effects observed include impacts on reproduction, population dynamics, oxidative stress, ingestion, physiology, feeding behaviour, metabolic and hepatic functions, as well as interactions with other contaminants (e.g. Anbumani and Kakkar 2018; Haegerbaeumer et al. 2019; Kögel et al. 2020). However, the extent to which the available data is useful to interpret consequences across different biological levels (cellular-organ-individual-population; Galloway et al. 2017) has been called into question (VKM 2019).

The potential risks of micro- and nanoplastics to the environment and biota health have been the subject of several recent reviews and risk assessments by international authorities including (i) the European Food Safety Authority (EFSA), Panel on Contaminants in the Food Chain (CONTAM) on the presence of nano- and microplastics in food (EFSA CONTAM Panel 2016); (ii) a technical paper from the Food and Agriculture Organization of the United Nations (FAO) on the status of knowledge on microplastics related to fisheries and aquaculture (Lusher et al. 2017); (iii) a scientific perspective on microplastics in nature and society (SAPEA 2019); (iv) an updated knowledge summary built on the foundations of the previous three reports (VKM 2019); and (v) an ecological and human health risk assessment conducted by the Joint Group of Experts on the Scientific Aspects of Marine Environmental Protection (GESAMP 2020). During the VKM systematic assessment (VKM 2019), publications were judged based on a set of criteria to assess their quality, and those with poor quality were excluded. The accepted papers were used to attempt conceptual human and environmental risk assessments; however, many uncertainties and knowledge gaps were identified. One of the most significant limitations was that nano- and microplastics were treated as one entity, ignoring their physicochemical heterogeneity (Rochman et al. 2019). There was also a disproportionate representation between different species and different environmental compartments (marine, brackish, freshwater, terrestrial), which hampered the understanding of impacts in specific ecosystems. Much of the information available focused on species which are routinely used in standard test guidelines developed by the Organization for Economic Cooperation and Development (OECD) and the International Organization for Standardization (ISO).

Here we provide an overview and synthesis of microplastic and nanoplastic ecotoxicology (2012 - August 2019) in the context of particle characteristics (e.g. polymer type, morphology, size fractions), interactive toxicological effects, taxonomic gradients and with a focus on other potential synergies with associated chemical compounds. The specific objectives of this chapter are to (1) synthesize the literature and scientific consensus regarding the ecotoxicity of microplastics and nanoplastics and their potential relationships with other chemical compounds; (2) evaluate the effects of microplastic and nanoplastic concentrations, polymer type, size and morphology, experimental design, exposure time and pathways on ecotoxicological endpoints; (3) identify critical data and knowledge gaps in microplastic and nanoplastic toxicity research; and (4) suggest approaches and guidelines for addressing the most pressing questions and for advancing microplastic and nanoplastic ecotoxicity research.

7.2 Methods Used for Review Process

7.2.1 Overall Review Process

A comprehensive assessment of available published peer-reviewed literature was conducted up to August 2019 using the Web of Science, ScienceDirect, Scopus, PubMed and Google Scholar databases. The search was based on a combination of keyword terms, such as microplastic, nanoplastic, effects, toxicity, specific phylum/sub-phylum and specific target organisms (e.g. fish, crustaceans, bivalves, etc.), in any topic, title or keywords. Additional targeted searches were conducted from references included in relevant peer-reviewed articles (including review papers), as well as relevant reports overlooked by the search engines used. Of the identified references, only those focusing on studies reporting ecotoxicological effects were retained for further analysis. Studies only describing ingestion and egestion of plastic particles without reporting toxicity assessment were excluded from the collected literature. The ingestion of nano- and microplastics by biota has been described in previous review articles (e.g. Collard et al. 2019; Wang et al. 2019b, 2020). Particles >5 mm were not included in this assessment. An overview of the review process can be found in Fig. 7.1.

7.2.2 Extraction and Compilation of Data

A total of 220 references containing relevant ecotoxicity data were selected for review, after which the following information was extracted and compiled in an EXCEL spreadsheet for subsequent analysis: (i) experimental design, (ii) group of organisms, (iii) particles used, (iv) ecotoxicological endpoints and (v) publication information.

In terms of experimental design, the information extracted was categorized according to (i) exposure time, as described by authors and converted into days; (ii) particle concentration, in mass and/or particle number; (iii) exposure regime, static, semi-static or flow-through; (iv) replication, as number of independent replicate experiments or number of replicate exposure vessels; (v) use of controls, negative

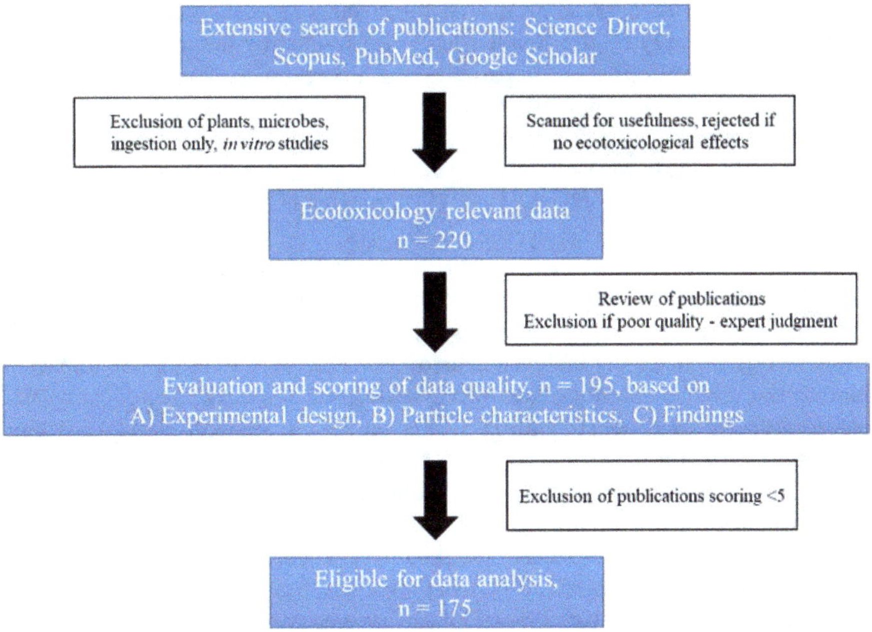

Fig. 7.1 Schematics on the literature review search of references containing relevant ecotoxicity data regarding micro- and nanoplastics

control (no plastic, only exposure media), additive/preservative control (e.g. tween 20, $NaNO_3$), particle control (kaolin, clay, etc.) or chemical control (co-exposure with other contaminants); (vi) confirmation of test concentration, nominal versus measured; (vii) exposure route, water, sediment/soil, food (e.g. inert pellets), prey (food chain) or others; and (viii) additional information, not included in the previous categories.

The types of organisms used in the studies reviewed were divided into the following taxonomic groups: Annelida, Arthropoda, Chordata, Cnidaria, Echinodermata, Mollusca, Nematoda, Phytoplankton and Rotifera. For each group, the following information was extracted: (i) taxonomic class; (ii) species, full Latin name; (iii) developmental stage, egg, embryo, larvae, juvenile, adult and others; (iv) feeding strategy, filter feeder, deposit feeder, scavenger, suspension feeder, predator or others; (v) supply of food during exposure; (vi) environmental compartment, freshwater, seawater or soil/sediment; (vii) replication, number of organisms per endpoint determination; and (viii) ingestion, checked, yes or no. Toxicity studies on higher plants, bacteria and *in vitro* were not included in this review.

For information on the particles used, the following categories were chosen as the most representative in terms of physicochemical characteristics: (i) polymer type; (ii) particle morphology, spheres, fibres, fragments (same as irregular), pellets or others if missing; (iii) surface modification, plain, COOH, NH_2, others or not specified; (iv) particle size; (v) co-exposure/mixture, yes or no in case of spiking with chemicals; (vi) chemical details, chemical name and concentration used; (vii) characterization, only by the supplier and/or additional by the authors; and (viii) others, additional information on particles, e.g. fluorescence, density, etc. In terms of particle type, the following list of polymer types was used to classify the particles used in the selected studies, which include the main groups of polymer materials reported in PlasticsEurope (2019): polyethylene (PE), polyethylene terephthalate (PET), polystyrene (PS), polypropylene (PP), polyvinylchloride (PVC), polyamide (PA), acrylonitrile butadiene styrene (ABS), nylon, polycarbonate (PC), polyhydroxy butyrate (PHB), polylactic acid (PLA), polymethylmethacrylate (PMMA), polyoxymethylene (POM), styrene acrylonitrile (SAN), phenylurea-formaldehyde (PUF), proprietary polymer as well as not specified (NS). High- and low-density PE were not differentiated but included in an overall PE group. To assess the impact of particle size (i.e. nanoplastic versus microplastic), one or more of the following size categories were used: < 0.05μm, 0.05–0.099μm, 0.1–0.99μm, 1–9μm, 10–19μm, 20–49μm, 50–99μm, 100–199μm, 200–500μm and > 500μm.

The effects reported were categorized following the levels of biological organization as suggested by Galloway et al. (2017): subcellular (e.g. enzyme activity, gene expression, oxidative damage), cellular (e.g. apoptosis, membrane stability), organ (e.g. histology, energetic reserves), individual (e.g. mortality, growth), population (e.g. reproduction, larval development) and ecosystem (e.g. behaviour, ecosystem function, community shifts). In cases where a large amount of data was generated in a specific study, detailed information on biological endpoints was also recorded, such as gene and protein expression data, enzymatic activities, histopathology effects, etc. Presence or absence of significant effects were recorded as yes or no, followed by the direction of the effect recorded as up (induction) and down (inhibition). Whenever disclosed, the ECx (concentration showing a x% effect), NOEC (no observed effect concentration) and LOEC (lowest observed effect concentration) values were also recorded.

Within the selected references, descriptions of experiments using different experimental conditions (e.g. time of exposure and concentration), two or more species (e.g. life stages and route of exposure) or particles with different characteristics (e.g. polymer type, size, morphology) were considered as individual records and added as separate entries in the data matrix. For example, whenever the size distribution for a given particle spanned more than one of the defined size categories, multiple entries were recorded, each corresponding to a size category. If a study included more than one species, a separate record was added for each species, each one with multiple entries dependent of the varying treatments used by the authors. Accordingly, the number of studies and corresponding entries presented in the results section represent the number of interactions of the classification criteria recorded for each reference, and not the total number of publications reviewed.

After revision of the 220 references collected, 25 were excluded due to poor quality in one or more of the classification criteria used. Examples were poor experimental design, lack of information on particles used or particle characterization, inadequate data representation or conclusions not supported by data. The exclusion of these 25 references was based on expert judgement, and data entries pertaining to these references were removed from the data matrix. The data matrix can be made available upon demand.

7.2.3 *Evaluation and Scoring of Data Quality*

The 195 references considered of acceptable quality were further evaluated and given a quality score based on the criteria presented in earlier publications. This was to ensure that the highest quality data generated through ecotoxicological studies was also the data that had the most impact in this analysis. Evaluation criteria were divided in three groups, experimental design, particle characterization and findings, as detailed in Table 7.1 (based on Connors et al. 2017; VKM 2019). Specifically:

– "Experimental design" included the use of reference controls and chemical controls, as well as replication within the test system. Maximum score = 3.

Table 7.1 Evaluation criteria used to score data quality of reviewed references (based on Connors et al. 2017; VKM 2019)

Criteria	Description	Scoring definition
Experimental design (0–3)	Use of reference controls	Use of reference particles other than plastic (e.g. kaolin, sand, etc.)
	Use of chemical controls	Applies to vector studies only, where the particles are spiked with one or more chemicals, or when further characterization was carried out and results indicate the presence of chemicals on the particles. Otherwise, 1 point should be automatically attributed
	Replication in test system	Exposure replication of minimum 3; total number of individuals: Depends on the endpoint
Characterization (0–5)	Particle size	Concentration range of particles used determined by authors (e.g. DLS, particle counter, etc.)
	Particle charge	Applies for nanoparticles only. If microparticles are used, 1 point should be automatically attributed
	Polymer confirmation	Confirmation of polymer used in exposure system (e.g. FT-IR)
	Chemical characterization	Applies for studies using spiked particles, particles obtained from the grinding of consumer goods, deployed particles, industrial particles (e.g. nurdles). Only in the case of particles obtained from a "trusted" supplier (e.g. Cospheric, sigma, etc.) and said to be "pristine", 1 point should be automatically attributed
	Test concentration confirmation	Test concentration measured in exposure system and not nominal concentration used
Findings (0–1)	Conclusions supported by the results	Accurate interpretation of the results without conjecture beyond experimental design

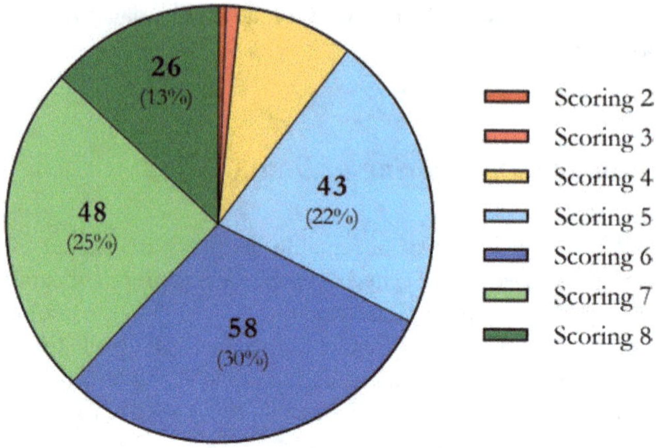

Fig. 7.2 Scoring of the 195 reviewed references. The number and % of references are only presented for those scored with 5 or more points

- "Particle characterization" included the reporting of particle size, particle charge, polymer confirmation, chemical characterization and confirmation of the test concentration. Maximum score = 5.
- "Findings" included the assessment of whether the conclusions were supported by the results. Maximum score = 1.

For each time a criterion was met, 1 point was attributed, and references were categorized based on a quality score out of 9. References that scored 4 or less were excluded from further analysis and corresponding data entries removed from the data matrix. Of the 195 references scored, 20 were eliminated due to low score, in which 17 papers scored 4 points, 2 papers scored 3 points and 1 paper scored 2 points. None of the papers scored either 1 or 9 points (Fig. 7.2).

7.2.4 Treatment of Extracted Data

Species sensitivity distributions (SSDs) were fitted for three relevant exposure routes: water exposure, sediment/soil exposure and food exposure. Ecotoxicity data for terrestrial, freshwater and marine compartments and species were extracted and summarized for use in the SSD model fitting. Information on polymer types and size classes were combined, and for this reason, studies using fibres were excluded from the SSDs. Ecotoxicity endpoints were limited to individual and population levels (Burns and Boxall 2018; Connors et al. 2017), and only NOECs and EC_{50} values were included. When only acute NOEC or EC_{50} data was available, chronic NOEC values were extrapolated as proposed by Posthuma et al. (2019). When multiple NOEC values were available for the same species, the geometric mean of the NOECs was used to summarize the information. To allow the comparison of ecotoxicological data from studies reporting different dose metrics, mass-based concentrations were converted to mg per litre (mg/L) and particle-based concentrations converted to particles per litre (particles/L). In the case of studies where particles were added via sediment/soil or via food, concentrations were converted to mg per kg (mg/kg) of sediment/soil or food and particles per kg (particle/kg) of sediment or food. As several studies only reported concentrations in either mass or particle number, two SSDs were created per exposure route. Studies where none of the above dose metrics were employed were excluded from the SSD fitting. The SSDs were realized as Bayesian distributional regression models assuming a log-normal data distribution (Ott 1990). All modelling was performed using statistical programming language R (R Core Team 2020) and its add-on package brms (Bürkner 2017, 2018). A total of 10,000 posterior draws were used to characterize each SSD. Where applicable, the value indicating the concentration at which 5% of the species are affected (hazard concentration, HC_5) was extracted from the posterior draws and summarized as average and 95% credible interval.

7.3 Results and Discussion

A key issue in understanding how microplastics and nanoplastics interact with the surrounding environment is their dynamic nature. The physicochemical properties of the parent material, including density, morphology, charge and size, are likely to influence particles' physical behaviour in the environment, fate (e.g. presence in the water column or in sediments), potential to adsorb environmental contaminants (e.g. Trojan horse effect), bioavailability and potential toxicological impacts on organism health (e.g. de Sá et al. 2018; Galloway et al. 2017; Haegerbaeumer et al. 2019; Kögel et al. 2020). The extensive literature review carried out showed that the responses of organisms to particle exposure were mostly dependent on particle characteristics as polymer type, size, morphology and surface alterations. However, it is possible that other factors were driving the observed impacts, as, for example, the presence of additive chemicals associated with the plastic particles, which are rarely considered in studies. A special emphasis has therefore been given to particle size, with a higher consensus in terms of increased internalization for smaller sized particles than larger ones and thus higher potential for toxic effects. A variety of experimental designs have been used to evaluate the effects of nanoplastics and microplastics in organisms, in which exposure time and particle concentration seem to be determinant for the induction of toxicity. Nonetheless, the observed effects were highly dependent on the environmental compartment assessed, in combination with specific morphological, physiological and behavioural traits of the species used, as, for example, developmental stage, trophic level and feeding strategy.

In terms of ecotoxicological effects, there is still little consensus regarding the biological impacts posed by plastic particles, as well as a limited understanding on the underlying toxic mechanisms causing the observed effects. This limited knowledge on mechanistic toxicity data also makes it difficult to understand and distinguish physical from chemical toxicological effects of plastic particles. And even though it is quite clear from wider literature that large particles (e.g. macroplastics) cause visible effects at the organism level (Kühn et al. 2015; Rochman 2015), the direct and indirect physiological effects of the smaller plastic particles remain elusive. Based on this review, effects were found at different levels of biological organization in a range of organisms. However, many of these studies used standard ecotoxicity approaches based on OECD or ISO guidelines that do not consider effects at the lower levels of biological organization such as cellular or subcellular mechanisms, which may be more sensitive and have a higher impact on the physiological traits of organisms, especially in the long term. To a small degree, some of the reviewed studies highlighted that the combination of nanoplastics and microplastics with organic and inorganic contaminants also modify and potentiate their toxicity towards biological systems. Nonetheless, the effects of chemical additives present in plastic particles are also understudied, and it is still not clear if the presence of these additives rather than the polymeric composition of particles are the main driver of the adverse effects reported in organisms. Based on the 175 publications reviewed, a more general and detailed report of the main factors influencing particle toxicity towards the different groups of organisms are presented in the sections below.

7.3.1 General Overview of Information Extracted from Reviewed Publications

7.3.1.1 Polymer Type, Morphology, Surface and Size

Within the 175 reviewed publications, the most commonly used polymer type was PS (90 studies, 51%), followed by PE (62 studies, 35%), PVC (17 studies, 10%) and PET (11 studies, 6%). The remaining polymer types (acrylonitrile butadiene styrene [ABS], nylon, polyamide [PA], polycarbonate [PC], polyhydroxybutyrate [PHB], polylactic acid [PLA], poly(methyl methacrylate) [PMMA], polyoxymethylene [POM], polypropylene [PP], styrene acrylonitrile resin [SAN]) were used in less than 5% in the reviewed studies. The use of PS and PE as polymers of choice in exposure studies is consistent with the most commonly found polymers in the environment, as PS, PE and PP are typically retrieved from surface waters and sediments (e.g. Koelmans et al. 2019 and references therein). Given that polymer type can influence the fate and behaviour of particles within test systems, in particular density and presence of chemical additives (e.g. Gallo et al. 2018), other polymers should be comprehensively assessed in order to build up knowledge regarding how their composition influence toxicity towards organisms.

Despite the prevalence of fragments, fibres and films in environmental samples due to degradation of larger pieces of plastic (see Burns and Boxall 2018; Kooi and Koelmans 2019; Phuong et al. 2016), the majority of studies focused on spherical particles (106 studies, 61%), with only 40 studies looking at the impacts of fragments/irregular particles (23%) and even less focusing on the effects of fibres (13 studies, 7%). The main reason for the use of spherical particles is that they are easier to produce than the other morphological types (e.g. fibres, fragments, foams), especially in terms of sufficient quantity within a certain size range. The irregular and non-standardized morphology of these particles also make them more difficult to characterize and track during exposure experiments, which results in poorly comparable ecotoxicity data. Nonetheless, irregularly shaped particles resulting from the fragmentation of larger plastic items or materials containing synthetic polymers as fibres have a higher environmental relevance and should be used more often in effects studies, especially in terms of increasing ecological relevance for advancing quantitative data to assess environmental risks.

Among the reported surface alterations, plain/pristine particles were used in 163 publications out of the 175 (93%) studies reviewed. Of all the particles reported with surface alterations, the majority was for PS, with PS-COOH and PS-NH_2 in the nano-size range being the most commonly used (10% and 9%, respectively). Particle surface chemistry, i.e. chemical groups and surface charge, was one of the main properties driving the behaviour of particles in the aquatic environment – this is particularly true for smaller sized particles – especially when it comes to stability, aggregation, mobility and sedimentation (e.g. Mudunkotuwa and Grassian 2011). In fact, particle surface charge, more so than polymer composition, has been suggested as the main driver behind behaviour and consequent toxicity of smaller sized plastics (Lowry et al. 2012; Nel et al. 2009). Even though functionalized particles are commonly used as surrogates for naturally altered particles, their prevalence in the environment has been questioned. The presence of negatively charged PS-COOH has been suggested as widespread in the environment, although there is very little information on its fate in different environmental compartments. Similarly, the presence of PS-NH_2 as a plastic degradation product in the environment has not yet been fully recognized/determined (Besseling et al. 2014).

An overview of the number of studies per particle type and size class is presented in Fig. 7.3. Of the size classes tested, most studies used particles smaller than those that can be detected with confidence in environmental matrices (<100μm, e.g. (de Ruijter et al. 2020)). Sixty-five of the reviewed studies used particles with sizes in the range 1–9μm (37%), followed by 43 studies with size in the range 20–49μm (25%), 36 studies with sizes in the range 50–99μm (21%) and 34 studies with sizes in the range 10–19μm (19%). As for smaller size ranges, 39% of the reviewed publications used particles <1μm (total 69 studies), with a predominance of particles within 0.1–0.99μm. Regarding fibres, the size ranges used were between 362 and 3000μm in length and 41 and 3000μm in diameter. In terms of size distribution per polymer type, for PS and PMMA a higher focus has been given to particles <10μm, especially for PS in the nano-range size, as seen in Fig. 7.3. This is the opposite of PE, as well as the remaining polymers reported, where most particles used have a size range > 1μm. Most of the studies comparing the effects of both nanoplastics and microplastics of the same polymer composition reported size-dependent effects, with an increase in toxicity with decreasing particle size (e.g. Jeong et al. 2016, 2017; Lee et al. 2013; Lei et al. 2018a; Snell and Hicks 2011). Nonetheless, this

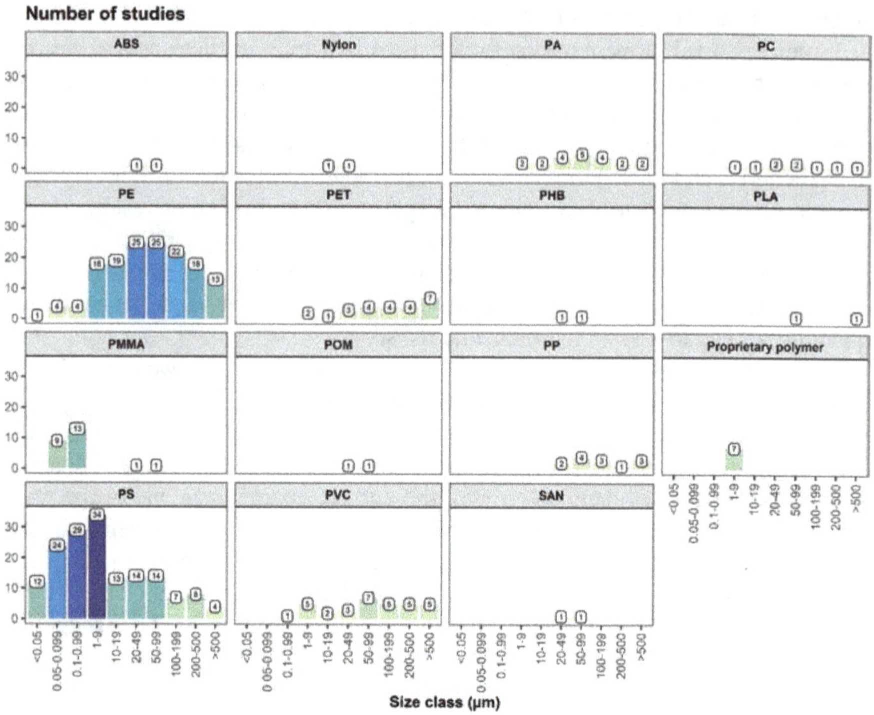

Fig. 7.3 Overview of the number of studies per particle type and size class. Note: There can be more than one size class within a study for a specific particle. See Methods Sect. 7.2.2. for more information on how particle size was categorized. *ABS* acrylonitrile butadiene styrene, *PA* polyamide, *PC* polycarbonate, *PE* polyethylene, *PET* polyethylene terephthalate, *PHB* polyhydroxy butyrate, *PLA* polylactic acid, *PMMA* polymethylmethacrylate, *POM* polyoxymethylene, *PP* polypropylene, *PS* polystyrene, *PVC* polyvinylchloride, *SAN* styrene acrylonitrile

size-toxicity correlation seems to be species and phyla dependent. Irrespective of the potentially higher adverse effects imposed by smaller sized particles in organisms, their detection in different environmental compartments and resulting uncertainties in terms of natural concentrations remain an ongoing analytical challenge. Nonetheless, their presence in the environment as a consequence of fragmentation and degradation of plastic debris is widely accepted, having been proven under laboratory conditions (e.g. Lambert and Wagner 2016) and where their occurrence in the North Atlantic subtropical gyre has also been suggested (Ter Halle et al. 2017).

Even though particle ingestion and egestion were not considered in this review chapter, the selective size ingestion of micro- and nanoplastics has been reported for a range of aquatic organisms (e.g. bivalves, Ward et al. 2019). Accordingly, the size distribution of microplastics and nanoplastics used in ecotoxicological studies need to be appropriate for the species used, as this may influence exposure and particle-organism interactions.

7.3.1.2 Experimental Conditions

Standard test protocol guidelines commonly used in toxicity testing of chemicals are not always suitable for testing of particles (e.g. Hermsen et al. 2018). Accordingly, ecotoxicity testing of nano- and microplastics often require modifications in experimental design to address specific particle behaviour and/or characteristics, leading to a lack of standardization. The lack of standardized test protocols for plastic particles results in a multiplicity of experimental conditions, which limits consistency and result comparison and interpretation (Connors et al. 2017; VKM 2019).

Considering the absence of consistent particle quantification in the environment in size ranges as small as those commonly used in ecotoxicological studies (Paul-Pont et al. 2018), the use of the so-called environmentally relevant doses of plastic particles also remains a challenge. Concentration range and units expressed in either mass or particle number are two of the main issues that have been highlighted related to the dosing of plastic particles in exposure systems. More than half of the publications reviewed reported particle concentrations in mass (minimum 7×10^{-7} mg/L to maximum 12,500 mg/L), with the most commonly used concentration range of 1–100 mg/L (organisms exposed via water, 72% of studies). As for particle mass used in exposures via food (17% of studies) or sediment/soil (10% and 7% of studies, respectively), concentrations varied from 7×10^{5} to 100 mg/kg food (most common 4000, 12,000, 100,000 mg/kg food) and 4×10^{5} to 1 mg/kg sediment/soil (most common 1000 to 50,000 mg/kg sediment/soil). Few studies reported concentrations in terms of particle number, with concentrations ranging from 1 to 8×10^{15} particles/L, 16 to 23×10^{7} particles/kg sediment/soil and 3×10^{5} to 1×10^{8} particles/kg food. Therefore, it seems that the nano- and microplastics used in the reviewed publications have been tested in numbers several orders of magnitude higher than those currently detected in the natural environment. This is particularly true for the small sized plastics within a wide range of polymer types, where realistic concentrations are rarely available for sizes >10μm and not available for sizes <10μm (for more information on environmental data on plastic contamination, check Litter Database webpage: http://litterbase.awi.de/litter). In addition, the failure to provide particle concentrations in both mass and number complicates the comparison of effect data across published studies, confounding the ability to reach precise conclusions over exposure and risk.

Exposure time is another important aspect related to varying experimental conditions used in nano- and microplastic ecotoxicological studies. The most commonly used exposure times in the reviewed studies were 48 h (27% studies), 24 h (18% studies), 96 h (17% studies) and 72 h (14% studies). These exposure durations are within those recommended in ecotoxicity guidelines for acute testing (e.g. OECD and ISO). In these tests, model organisms are normally exposed to high concentrations of a test compound over a short period of time, after which effect endpoints such as mortality or development are commonly assessed. Even though several of these studies showed evidence of deleterious effects at high concentrations, there are still knowledge gaps – which are hidden by the present focus in acute ecotoxicological testing, relating to limited environmental relevance. As exposure concentration and duration are two major parameters influencing toxicity, results based on short-term and high exposure concentrations make it difficult to extrapolate data to a more realistic scenario of exposure to low concentrations over a long period of time. One of the main gaps in the reviewed studies was the underrepresentation of long-term exposures at environmentally relevant concentrations and their consequent long-term effects at the organism and ecosystem levels (e.g. chronic exposure, whole life cycle, multi-generational effects). Long-term (or chronic) studies on the effects of nano- and microplastics were mostly carried out for 28 and 21 days (11% studies each), followed by 14 days (10% studies). Only a very small percentage of studies have used an exposure period higher than 28 days, with only 4 studies looking at ecotoxicological effects above 3 months of exposure (maximum 240 days, i.e. 8 months). Long-term exposures carried out over more than 1 life stage or whole organism's lifespan allow to focus on population-relevant adverse endpoints (e.g. reproduction), as well as other sublethal effects that might constitute more reliable endpoints for risk assessment and are therefore urgently needed.

7.3.1.3 Organisms Used in Ecotoxicological Studies

When it comes to environmental compartments, most test organisms used were from the marine environment (61%), followed by freshwater (31%) and terrestrial (8%) compartments, as presented in Fig. 7.4. Only 1 study reported the use of brackish organisms (1%). This highlights that the effects of nano- and microplastics on terrestrial and freshwater ecosystems have been understudied and deserve further attention (e.g. Adam et al. 2019; Haegerbaeumer et al. 2019; Horton et al. 2017; Strungaru et al. 2019). These knowledge gaps are of particular concern, especially when terrestrial and freshwater environments are considered the main sources and transport pathways of plastic particles to the marine environment. Given that many

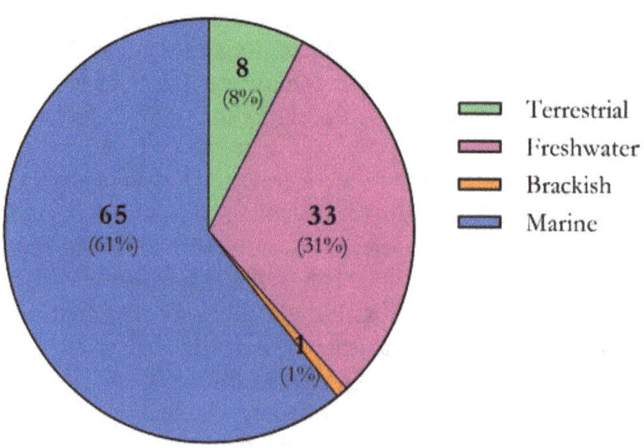

Fig. 7.4 Number of species (total of 107) from each environmental compartment used in the reviewed references

plastic particles are used and disposed on land, terrestrial environments will be subject to extensive pollution by particles of varying characteristics at high concentrations, making terrestrial organisms at high risk of exposure. As for freshwater organisms, these will be directly affected by terrestrial runoff and other anthropogenic sources (e.g. wastewater treatment discharge, sewage sludge application), potentially containing high levels of plastic particles, as well as other associated contaminants (Adam et al. 2019; Horton et al. 2017 and references therein).

At the phylum level, Arthropoda was the most studied (34%, 59 publications), followed by Chordata (23%, 41 publications), Mollusca (21%, 36 publications), Phytoplankton (14%, 25 publications), Annelida (9%, 16 publications), Cnidaria and Echinodermata (2% each, 4 publications), Rotifera (2%, 3 publications) and finally Nematoda (1%, 1 publication). The freshwater crustacean *Daphnia magna* (17% overall studies) was the most studied species, followed by the marine mussel *Mytilus galloprovincialis* and the freshwater zebrafish *Danio rerio* (both with 6% of overall studies). In terms of developmental stage, most of the studies assessed effects in adult organisms (42%, 73 studies total) and a small percentage used juveniles or neonates (both with 14%, 25 studies). Very few studies have looked at whole cycle assessments, 3% of the total of reviewed publications, and those that did were only directed towards arthropods. In terms of feeding strategy, 32% of the species used were filter feeders, followed by photosynthetic organisms (21%), predators (17%), detritivores (10%), grazers (9%), scavengers (8%) and deposit feeders (5%). Only one herbivore and one microbivore were used.

Even though the organisms used in the reviewed publications have different roles in terrestrial and aquatic food webs, there is still a lack of studies conducted on organisms other than fish, small crustaceans and bivalves. Specifically, more studies on the effects of nano- and microplastics on organisms that are the basis of aquatic food chains should be conducted (e.g. planktonic species). These species have critical roles in ecosystem balance and might be at highest risk of exposure due to their feeding strategies and relative position in the water column. Moreover, small plastic particles are easily confused as food and ingested by planktonic species, thus serving as a route of transfer to secondary and tertiary consumers in food chains (Botterell et al. 2018). In addition, soil- and sediment-dwelling organisms are of major importance, as soil/sediment is considered the main sink for contaminants in the environment, increasing the likelihood of synergistic effects of plastic particles with other environmental contaminants (Adam et al. 2019; Horton et al. 2017 and references therein). Furthermore, targeted studies on species other than those commonly used in OECD and ISO guidelines should also be conducted, as the toxicological and mechanistic effect data on these species might not provide sufficient information into impacts on other ecologically relevant species. The same can be said in terms of transferring knowledge from marine to freshwater or terrestrial environment. Given the differences in habitat, physiological traits and feeding mechanisms, it is not clear as to what extent ecotoxicological effects on marine organisms can be applied to freshwater and terrestrial species within the same taxonomical group and vice versa.

7.3.1.4 Levels of Biological Organization

Most of the reviewed studies focused on the effects of nano- and microplastics at the individual level (133 studies, 40%), followed by the subcellular level (78 studies, 23%). The population level has been addressed in 45 studies (14%), ecosystem in 33 (10%), closely followed by the organ level with 30 studies (9%). Only 13 studies (4%) analysed effects at the cellular level. Within the individual endpoints, growth and mortality were the most studied (74 and 73 studies, respectively), while at the subcellular level, effects looking at alterations in gene expression (41 studies) were the most frequent, followed by oxidative stress (26 studies) and enzymatic activities (24 studies). Within population-related endpoints, the most determined were reproduction (21 studies) and larval development (16 studies). Within ecosystem, 29 studies looked at behaviour and 22 looked at community shifts. As for organ level, most studies (17) looked at histopathological alterations, followed by nine studies looking at energy reserves. At the cellular level, eight studies looked at membrane stability, five at cell size and four at both cell number and cell complexity. When looking at the number of studies categorized by environmental compartment (Fig. 7.5), the majority of the studies for both freshwater and marine environments

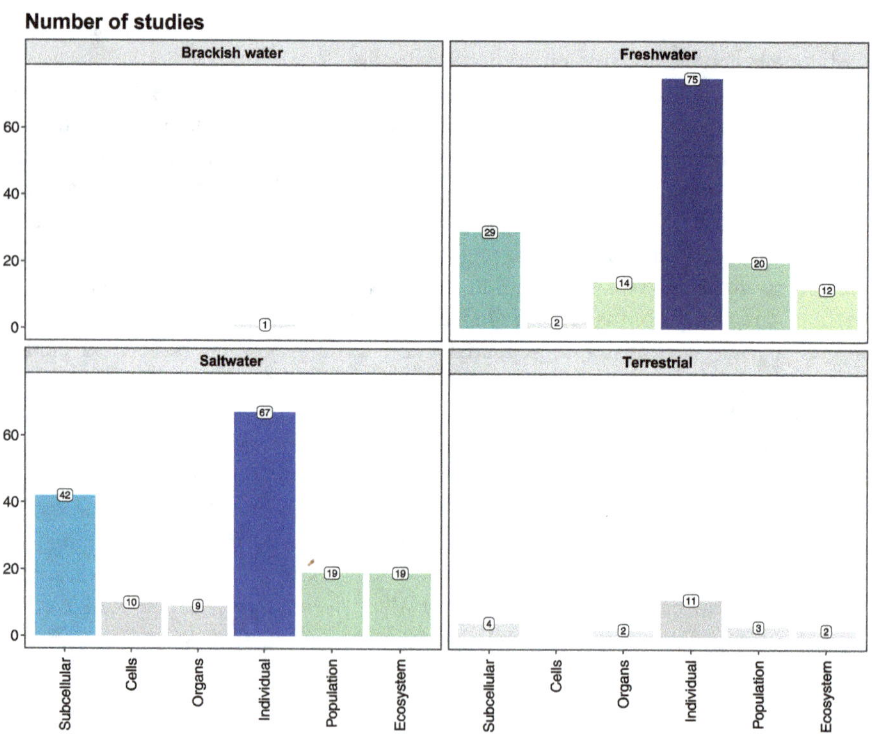

Fig. 7.5 Number of studies categorized by level of biological organization per environmental compartment

covered endpoints at the individual level (75 and 72 studies, respectively), followed by effects at the subcellular level (29 and 42 studies, respectively). Impacts at the individual and cellular levels were also the most determined in terrestrial organisms (ten and 4 studies, respectively), while only one study covered individual endpoints in the brackish environment. Studies on effects at the cellular level were less common in freshwater and marine environments (two and ten studies, respectively), while no studies addressed this level of biological organization in terrestrial and brackish environments.

7.3.2 *Ecotoxicological Effects*

While a range of ecotoxicological effects caused by plastic particle exposure have been documented across several groups of organisms, there are still distinct research gaps concerning effects of both nano- and microplastics in specific taxonomical groups. In the following paragraphs, particle characteristics, exposure conditions and consequent ecotoxicological effects will be described for each taxonomical group considered in the present review: Phytoplankton, Cnidaria, Nematoda, Rotifera, Arthropoda, Annelida, Mollusca, Echinodermata and Chordata.

7.3.2.1 Phytoplankton

Phytoplankton include unicellular organisms such as microalgae that are at the bottom of the aquatic food chain. Small disruptions of microalgae populations due to exposure to plastic particles may lead to serious repercussions at the ecosystem level, being thus imperative to characterize the risks/effects of plastic particles on this taxonomical group (Prata et al. 2019). Phytoplankton were evenly represented from marine and freshwater environments in the reviewed studies (12 and 13 studies, respectively). Exposure studies included 21 different species belonging to 8 different classes (Bacillariophyceae, Chlorodendrophyceae, Chlorophyceae, Coscinodiscophyceae, *Cyanophyceae*, Dinophyceae, Prymnesiophyceae and Trebouxiophyceae). The most used class was Chlorophyceae (14 studies). *Raphidocelis subcapitata*, previously named as *Pseudokirchneriella subcapitata*, was the most used species with four studies. Six other species (*Chaetoceros neogracile*, *Chlamydomonas reinhardtii*, *Chlorella pyrenoidosa*, *Dunaliella tertiolecta*, *Scenedesmus obliquus* and *Skeletonema costatum*) had two studies each, while the remaining had only one publication.

A total of 7 different polymers were used across the 25 reviewed studies, with PS as the most studied polymer (15 studies). Five studies used PE, four used PVC, two used PP, while PMMA, proprietary polymer and PUF were represented by one study each. Most studied PS spheres (n = 12), while only two used PVC spheres. Regarding size, eight studies used PS particles ranging between 0.05 and 0.099µm, and four used PS particles between 1 to 9µm and 0.1 to 0.99µm. There were two studies on PE particles between 50 and 99μm and PVC particles between 1 and 9μm. In terms of particle surfaces, plain PS particles (n = 7 studies) were the most used, followed by PS-COOH (n = 6) and PS-NH_2 (n = 5).

All phytoplankton publications addressed effects at the individual level, with 60% reporting effects. Growth was the most common endpoint (24 studies, 21 with effects), followed by pigment content (9 studies, 7 with observed effects), photosynthesis and photosynthetic performance (8 studies, 7 with effects) and chlorophyll a content (1 study with significant effects) (Baudrimont et al. 2020; Bellingeri et al. 2019; Bergami et al. 2017; Besseling et al. 2014; Bhargava et al. 2018; Canniff and Hoang 2018; Casado et al. 2013; Chae et al. 2018; Gambardella et al. 2018; Garrido et al. 2019; González-Fernández et al. 2019; Lagarde et al. 2016; Liu et al. 2019; Long et al. 2017; Luo et al. 2019; Mao et al. 2018; Nolte et al. 2017; Prata et al. 2018; Sendra et al. 2019; Seoane et al. 2019; Thiagarajan et al. 2019; Zhang et al. 2017; Zhao et al. 2019; Zhu et al. 2019). At the cellular level, effects on membrane stability (four studies, three with effects), cell complexity (three studies, all with effects) and cell size (four studies, three with effects) were addressed in marine and freshwater species (González-Fernández et al. 2019; Liu et al. 2019; Mao et al. 2018; Sendra et al. 2019; Seoane et al. 2019). Nine studies looked at several effects at the subcellular level, including oxidative stress (six studies, all observing effects), lipid peroxidation (three studies, two with effects), reactive oxygen species (ROS) formation (one study, no effects), neutral lipid content (one study with effects), protein content (two studies with effects), DNA damage (one study with effects) and gene expression (one study with effects) (Bellingeri et al. 2019; González-Fernández et al. 2019; Lagarde et al. 2016; Liu et al. 2019; Mao et al. 2018; Sendra et al. 2019; Seoane et al. 2019; Thiagarajan et al. 2019; Zhu et al. 2019). Only one publication studied effects at the ecosystem level, such as bacteria concentration and community shifts, with effects only reported for the latter (González-Fernández et al. 2019).

Overall, phytoplankton growth does not seem to be greatly impacted by micro- or nanoplastic exposure, for which little or no effects were reported for both freshwater and marine species. However, deleterious effects were seen at concentrations considered high. The lowest concentration at which effects on growth were reported was 0.001 mg/L for *D. tertiolecta* exposed to PS spheres (72 hrs, size range 0.1 to 0.99μm), even though complete growth inhibition was not achieved (Gambardella et al. 2018). In this study, a dose-dependent growth inhibition was observed in exposed microalgae and associated with the use of energy sources in detoxification processes, such as the generation of extracellular polysaccharides (Gambardella et al. 2018). Of the 25 reviewed studies, only 2 reported EC_{50} values for PS nanoplastics: an EC_{50} value of 12.97 mg/L was recorded for the marine microalgae *D. tertiolecta* (size range 0.05–0.099μm) (Bergami et al. 2017), while EC_{50} of 0.58 mg/L and 0.54 mg/L were obtained for freshwater microalga *P. subcapitata* (polyethyleneimine PS with different size ranges of 0.05–0.099 and 0.1–0.99μm, respectively) (Casado et al. 2013). For sublethal effects, the consensus is that toxicity in microalgae was influenced by size and surface chemistry of particles, with nanoplastics exerting stronger impairment than their micro-sized counterparts (e.g.

Bergami et al. 2017; Seoane et al. 2019; Zhang et al. 2017). PS nanoplastics, size range 0.05–0.99µm, were found to induce oxidative stress in the form of ROS formation (PS-NH_2 and plain PS (González-Fernández et al. 2019; Sendra et al. 2019)), result in effects on protein and neutral lipid content, affect membrane stability, cause DNA damage (plain PS (Sendra et al. 2019)), decrease pigment content including chlorophyll a (PS, PS-NH_2 and PS-COOH (Besseling et al. 2014; González-Fernández et al. 2019; Sendra et al. 2019)), alter cell size and complexity (PS-NH_2 and plain PS (González-Fernández et al. 2019; Sendra et al. 2019)) as well as cause community shifts (PS-NH_2 (González-Fernández et al. 2019)) in both freshwater and marine microalgae. Furthermore, positively charged PS-NH_2 have been shown to have higher interaction and toxicity than negatively charged PS-COOH and plain PS due to increased adhesion onto algal surfaces, with particle charge being recognized as the cause for the increased severity (Bergami et al. 2017; Chae et al. 2018; Nolte et al. 2017).

Overall, ecotoxicological data obtained for microalgae demonstrated that exposure to nano- or microplastics caused a variety of cellular and biochemical effects, from altered expression of genes involved in metabolic pathways, to photosynthetic impairment and growth inhibition (e.g. Lagarde et al. 2016; Mao et al. 2018). The toxicity observed to microalgae seems to be dependent of many factors including particle size (Zhang et al. 2017), polymer type (Lagarde et al. 2016), surface chemistry (González-Fernández et al. 2019; Seoane et al. 2019), particle concentration (Mao et al. 2018), exposure time, as well as targeted species (Long et al. 2017). Nonetheless, the environmental relevance and toxicity mechanisms of nano- and microplastics in microalgae remain unclear. This is mostly due to the determination of growth inhibition as the most common toxicological endpoint, in which the exposure duration is too short, and it is not possible to clearly discriminate between direct toxic effects and indirect physical effects caused by particles. Limitations in the use of this method have also been highlighted in studies using nanomaterials, mostly related to particle interference with algal growth quantification techniques (i.e. measurement chlorophyll a fluorescence) due to a shading effect (Handy et al. 2012). The presence of particles in suspension can cause shading either by reducing the access of algae to light or by obstructing the fluorescence signal from the algae to the fluorescent detector. This shading effect will impact the accuracy of the measured fluorescence response, leading to an underestimation of chlorophyll a quantification, thereby overestimating the overall toxic effect (Farkas and Booth 2017). In view of the important role that phytoplankton have in aquatic food webs, there is a need to develop better toxicological assays/endpoints with increased sensitivity that are able to reveal underlying toxic effects of plastic particles.

7.3.2.2 Cnidaria

The group Cnidaria is composed of aquatic organisms with basic body forms, swimming medusae or sessile polyps, that inhabit both the freshwater and the marine environments, even though more predominant in the latter. Examples of cnidarians are sea anemones, corals and jellyfish. The cnidarians used in the reviewed publications were all coral species and exclusively from the marine environment. Nine species were represented across four studies, all from the class Anthozoa. *Pocillopora damicornis* was the only species used in more than one study. The eight other species (*Acropora formosa, A. humilis, A. millepora, Montastraea cavernosa, Orbicella faveolata, Pocillopora verrucosa, Porites cylindrica, P. lutea*) were all used in single studies. All Cnidaria species investigated were filter feeders and were exposed to particles via water. Most studies were carried out on polyps (two studies).

Four studies have been carried out on Cnidaria investigating irregular fragments and beads composed of two polymer types. PE was used in three of the four studies (Hankins et al. 2018; Reichert et al. 2018; Syakti et al. 2019), while only one study used PS (Tang et al. 2018). Two studies used PE fragments (Reichert et al. 2018; Syakti et al. 2019), one used PE beads (Hankins et al. 2018), and the remaining study did not specify the morphology of PS particles used (Tang et al. 2018). In terms of size, one study focused on the smallest size category, 0.1 to 0.99µm (Jia Tang et al. 2018); PE fragments were studied in the size range 20–49µm (Reichert et al. 2018), 50–99µm (Reichert et al. 2018; Syakti et al. 2019) and 100–199µm (Reichert et al. 2018; Syakti et al. 2019); and one study used the size range 200–500µm (Syakti et al. 2019). PE beads were investigated in the size ranges 50–99µm, 100–199µm, 200–500µm and > 500µm during a single study (Hankins et al. 2018).

The subcellular level was studied in one publication reporting effects on enzymatic activity and gene expression (Tang et al. 2018). At the individual level, two studies investigated and reported bleaching (Reichert et al. 2018; Syakti et al. 2019); one study investigated and reported effects on mucus production, tissue necrosis and growth (Reichert et al. 2018); one study investigated and reported mortality and tissue necrosis (Syakti et al. 2019); and one study investigated calcification but did not observe any effects (Hankins et al. 2018). Only one publication studied community shifts, although no effects were observed on symbiont density or symbiont chlorophyll content (Tang et al. 2018). Bleaching was the most common endpoint, with both studies detecting effects. No studies were found at the population level.

Regarding concentrations and particle size, only a single concentration (50 mg/L) and size (1–9µm) was used to investigate subcellular-level effects (Tang et al. 2018). The effects of PS on enzymatic activity were investigated, where alterations in superoxide dismutase, alkaline phosphatase, catalase and glutathione S-transferase activity were observed throughout exposure. No effects were observed for phenoloxidase activity.

The reported effects at the individual level ranged from exposure to 50 mg/L to 150 mg/L. Exposure to PE fragments increased mortality, bleaching and necrosis in *A. formosa* after 2 days of exposure at 50, 100 and 150 mg/L (size range 50 to 500µm (Syakti et al. 2019)), as well as in *A. humilis, A. millepora, P. cylindrica, A. humilis, P. verrucosa* and *P. damicornis* after 28-day exposure at 100 mg/L (size range 20 to 100µm (Reichert et al. 2018)). Growth was also impaired across these species, but this was dependent on the size of particles used in the exposure. Mucus production only appeared to be affected in *P. lutea* also exposed to PE fragments

(100 mg/L, size range 20–100μm) (Reichert et al. 2018). At the ecosystem level, the only observed effect was a community shift in chlorophyll content symbiont at 12-hr exposure to PS 50 mg/L (Tang et al. 2018).

7.3.2.3 Nematoda

Nematodes, also called roundworms, are unsegmented worms found in almost every terrestrial and aquatic habitat. Only a single study addressed the effect of microplastics on nematodes (Judy et al. 2019). The nematode *Caenorhabditis elegans*, which lives in the pore water of soils, was exposed to fragments larger than 500μm, produced by shredding consumer products (Judy et al. 2019). The exposure scenarios used organisms at the adult stage, exposed through contact with the soil solution, implying both dermal and trophic exposure to microplastics.

The effects of a single high concentration (5 g/kg soil dry weight) of three polymer types (PE, PET, PVC) were assessed at the individual level (mortality and reproduction), after various contact time between soil and plastics (0, 3 and 9 months). Increased mortality was only observed for PET incubated in soil for 3 months, while decreased reproduction was only observed for PVC incubated in soil for 9 months (Judy et al. 2019).

7.3.2.4 Rotifera

Rotifers are organisms that are bilaterally symmetrical and have a microscopic size and unsegmented soft body, with a common distribution in both the freshwater and marine environments. As main components of zooplankton, these small organisms have an important ecological role in aquatic ecosystems. This taxonomic group was only represented by a single marine species, *Brachionus plicatilis*. Two developmental stages of *B. plicatilis* were used in exposure studies, neonates (Gambardella et al. 2018; Manfra et al. 2017) and nauplii (Beiras et al. 2018), both exposed via water. All studies investigated the effect of microplastic spheres, either composed of PS (Beiras et al. 2018; Gambardella et al. 2018) or PE (Manfra et al. 2017). In terms of size, two studies looked at particles <0.05μm (Gambardella et al. 2018; Manfra et al. 2017), one study looked at particles 0.05–0.099μm (Manfra et al. 2017), and one study looked at 1–9μm sized particles (Beiras et al. 2018). Two studies described the surface of the particles, Gambardella et al. (2018) used plain PS spheres, and Manfra et al. (2017) looked at both COOH and NH_2 coated PS spheres.

All publications looked at individual-level effects, specifically mortality. No studies assessed subcellular or population-level effects and only one study considered ecosystem-level effects, specifically alterations in swimming speed (Gambardella et al. 2018). Neonates exposed to PS-NH_2 spheres (0.001–50 mg/L) exhibited significant mortality only when concentrations exceeded 10 mg/L (Manfra et al. 2017). On the other hand, PS-COOH spheres did not induce any effect at the same concentrations (Manfra et al. 2017). In another study, nauplii exposed to PE spheres were only significantly affected after 48 hrs of exposure, when concentrations exceeded 1 mg/L (Beiras et al. 2018). Finally, PS spheres (<0.05μm) only affected the swimming speed of neonates after 48-hr exposure (0.001–10 mg/L) (Gambardella et al. 2018).

7.3.2.5 Arthropoda

Arthropoda is the largest group of the animal kingdom, which includes invertebrate organisms that have an exoskeleton, a segmented body and jointed appendages. Arthropods are widely represented in every environmental compartment and include crustaceans, insects, isopods and amphipods, among others. Most of the studies conducted with Arthropoda (39 of 57) were in the freshwater environment, followed by 16 studies in the marine environment, 3 studies in terrestrial and only 1 in the brackish environment. Twenty-nine Arthropoda species from 5 classes, Branchiopoda, Entognatha, Hexanauplia, Insecta and Malacostraca, were studied: *Acartia tonsa*, *Amphibalanus amphitrite*, *Artemia franciscana*, *Asellus aquaticus*, *Calanus finmarchicus*, *Calanus helgolandicus*, *Carcinus maenas*, *Centropages typicus*, *Ceriodaphnia dubia*, *Chironomus tepperi*, *Corophium volutator*, *Daphnia galeata*, *Daphnia magna*, *Daphnia pulex*, *Echinogammarus marinus*, *Eriocheir sinensis*, *Folsomia candida*, *Gammarus fossarum*, *Gammarus pulex*, *Hyalella azteca*, *Idotea emarginata*, *Lobella sokamensis*, *Nephrops norvegicus*, *Palaemonetes pugio*, *Parvocalanus crassirostris*, *Platorchestia smithi*, *Porcellio scaber*, *Talitrus saltator* and *Tigriopus fulvus*. Fifteen of the species were Malacostraca, while there was only one study on Insecta (*Chironomus tepperi*; Ziajahromi et al. 2018). *Daphnia magna* was by far the most used species (n = 29 publications), followed by *Artemia franciscana* (n = 4 publications). Overall, 14 species were from the marine environment, 11 from freshwater, 3 terrestrial and 1 from brackish water.

Most of the Arthropoda species were filter/suspension feeders (6 species in 35 studies). Nine studies used eight detritivores species, seven studies included seven grazer species, and four studies used four scavenger species. Deposit feeders (two species), filter feeders (one species) and grazer and detritivores (one species) were represented by two publications each. Only one publication studied a predator species, *Eriocheir sinensis*. Most studies were carried out on adults (27 studies) and neonates (23 studies), while juveniles (7 studies), nauplii (6 studies), larvae (2 studies) and 1-week-old organisms (1 study) were less studied. Five publications studied the whole cycle of *D. magna* and *D. pulex*. Filter/suspension feeders and predators were exposed via water (37 studies). On the other hand, detritivores were exposed via water (three studies), sediment (two studies), soil (two studies) and food (two studies). Grazers were also exposed via water (five studies), sediment and food, and deposit feeders were exposed via water and sediment. Lastly, scavenger organisms were only exposed via food (four studies).

Fourteen polymer types were studied using Arthropoda, in a total of 57 publications. PS was the most studied polymer, followed by PE (31 and 14 studies, respectively). PET was represented by five publications, while PA, PMMA and PP had four each. Proprietary polymer and PVC had three and two studies, respectively. All the other particle types (ABS, nylon, PC, PHB, POM and SAN) were represented by one study each. Most of the studies used spheres (30 and 11 using PS and PE, respectively), while the remaining particle shapes had less than 5 studies each. Regarding size, PS particles between 1–9μm, 0.1–0.99μm and 0.05–0.099 were used in 13, 12 and 10 publications, respectively. Seven studies used PE particles between 20 and 49μm. The remaining size classes were used in five or less studies. ABS, PC, PHB, POM and SAN were only studied within the size range 20 to 49μm. Regarding particle surface, PS-COOH was the most studied with seven publications, followed by PS plain and PS-NH_2 with six studies each, all particles within the nano-scale. Particles with other surface modifications were used in five or less publications each.

Effects at the individual level (51 studies, corresponding to 89% of studies) were the most commonly determined in arthropods, followed by effects at the population (18 studies, 32% of studies) and subcellular, ecosystem and organ levels (11, 7 and 5 studies, corresponding to 19%, 12% and 9% of studies). When comparing the different levels of biological organization, the percentage of reported effects was comparable to those reporting no effects. Gene expression was the most common endpoint determined within the subcellular level (Bergami et al. 2017; Fadare et al. 2019; Gambardella et al. 2017; Heindler et al. 2017; Imhof et al. 2017; Lin et al. 2019b; Liu et al. 2018, 2019; Tang et al. 2019; Yu et al. 2018; Zhang et al. 2019), followed by enzymatic activity and neurotoxicity (Gambardella et al. 2017; Lin et al. 2019b; Yu et al. 2018) as well as oxidative stress (Lin et al. 2019b; Yu et al. 2018; Zhang et al. 2019). Energy reserves (Cole et al. 2019; Cui et al. 2017; Kokalj et al. 2018; Weber et al. 2018) and alterations in hepatosomatic index (Yu et al. 2018) were the endpoints targeted at the organ level. At the individual level, mortality (Au et al. 2015; Beiras et al. 2018; Bergami et al. 2016, 2017; Besseling et al. 2014; Bhargava et al. 2018; Blarer and Burkhardt-Holm 2016; Booth et al. 2016; Bosker et al. 2019; Bruck and Ford 2018; Canniff and Hoang 2018; Casado et al. 2013; Cole et al. 2015; Cui et al. 2017; Fadare et al. 2019; Gambardella et al. 2017; Gerdes et al. 2019; Gray and Weinstein 2017; Hämer et al. 2014; Horton et al. 2018; Imhof et al. 2017; Jemec et al. 2016; Kim et al. 2017; Kokalj et al. 2018; Lin et al. 2019b; Liu et al. 2018; Ma et al. 2016; Mattsson et al. 2017; Nasser and Lynch 2016; Ogonowski et al. 2016; Pacheco et al. 2018; Redondo-Hasselerharm et al. 2018; Rehse et al. 2016, 2018; Rist et al. 2017; Tang et al. 2019; Tosetto et al. 2016; Ugolini et al. 2013; Vicentini et al. 2019; Weber et al. 2018; Wu et al. 2019a; Yu et al. 2018; Zhang et al. 2019, p. 201; Ziajahromi et al. 2017) and growth (Au et al. 2015; Bergami et al. 2016; Besseling et al. 2014; Bruck and Ford 2018; Cole et al. 2019; Gerdes et al. 2019; Hämer et al. 2014; Imhof et al. 2017; Jemec et al. 2016; Kokalj et al. 2018; Liu et al. 2019; Ogonowski et al. 2016; Pacheco et al. 2018; Redondo-Hasselerharm et al. 2018; Rist et al. 2017; Jinghong Tang et al. 2019; Vicentini et al. 2019; Weber et al. 2018; Welden and Cowie 2016; Yu et al. 2018; Zhao et al. 2015; Zhu et al. 2018; Ziajahromi et al. 2017) were the most studied, alongside feeding behaviour (Blarer and Burkhardt-Holm 2016; Bruck and Ford

2018; Cole et al. 2013, 2019; Hämer et al. 2014; Kokalj et al. 2018; Ogonowski et al. 2016; Redondo-Hasselerharm et al. 2018; Rist et al. 2017; Straub et al. 2017; Watts et al. 2015; Weber et al. 2018; Welden and Cowie 2016; Zhu et al. 2018), development (Blarer and Burkhardt-Holm 2016; Ma et al. 2016; Straub et al. 2017), energy reserves (Watts et al. 2015; Welden and Cowie 2016), respiration rate (Cole et al. 2015) and gut microbial diversity (Zhu et al. 2018). Endpoints related to population level included alterations in reproductive output (Au et al. 2015; Besseling et al. 2014; Bosker et al. 2019; Canniff and Hoang 2018; Cole et al. 2015; Cui et al. 2017; de Felice et al. 2019; Heindler et al. 2017; Imhof et al. 2017; Liu et al. 2019; Ogonowski et al. 2016; Pacheco et al. 2018; Rist et al. 2017; Vicentini et al. 2019; Zhu et al. 2018; Ziajahromi et al. 2017, 2018), followed by larval development (Ziajahromi et al. 2018) and population size (Heindler et al. 2017). At the ecosystem level, only alterations in behaviour (e.g. swimming activity, phototactic response, distance and acceleration) were recorded upon exposure (Booth et al. 2016; Chae et al. 2018; de Felice et al. 2019; Frydkjær et al. 2017; Gambardella et al. 2017; Kim and An 2019; Lin et al. 2019b; Tosetto et al. 2016).

From the terrestrial species included in the ecotoxicological assessments reviewed, effects on feeding behaviour, growth, gut microbial diversity and reproduction were seen for *F. candida* in response to PVC (1000 mg/kg soil, size range 80–250µm) (Zhu et al. 2018). These effects were attributed to changes in soil structure due to the presence of microplastics that led to alterations in feeding behaviour and capacity to find high-quality food, thus influencing nutrient absorption (Zhu et al. 2018). Similar findings were found for *L. sokamensis* exposed to PE (1000 mg/kg soil, size range 20–49µm) and PS (4, 8 and 1000 mg/kg soil, size ranges 0.1–0.99, 20–49 and 200–500µm) (Kim and An 2019). In this study, springtails showed altered behaviour in response to microplastic movement into soil bio-pores, at lower concentrations and size ranges than those reported for *F. candida* (4 and 8 mg/kg soil for PS 0.1–0.99µm compared to 1000 mg/kg soil PVC 80–250µm). Both studies highlight that the behaviour of plastic particles in soil does not only affect the behaviour of soil-dwelling organisms and lead to high adverse effects (e.g. impaired growth and reproduction), but their presence can also have wider implications for effective management of soils (Kim and An 2019; Zhu et al. 2018).

Several biological endpoints have been determined in freshwater arthropods in response to both nano- and microplastics, with toxicity being dependent on polymer type (e.g. Au et al. 2015), particle size (e.g. de Felice et al. 2019), surface chemistry (e.g. Lin et al. 2019b) and time of exposure (e.g. Liu et al. 2019). As mentioned previously, the crustacean *Daphnia* sp. was the most used organism to assess the ecotoxicological effects of plastic particles via water exposure, for which acute and chronic toxicity has been reported for different particles. Adverse effects including mortality (LOEC 0.005 mg/L, PS spheres 10–19µm (P. Zhang et al. 2019)), abnormal development (adults LOEC 0.1 mg/L and offsprings LOEC 5 mg/L, PS spheres 0.05–0.099µm (Liu et al. 2019 and Cui et al. 2017, respectively)), swimming behaviour (LOEC 1 mg/L for PE fragments 10–19µm, PS spheres 0.1–0.99µm and PS-NH_2 0.05–0.099µm (Frydkjær et al. 2017; Lin et al. 2019b)) and reproductive output (LOEC 0.02 mg/L, proprietary polymer 1–9µm (Pacheco et al. 2018)) were the most commonly described. In terms of sediment exposure, the effects of PE at environmentally relevant concentration (500 particles/kg sediment, size range 1–49μm) were evaluated using the chironomid *C. tepperi* (Ziajahromi et al. 2018) after 5 and 10 days of exposure. The authors reported that exposure to PE negatively affected the survival, growth (i.e. body length and head capsule) and emergence of chironomids, with the observed effects being strongly dependent on particle size.

Ecotoxicological studies of marine arthropods showed that smaller sized plastic particles had a stronger impact, with surface chemistry playing a significant role for the effects seen. This is the case of *A. franciscana* exposed to PS nanoplastics with different surface alterations, for which the lowest LOECs for different endpoints were recorded. Also, when comparing the long-term toxicity of PS-COOH and PS-NH_2 (size range 0.05–0.099μm), Bergami et al. (2017) observed a concentration-dependent mortality in brine shrimp after 14 days, with the latter showing a higher impact (EC_{50} = 0.83 mg/L). In addition, alteration in genes involved in moulting were also recorded at the lowest concentration tested of 0.01 mg/L, further suggesting that the disruption of larval moulting and energy metabolism may play a role in the toxicity of nanoplastics towards arthropods. In another study by Gambardella et al. (2017), short-term exposure of *A. franciscana* and *A. amphitrite* to PS nanoplastics (size range 0.1–0.99μm) at low concentrations (0.001 to 10 mg/L) did not affect survival but impacted swimming behaviour, increased expression of catalase and inhibited acetylcholinesterase activity in exposed organisms. As only sublethal effects were observed, the authors highlight that behavioural responses seem to be more sensitive than mortality in plastic toxicity assessments, especially after short-term exposure.

Arthropoda was the most heterogeneous of the taxonomical groups assessed, including a wide range of species belonging to the terrestrial and aquatic compartments with different developmental stages and feeding strategies. Several effects covering different levels of biological organization were reported, with impacts on feeding behaviour, growth, development, reproduction and lifespan being highlighted as the most significant. These findings emphasize the need to perform long-term exposures covering whole cycle assessments to fully understand the magnitude and consequences of plastic particles to the aquatic environment. This is particularly important for species belonging to zooplankton, an important food source for secondary consumers, as these represent a possible route by which plastic particles could enter food chains and be transferred up the trophic levels. In addition, a significant impact on the lifespan of these organisms might have serious consequences in the balance of aquatic ecosystems (Botterell et al. 2018).

7.3.2.6 Annelida

The Annelida group is composed of segmented worms, such as earthworms, lugworms and leeches. Annelids can be found in all types of habitat, and one of their most important ecological roles is reworking of soils and sediments. The terrestrial environment was represented by nine studies (covering three species) and the marine environment by seven studies (also covering three species). The marine environment was represented by three species belonging to the Polychaeta class: *Arenicola marina* (five studies), *Hediste diversicolor* (one study) and *Perinereis aibuhitensis* (one study). The terrestrial environment was represented by three species of the Clitellata class: *Eisenia fetida* (five studies), *Lumbricus terrestris* (three studies) and *Eisenia andrei* (one study). All but one of the studies (where life stage was not specified) used adult organisms. In the terrestrial environment, soil was spiked with microplastics in eight out of nine studies, the remaining study using spiked food (leaf litter). However, both dermal and trophic exposure can be expected from these two exposure scenarios, due to constant burrowing and feeding activity of the earthworms. For the aquatic environment, spiked sediment was also the main exposure scenario (six out of seven studies), with only one study using spiked water.

The most studied polymer type was PE (nine studies, Besseling et al. 2017; Huerta Lwanga et al. 2016; Judy et al. 2019; Prendergast-Miller et al. 2019; Rillig et al. 2017; Rodríguez-Seijo et al. 2017; Rodríguez-Seijo et al. 2018a, b; Wang et al. 2019a), followed by PS (five studies, Besseling et al. 2013; Cao et al. 2017; Leung and Chan 2018; Van Cauwenberghe et al. 2015; Wang et al. 2019a), PVC (four studies, Browne et al. 2013; Gomiero et al. 2018; Judy et al. 2019; Wright et al. 2013) and PET (one study, Judy et al. 2019). The morphology of the particles was not always provided by the authors, but when it was the case, spheres and fragments were the most common shapes, each covered by six studies. Interestingly in one study, characterization by scanning electron microscopy revealed that particles sold as spheres were in fact flakes (Cao et al. 2017). Overall, particles ranging from below 1μm to 5 mm were studied, with most studies focusing on particles above 100μm (12 out of 16 studies). When particles were prepared in the laboratory, the lowest and largest particle sizes were not always provided (e.g. Huerta Lwanga et al. 2016). None of the 16 studies on Annelida reported any surface characterization or functionalization.

The individual level was assessed in all 16 studies on annelids, followed by subcellular (9 studies), ecosystem (6 studies) and population (3 studies) levels. Only one study covered effects at the cellular and organ level. At the individual level, mortality and growth were the most studied endpoints (both covered by 10 studies), although being the least affected endpoints across species, environmental compartments, polymer types and sizes. Mortality was never observed, except in one study with PS flakes at environmentally irrelevant concentrations (5 and 20 g/kg soil dry weight). Growth was rarely affected, and only at environmentally irrelevant concentrations for pristine plastic particles (from 10 g/kg PS flakes and from 4 g/kg PE spheres).

The lowest concentrations inducing effects at the subcellular level were observed for exposure to PE fragments (size classes 200–500 and > 500μm), which increased protein, lipid and polysaccharide contents in earthworms at 62 mg/kg, decreased catalase activity at 125 mg/kg and increased lipid peroxidation at 250 mg/kg (Rodríguez-Seijo et al. 2017, 2018a). PS fragments of similar size (200–500μm) were found to increase peroxidase activity in earthworms at 10 g/kg (the lowest concentration tested by Wang et al. 2019a). In marine annelids, PVC fragments

(100–199μm) induced inflammation at 5 g/kg (the lowest concentration tested by Wright et al. (2013)).

At the ecosystem level, negative results were most frequently reported, e.g. no avoidance of PE fibres (40 × 400μm) at up to 10 g/kg (Prendergast-Miller et al. 2019) and PE, PET and PVC fragments (>500μm) at 5 g/kg (Judy et al. 2019) by earthworms and no effect of PE spheres (particle size distribution ranging from <50μm to >100μm) at up to 12 g/kg on burrow formation by earthworms (Huerta Lwanga et al. 2016). The only effects seen were on the feeding activity of marine annelids, where PVC fragments (100–199μm) at 10 and 50 g/kg increased the feeding activity of *Arenicola marina* (Wright et al. 2013).

Overall, the data on the ecotoxicological effects of plastic particles on Annelida is very limited but seem to suggest a moderate to low risk to these organisms. One of the reasons could be linked to the ecological traits of annelids, adapted to continuously ingest vast amounts of non-nutritious particles, through their burrowing and feeding activities. It should also be noted that the absence of avoidance behaviour and detrimental effects on annelids make them efficient vectors of plastic particles not only to their predators but also to the whole ecological compartment, due to their intense bioturbation activity.

7.3.2.7 Mollusca

The Mollusca group includes several ecologically and commercially important filter feeders (e.g. mussels and clams) that due to their habitat and feeding behaviour are likely to encounter plastic particles of varying sizes. Most of the studies for Mollusca focused on marine species (29 studies, 13 species), followed by freshwater (6 studies, 4 species) and terrestrial species (a single study, 1 species). The 17 species belonged to 2 classes, Bivalvia and Gastropoda: *Abra nitida*, *Achatina fulica*, *Corbicula fluminea*, *Crassostrea gigas*, *Dreissena polymorpha*, *Ennucula tenuis*, *Meretrix meretrix*, *Mytilus edulis*, *Mytilus galloprovincialis*, *Mytilus* sp., *Ostrea edulis*, *Perna perna*, *Perna viridis*, *Pinctada margaritifera*, *Potamopyrgus antipodarum*, *Scrobicularia plana* and *Sphaerium corneum*. The most commonly studied species was the mussel *M. galloprovincialis* (in 11 studies). Most of the species used were filter feeding (13 species in 33 studies), followed by grazer species (2 species in 2 studies), while only 1 study used deposit feeders (2 species). Most studies were carried out on adults (28 studies), with 7 studies using larvae, 4 studies embryos, 2 studies gametes and 1 study juveniles. Filter-feeding organisms were exposed mainly via water (28 studies) and 1 via water plus muddy sediment. For these organisms, two studies used exposure via food and two studies via sediment. The deposit feeders were exposed via sediment, while the grazers via food and soil.

For Mollusca, 36 studies looked at the effects of 9 different polymers, with PS being the most studied polymeric material (total 20 studies). Overall, 12 studies used PE and 4 studies used PVC and PET. There were two studies for PLA and two for proprietary polymer, while all the other polymers (PA, PC and PP) only had one study each. Most of the studies were performed with PS spheres (n = 14), followed by PE and PS fragments (eight and three studies, respectively). Two studies used PET fibres and spheres of proprietary polymer, while the remaining morphologies only had one study each. Regarding size, the highest number of studies (12 in total) used PS particles between 1 and 9μm. Studies with PE particles used size ranges of 20–49μm and 50–99μm with five studies each, along with PS particles with sizes 0.1–0.99μm, 20–49μm and 20–49μm. All the other particle size distributions had less than five studies each. Only studies using PS particles reported particle surface information, for which four studies used PS-NH_2, three studies used plain and COOH and one used PS with sulphate groups, where all particles were within the nano-scale. Most of the reviewed studies only reported effects for particles above 1μm, with only a small number showing impacts with particles within nano-range, more specifically PS and PE. This is the reflection of the size-dependent threshold commonly associated with the particle-selection feeding behaviour characteristic of most of the species included in this taxonomical group (Van Cauwenberghe and Janssen 2014; Wegner et al. 2012).

In terms of levels of biological organization, effects at the subcellular (23 studies, with 18 reporting effects) and individual level (22 studies, with 12 reporting effects) were the most studied. There was only one study at an ecosystem level (reporting effects) but 11 analysing effects at the population level (7 with observed effects). Overall, 11 studies analysed effects on organs (with 6 reporting effects) and 7 in cells (6 reporting effects). The most studied endpoint was related to impacts in feeding behaviour (15 studies), with 9 reporting significant effects related to filtration and ingestion rate, absorption and assimilation efficiency (Capolupo et al. 2018; Cole and Galloway 2015; Gardon et al. 2018; Green 2016; Guilhermino et al. 2018; Oliveira et al. 2018; Revel et al. 2019; Rist et al. 2016, 2019; Rochman et al. 2017; Santana et al. 2018; Song et al. 2019; Sussarellu et al. 2016; Wegner et al. 2012; Woods et al. 2018). Endpoints related to oxidative stress were the second most common endpoint, with 14 studies, 8 of which showing impacts on lipid peroxidation, formation of reactive oxygen species and total oxyradical scavenging capacity (Avio et al. 2015; Brandts et al. 2018b; Gonçalves et al. 2019; González-Fernández et al. 2018; Guilhermino et al. 2018; Magni et al. 2018; Oliveira et al. 2018; Paul-Pont et al. 2016; Revel et al. 2019; Ribeiro et al. 2017; Santana et al. 2018; Song et al. 2019; Sussarellu et al. 2016; von Moos et al. 2012). In combination with oxidative stress, alteration in enzymatic activity was also one of the main endpoints determined in molluscs (reported in 12 studies), with 10 studies showing alterations to antioxidant enzymes (Avio et al. 2015; Brandts et al. 2018b; Franzellitti et al. 2019; Gonçalves et al. 2019; Guilhermino et al. 2018; Magni et al. 2018; Oliveira et al. 2018; Paul-Pont et al. 2016; Pittura et al. 2018; Revel et al. 2019; Ribeiro et al. 2017; Song et al. 2019). Alterations in gene expression were also a common endpoint in most of the reviewed studies (12 studies), with 10 reporting up- and downregulation of genes involved in different metabolic pathways as detoxification, immunity, apoptosis, energy reserves, etc. (Avio et al. 2015; Balbi et al. 2017; Brandts et al. 2018a; Capolupo et al. 2018; Détrée and Gallardo-Escárate 2017, 2018; Franzellitti et al. 2019; Paul-Pont et al. 2016; Pittura et al. 2018; Revel et al. 2019; Rochman et al. 2017; Sussarellu et al. 2016). Histopathological alterations were also included in some of these studies to understand the effects of particle ingestion in different organs (total nine studies), with five studies reporting alterations in the gills and digestive glands of exposed organisms (Bråte et al. 2018; Gardon et al. 2018; Gonçalves et al. 2019; Guilhermino et al. 2018; Paul-Pont et al. 2016; Revel et al. 2019; Rochman et al. 2017; Song et al. 2019; von Moos et al. 2012). Five out of eight studies reported significant genotoxicity of the plastic particles used, expressed as DNA damage or micronuclei formation (Avio et al. 2015; Brandts et al. 2018a; Bråte et al. 2018; Magni et al. 2018; Pittura et al. 2018; Revel et al. 2019; Ribeiro et al. 2017; Santana et al. 2018). Seven studies also analysed the neurotoxicity of particles, with six reporting significant alterations in acetylcholinesterase activity (Avio et al. 2015; Brandts et al. 2018a; Guilhermino et al. 2018; Magni et al. 2018; Oliveira et al. 2018; Pittura et al. 2018; Ribeiro et al. 2017). Several endpoints related to population effects were determined in molluscs, most of which related to fecundity (six studies, Gardon et al. 2018; González-Fernández et al. 2018; Imhof and Laforsch 2016; Luan et al. 2019; Sussarellu et al. 2016; Tallec et al. 2018), offspring viability (one study, Capolupo et al. 2018), larval development (seven studies, Balbi et al. 2017; Beiras et al. 2018; Cole and Galloway 2015; Luan et al. 2019; Rist et al. 2019; Sussarellu et al. 2016; Tallec et al. 2018) and juvenile development (one study, Imhof and Laforsch 2016). Of these endpoints, only those related to fecundity (e.g. fertilization yield, gamete quality hatching rate, etc.) and larval development showed a significant effect. General health endpoints including growth (eight studies, Détrée and Gallardo-Escárate 2018; Gardon et al. 2018; Green 2016; Imhof and Laforsch 2016; Redondo-Hasselerharm et al. 2018; Rist et al. 2019; Santana et al. 2018; Song et al. 2019), energy reserves (five studies, Avio et al. 2015; Bour et al. 2018; Brandts et al. 2018a; Pittura et al. 2018; von Moos et al. 2012), condition index (six studies, Bour et al. 2018; Revel et al. 2019; Ribeiro et al. 2017; Santana et al. 2018; Sussarellu et al. 2016; von Moos et al. 2012), respiration rate (three studies, Gardon et al. 2018; Green 2016; Rist et al. 2016) and scope for growth (one study, Gardon et al. 2018) were also included in several studies; however, these were the less sensitive endpoints, where only one to two studies reported a significant effect.

Of the four freshwater species used in the studies reviewed, significant impacts were only recorded for *D. polymorpha* exposed to PS (1–9μm, LOEC 50000 particles/L) (Magni et al. 2018) and *C. fluminea* following exposure to a proprietary polymer (1–9μm, LOEC 0.13 mg/L) (Guilhermino et al. 2018; Oliveira et al. 2018), as well as PET, PE, PVC and PS fragments (Rochman et al. 2017). In the study by Rochman et al. (2017), *C. fluminea* was exposed to environmental concentrations and sizes of PET, PE, PVC and PS fragments (sizes range 50 to >500μm) for 28 days, after which histopathological alterations were recorded (LOEC 2.8 mg/L). The authors highlight that the effects observed in exposed clams were specific to the polymer type used.

Several ecotoxicological effects across the different levels of biological organization were recorded for marine molluscs. Interestingly, mortality was one of the least sensitive endpoints in organisms exposed either via sediment or water, even at very high concentrations. Only Rist et al. (2016) reported substantial mortality in *P. viridis* exposed to PVC after 91 days of exposure (size range 1–49μm, 2160 mg/L); however, no significant statistical differences were found compared to the control condition. Mussels belonging to the genus *Mytilus* were the most used marine species used in the reviewed studies, for which a wide range of biological endpoints were determined. The biological endpoints for which significant effects were recorded included byssus production and immunity deficiency (LOEC 0.025 mg/L, PE fragments >500μm) (Green et al. 2019), mortality, concentration and phagocytic activity of circulation haemocytes, histopathological alterations, ROS production and lipid peroxidation (LOEC 0.032 mg/L, PS spheres 1–9μm) (Paul-Pont et al. 2016), antioxidant enzymatic activity and genotoxicity (LOECs of 0.000008 mg/L and 0.01 mg/L, respectively, mixture PE and PP fragments, 200–500μm) (Revel et al. 2019), feeding behaviour (LOEC 3000 particles/L, PET fibres 200 to >500μm) (Woods et al. 2018), alterations in gene and protein expression, growth (LOEC 0.03 mg/L, PE and PLA fragments 1 to 50μm) (Détrée and Gallardo-Escárate 2018), larval malformations (LOEC 0.00042 mg/L, PS spheres, 1–9μm) (Rist et al. 2019), lysosomal membrane stability (LOEC 1500 mg/L, PE and PS fragments size range from <0.05 to 99μm) (Avio et al. 2015) and neurotoxicity (LOEC 0.05 mg/L, PS spheres 0.1–0.99μm) (Brandts et al. 2018b).

The gastropod *A. fulica* was the only terrestrial species in the ecotoxicological studies reviewed, for which effects were recorded following 28 days of exposure to PET fibres (length 1260μm, diameter 76μm) at concentrations ranging from 14 to 710 mg/kg sediment (Song et al. 2019). The authors reported alterations in feeding behaviour (LOEC 14 mg/kg sediment) upon exposure that resulted in histopathological alterations in the gastrointestinal tract (LOEC 140 mg/kg sediment) and oxidative stress in the liver (LOEC 710 mg/L).

Mollusca was the taxonomical group for which a wider range of biological endpoints were determined. Overall, the reviewed data highlighted that acute and chronic toxicity of plastic particles in molluscs seem to be dependent not only on particle characteristics such as polymer type (Avio et al. 2015; Rochman et al. 2017), concentration range (Gardon et al. 2018; Rochman et al. 2017), particle size (Tallec et al. 2018) and surface chemistry (Cole and Galloway 2015; Luan et al. 2019), but also on organism-specific traits such as developmental stage (Balbi et al. 2017; Rist et al. 2019) and tissue analysed (Brandts et al. 2018b; Revel et al. 2019; Ribeiro et al. 2017). Furthermore, the reviewed findings further emphasize the need to conduct studies with freshwater and terrestrial species, especially when considering their higher risk of exposure to plastic particles. It is also worth mentioning that this taxonomical group includes many filter-feeding species with a high tendency for particle retention, thus representing a possible source of transfer across higher trophic levels and potentially to humans.

7.3.2.8 Echinodermata

Echinoderms are exclusively marine invertebrate species that have a widespread distribution throughout the ocean. These organisms inhabit a diverse array of cold water and tropical ecosystems including habitats from coastal, intertidal zones to offshore, as well as deep water areas. Common echinoderms include sea cucumbers, starfish and sea urchins. Four microplastic ecotoxicology studies were reviewed for echinoderms representing the marine environment. Sea urchin species were used in all studies: *Paracentrotus lividus* was used in three studies (Beiras et al. 2018; Della Torre et al. 2014; Messinetti et al. 2018), while *Tripneustes gratilla* was used in one study (Kaposi et al. 2014). Early life stages of sea urchins were used for all studies (larvae/embryo (Beiras et al. 2018; Della Torre et al. 2014; Messinetti et al. 2018)). All studies with echinoderms were performed via water exposure. Reviewed studies used PS (two studies) and PE (two studies) microparticles. Experimental studies on echinoderms varied with PS with two different surface charges being used at the 40–50 nm size range and 10μm PS spherical microparticles. PE of similar size ranges similar as natural food of zooplankton organisms (1–500μm) were also used, as well commercial PE ranging from 10 to 45μm.

The individual level was studied in all four studies and one study included endpoints at the cellular level (Della Torre et al. 2014). The effects of carboxylated PS (PS-COOH) and amine PS (PS-NH_2) nanoplastics were used to evaluate embryotoxicity in *P. lividus*, specifically disposition, embryo development and gene expression. No embryotoxicity was observed for PS-COOH which formed microaggregates and was anionic up to 50μg/mL. However, PS-NH_2, which was better dispersed and cationic, caused developmental defects (EC_{50} 3.85μg/mL 24 hours post fertilization and EC_{50} 2.61μg/mL 48 hours post fertilization). These findings suggest that surface charge and particle aggregation dynamics in seawater influence embryotoxicity. Collectively, the findings of Della Torre et al. (2014) highlight the importance of different aggregation states and surface properties of nanoplastics and how they lead to differences in uptake, exposure and disposition routes and overall impacts.

The effects of ingesting microplastics in larval *T. gratilla* were proportionally related to the concentration of PE microspheres and ingestion was reduced in the presence of biological fouling and phytoplankton food. An unrealistically high concentration of PE microspheres (300 spheres/mL) affected larval growth with no significant effect on survival observed. Conversely, at environmentally realistic concentrations, there was little effect observed on growth or survival (Kaposi et al. 2014).

The planktotrophic larvae of *P. lividus* were utilized to evaluate the effects of PS microbeads on juvenile development. *P. lividus* larvae were able to ingest microplastics, albeit at a lower rate, in comparison to the sessile filter-feeding ascidian (*Ciona robusta*) juveniles. No effect of PS microbeads, at any concentration (control vs. 0.125, 1.25, 12.5 and 25μg/mL), was observed on larval survival, whereas growth was negatively affected, with shorter larvae observed in the 25μg/mL treatment (Messinetti et al. 2018).

7.3.2.9 Chordata: Fish

Marine and freshwater environments are evenly represented in fish studies, with 19 and 20 studies, respectively. Overall, 18 different species were used in fish studies (*Acanthochromis polyacanthus*, *Acanthurus triostegus*, *Bathygobius krefftii*, *Carassius carassius*, *Clarias gariepinus*, *Cyprinodon variegatus*, *Danio rerio*, *Dicentrarchus labrax*, *Lates calcarifer*, *Oncorhynchus mykiss*, *Oreochromis niloticus*, *Oryzias latipes*, *Oryzias melastigma*, *Pimephales promelas*, *Pomatoschistus microps*, *Sparus aurata*, *Symphysodon aequifasciatus*). The most commonly studied species is the zebrafish *D. rerio* (12 studies, corresponding to 26% of studies). The European seabass (*D. labrax*) and the common goby (*P. microps*) are the most commonly studied marine species (six studies, 13% of studies each). Most studies were carried out on embryo/larvae (11 studies, 28% of studies) or juvenile (16 studies, 41% of studies) fish, while studies on adult fish only represent 18% of the studies (7 studies). Six studies did not report the developmental stage of the test species.

Fish exposure to microplastics was performed either directly via water (27 studies, 69% of studies) or via the trophic route (13 studies, 33% of studies). For the later, two main methods are found in the literature. The first method consists in exposing living prey to microplastics then feeding them to fish (Cedervall et al. 2012; Mattsson et al. 2015, 2017; Skjolding et al. 2017; Tosetto et al. 2017). The second method consists in spiking artificial food with known concentrations of microplastics and feed it to fish (Ašmonaitė et al. 2018a, b; Caruso et al. 2018; Granby et al. 2018; Jovanović et al. 2018; Mak et al. 2019; Mazurais et al. 2015; Rochman et al. 2013). While the first method is more representative of trophic interactions in the environment, microplastic ingestion by living prey is not a controlled parameter, and spiking artificial food therefore offers better control of exposure concentrations. The numbers of studies reporting adverse effects, as well those reporting an absence of effect, are similar for marine and freshwater environments and for the different exposure routes. This suggests that these parameters are not likely to influence the occurrence of effects in fish following exposure to microplastics.

More than 92% of studies conducted on fish species used PS (45% = 18 studies) or PE (47.5% =15 studies) microplastics. Commercially available (micro)spheres are the most represented particle morphology and are used in 56% of the studies (22 studies). Undetermined fragments are used in 46% of the studies (18 studies), and close to 13% of the studies (5 studies) did not disclose particle morphology. Four studies used microplastics produced by grinding larger plastic items (Caruso et al. 2018; Choi et al. 2018; Lei et al. 2018b). A broad range of particle sizes have been tested, with the vast majority of studies using microplastics comprised between 0.1 and 500μm. Most studies investigating the effects of microplastics presenting different properties compared different particle sizes: 49% (19 studies) studied microplastics presenting different sizes, while only one and two studies compared microplastic morphology and polymer type, respectively.

In fish studies, the subcellular level is the most frequently studied level of biological organization (23 studies, 59% studies), followed by the individual, ecosystem, organ and population levels, respectively (16, 16, 13 and 8 studies, respectively, corresponding to 41%, 41%, 33% and 21% of studies). For each organization level, all the studied endpoints were listed and sorted as "impacted" or "not impacted" following exposure to microplastics. For most organization levels, the numbers of endpoints not impacted are very close to the numbers of impacted endpoints. At cellular and subcellular levels, oxidative stress is the main endpoint studied (Ašmonaitė et al. 2018a; Chen et al. 2017; Choi et al. 2018; Ding et al. 2018; Ferreira et al. 2016; Karami et al. 2017; LeMoine et al. 2018; Luís et al. 2015; Mak et al. 2019; Oliveira et al. 2013; Rochman et al. 2013; Wang et al. 2019c), as well as lipid peroxidation (Barboza et al. 2018; Ding et al. 2018; Ferreira et al. 2016; Fonte et al. 2016; Oliveira et al. 2013; Wen et al. 2018a), immune and/or inflammatory responses (Brandts et al. 2018a; Choi et al. 2018; Granby et al. 2018; Mazurais et al. 2015), neurotoxicity (Barboza et al. 2018; Ding et al. 2018; Ferreira et al. 2016; Fonte et al. 2016; Luís et al. 2015; Oliveira et al. 2013; Rainieri et al. 2018), energy production (Barboza et al. 2018; Oliveira et al. 2013; Wen et al. 2018a), endocrine disruption (Wang et al. 2019c) and gut tight junctions proteins, as well as active transport through gut (Ašmonaitė et al. 2018b). At the organ level, most studies focus on histological changes (Ašmonaitė et al. 2018b; Choi et al. 2018; Jovanović et al. 2018; Karami et al. 2016, 2017; Lei et al. 2018b; Mak et al. 2019; Rainieri et al. 2018; Rochman et al. 2013; Wang et al. 2019c), but other endpoints were also studied, such as intestine permeability (Ašmonaitė et al. 2018b; Jovanović et al. 2018), blood and plasma chemistry and metabolite concentrations (Jovanović et al. 2018; Mattsson et al. 2015, 2017), brain weight and water content (Mattsson et al. 2015, 2017), liver glycogen (Karami et al. 2016; Rochman et al. 2013), lipid metabolism (Cedervall et al. 2012) and gut microbiota (Caruso et al. 2018; Jin et al. 2018). Endpoints studied at the population level comprise fish fecundity (e.g. number of eggs laid and hatching rate) (Cong et al. 2019; LeMoine et al. 2018; Wang et al. 2019c), embryo survival and development (Batel et al. 2018; Pitt et al. 2018) and larval survival, development and behaviour (Chen et al. 2017; Choi et al. 2018; Malinich et al. 2018). Endpoints at the ecosystem levels relate to behaviour and include feeding behaviour (e.g. feeding time, foraging, predatory performance), environment exploration and fish locomotion (Cedervall et al. 2012; Choi et al. 2018; Critchell and Hoogenboom 2018; de Sá et al. 2015; Ferreira et al. 2016; Fonte et al. 2016; Guven et al. 2018; Jacob et al. 2019; Luís et al. 2015; Mak et al. 2019; Malinich et al. 2018; Mattsson et al. 2017; Pitt et al. 2018; Skjolding et al. 2017; Tosetto et al. 2017; Wen et al. 2018a). Contrary to the above-described levels of biological organization, for which the numbers of impacted and non-impacted endpoints are similar, at the individual level more studies report an absence of effects (11 studies) than the observation of adverse effects (3 studies) following microplastic exposure. Mortality was reported for medaka larvae exposed to PS sphere (10µm, 100,000 part./L) for 14 days (Cong et al. 2019) and for juvenile goby exposed to PE spheres (1–5µm, 184µg/L) for 4 days (Fonte et al. 2016), and weight loss was observed in crucian carp exposed to PS nano-spheres via trophic chain for 42 days (Cedervall et al. 2012). Other studies investigating fish mortality, growth or body condition reported an absence of effect (Critchell and Hoogenboom 2018; Ding et al. 2018; Granby et al. 2018; Jovanović et al. 2018; Karami et al. 2017; Lei et al. 2018b; LeMoine et al. 2018; Mazurais et al. 2015; Oliveira et al. 2013; Wen et al. 2018a, b;), and in one case reported mortality only at the highest concentration test (PMMA nano-spheres, 20 mg/L) (Brandts et al. 2018a).

7.3.3 Species Sensitivity Distributions

Species sensitivity distributions (SSDs) are a common approach used in environmental protection, risk assessment and management practices to describe interspecies sensitivity and estimate community-level risks for a specific stressor. An SSD is derived by fitting a selected statistical model, in this case a lognormal distribution, to available ecotoxicity effect data for species from different taxonomical groups, after which predictions of the % of species affected can be calculated (Posthuma et al. 2019). The SSD captures the interspecies variability, which can then be used to derive key risk assessment components, such as the concentration at which 5% of the species in an ecosystem can be affected. This key regulatory parameter is commonly known as the "hazardous concentration for 5% of the species" or HC_5 and is normally used to derive environmental quality criteria standards (Besseling et al. 2019; Burns and Boxall 2018 and references therein). Even though this approach is commonly used to assess the risk of other environmental chemicals, only recently it has been applied to both microplastic and nanoplastic data (Adam et al. 2019; Besseling et al. 2019; Burns and Boxall 2018; Everaert et al. 2018; VKM 2019).

With the ecotoxicological data collected from the reviewed publications, three SSDs for microplastic were investigated for water, sediment/soil and food exposure routes, after which the HC_5 corresponding to concentrations expressed in mass and particle number when available were estimated (Fig. 7.6). However, the lack of ecotoxicological data for species covering the different environmental compartments limited the applicability of SSDs in this case, thus decreasing the overall success of the hazard assessment of microplastics and nanoplastics. SSDs are as robust as the quality of their ecotoxicological data, and usually at least 12 different species are considered a minimum for fitting an SSD (Posthuma et al. 2019). Accordingly, even though a total of 107 species covering key taxonomical groups were comprehensibly assessed in the 175 publications reviewed, only 12–58 were used to build the SSDs. This represents a subset of the total data, depending on the availability of data for the exposure matrix (water or sediment/soil) and the exposure quantification (mass or particles).

As the total microplastic toxicity data on freshwater and marine environments is still limited, information collected on marine, freshwater and terrestrial species were combined according to exposure route (water, sediment/soil and food) to increase the number of feeding strategies and trophic levels included in the SSDs, thus increasing statistical power. No distinction was made between particle characteristics due to insufficient data within a certain particle size and polymer type. In

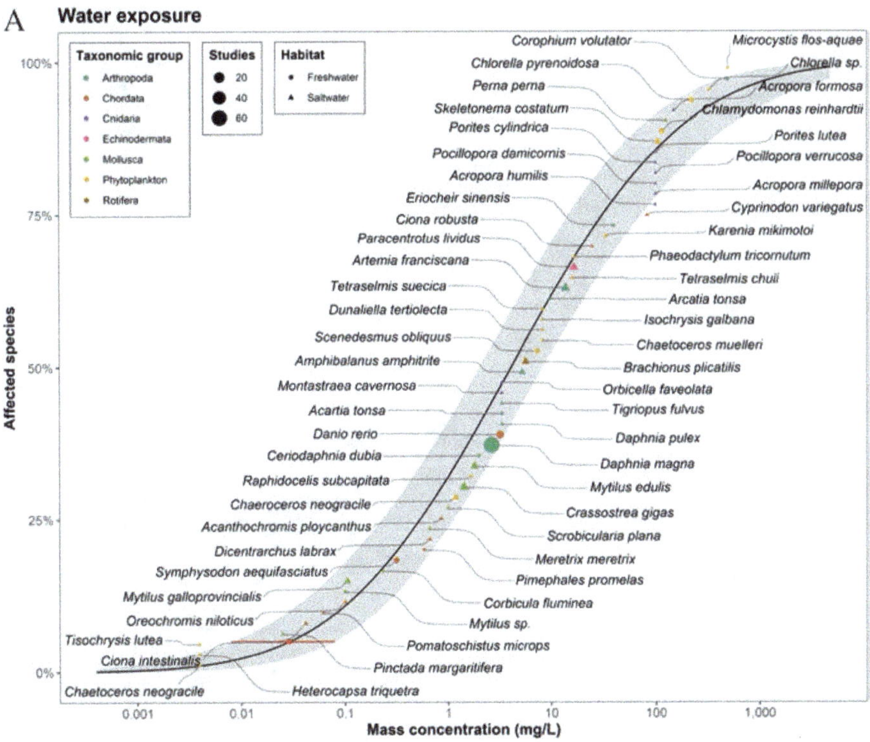

Fig. 7.6 Species sensitivity distributions (SSDs) for (**a**) species exposed via the water phase with data divided by particle concentration expressed as mass (mg/L) ($n = 58$); (**b**) species exposed via the water phase with data divided by particle concentration expressed as particle number (*million* particles/L) ($n = 31$); and (**c**) species exposed via the sediment and soil phase with data shown only for particle concentration as mass number (mg/kg) ($n = 12$). The average SSDs are plotted as solid black lines, and the 95% credible interval as grey ribbon. The HC_5 (concentration at which 5% of the species are affected) is represented as a red point in combination with the 95% credible intervals. Taxonomic groups are represented in different colours, with the different habitats divided by shape and where size reflects the number of studies included

addition, only data pertaining to individual and population levels were considered (e.g. mortality, growth, reproduction), for which both NOECs and EC_{50}/LC_{50} values were used.

The poor standardization in terms of reporting of experimental conditions was another factor influencing the construction of SSDs. For example, the lack of information on exposure concentrations expressed in mass and particle number further limited the usable data sets. Dose metrics were standardized to either mass- or particle-based concentrations. When it was not possible to perform this conversion, the studies were excluded from the SSD fitting. Most of the excluded studies were for exposure via food (e.g. fish), leaving insufficient data available to construct SSDs, as only 6 and 3 data points were available (for mass concentration and

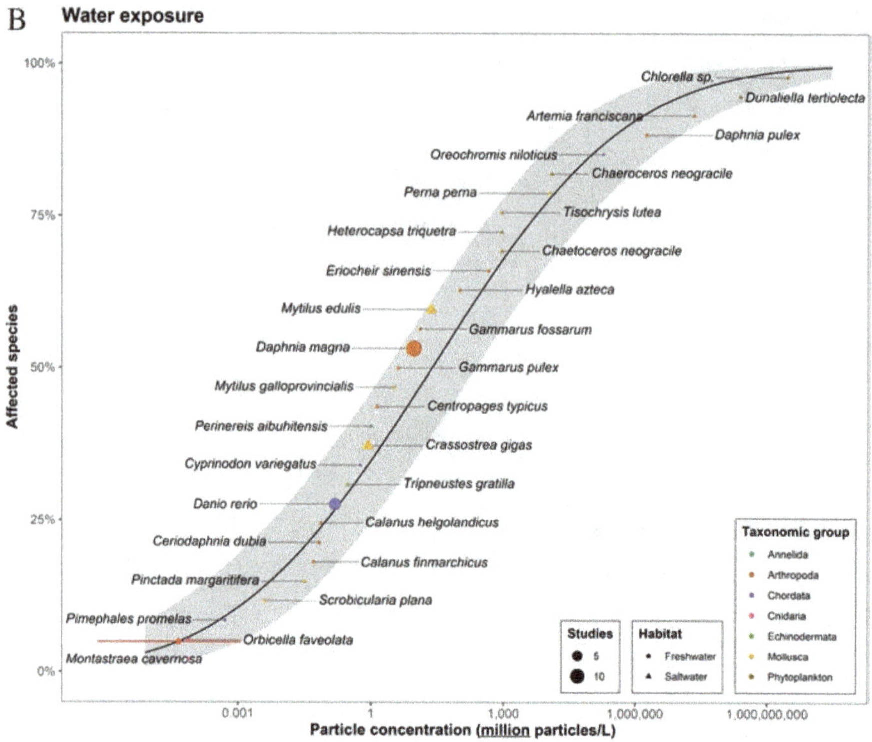

Fig. 7.6 (continued)

particle concentration, respectively). Overall, tentative SSDs reflecting the combined variability of species sensitivity, plastic properties and effect mechanisms were only constructed for water exposure as a function of particle dosage (both mass and number) and sediment/soil exposures as a function of particle dosage (mass only). Due to insufficient data, the particle-based sediment exposure route and the entire dietary exposure route were excluded from the SSD analyses. The SSD for mass-based water exposure was fitted to data from 101 studies, covering 58 species across 7 taxonomic groups and 2 habitats. Its particle-based counterpart was fitted to data from 39 studies, covering 31 species across 7 taxonomic groups and 2 habitats. For the mass-based sediment exposure route, the SSD was fitted to data from 17 studies, covering 12 species across 4 taxonomic groups and 3 habitats; note that in terms of species coverage, this is considered a minimum acceptable coverage.

The separately constructed SSDs for organisms exposed via water and sediment/soil (expressed in mass and particle number) are shown in Fig. 7.6. Of the studies where concentrations were expressed by particle mass, microalgae species were the most and least sensitive species to exposure via the water phase (Fig. 7.6a). The most sensitive species was the marine microalgae *C. neogracile* (PS-NH_2 spheres, <1μm), (González-Fernández et al. 2019), while the most sensitive freshwater species was the clam *C. fluminea* (proprietary polymer, 1–9μm) (Oliveira et al. 2018).

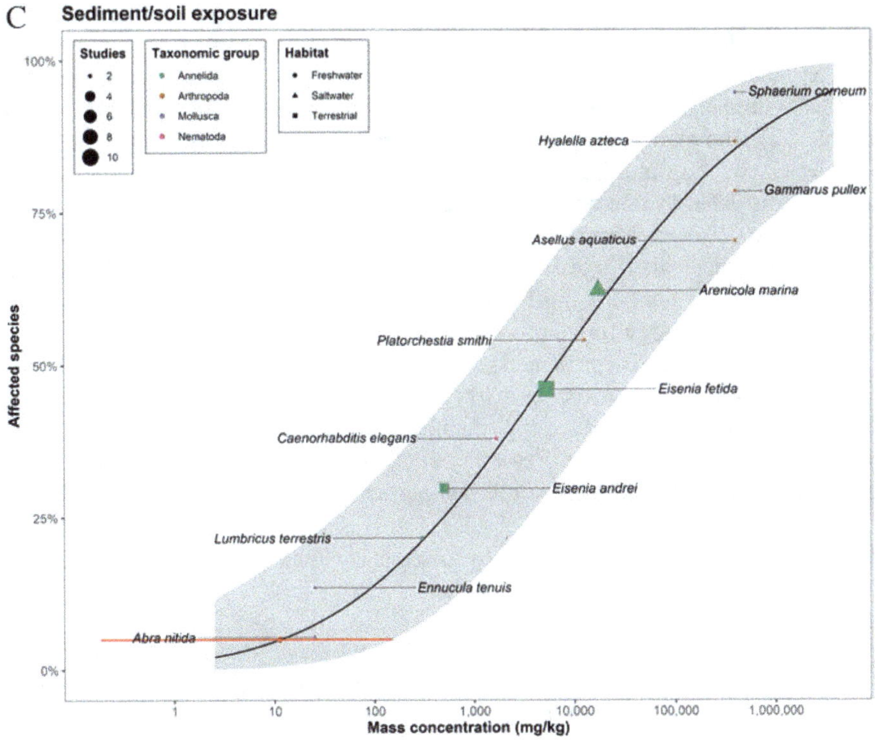

Fig. 7.6 (continued)

The least sensitive freshwater species was *M. flos-aquae* (PVC and PP, 100–199µm) (Wu et al. 2019b), while the cnidarian *A. formosa* was the least sensitive marine species (PE fragments, size range 50 to 500µm (Syakti et al. 2019). The derived HC_5 for this SSD was 28.9µg/L (95% CI 7.94–79.1µg/L). For the water exposure SSD built with data expressed in terms of particle number (Fig. 7.6b), the cnidarians *M. cavernosa* and *O. faveolata* were the most sensitive species (PE beads, >50µm (Hankins et al. 2018)), while the least sensitive was the freshwater microalgae *Chlorella* sp. (Thiagarajan et al. 2019). The derived HC_5 for this SSD was 41.6 particles/L (95% CI 0.58–1176 particles/L). For exposures either via sediment or soil (Fig. 7.6c), the SSDs obtained for particle concentration in mass showed that the most sensitive species were the marine clams *A. nitida* and *E. tenuis* (PE fragments >1µm) (Bour et al. 2018), followed by the terrestrial annelid *L. terrestris* (PE spheres <1 to >500µm) (Huerta Lwanga et al. 2016). The least sensitive species were the freshwater snail *S. corneum* (PS fragments >20µm (Redondo-Hasselerharm et al. 2018)) and the freshwater arthropod *H. azteca* (PE and PS fragments 10–500µm) (Au et al. 2015; Redondo-Hasselerharm et al. 2018). The derived HC_5 for this SSD was 11.3 mg/kg (95% CI 0.18–151 mg/kg). As mentioned above, construction of an SSD for particle-based sediment exposure was not possible due to lack of sufficient data.

The mass-based water exposure HC_5 value (28.9μg/L) obtained in the present review is higher than that previously reported for microplastics (0.08–5.4μg/L) (Table 7.2). The main reason for this difference is the inclusion of a higher number of species covering multiple taxonomical groups. On the other hand, the particle number-based HC_5 value was 41.6 particles/L, which is within the range provided by the VKM (2019) assessment. Even though this estimate included a larger data set (31 species) than other assessments, the number of studies that provide particle concentrations in number is still quite limited. No other HC_5 values expressed in mg/kg exist in literature for comparison.

Even though the SSDs presented here are more robust as they are based on larger data sets and add to the existing SSDs in literature, several knowledge gaps still need to be addressed to reduce uncertainties and improve the robustness and relevance of the obtained results (Besseling et al. 2019; Burns and Boxall 2018). For this reason, ecotoxicity testing of relevant particle sizes, shapes and polymer types,

Table 7.2 – HC_5 values obtained from species sensitivity distribution analysis collected from literature

HC_5 (μg/L)	HC_5 (particles/L)	HC_5 (mg/kg)	Notes	References
28.9 (7.94–79.1)	41.6 (0.58–1176)	11.3[a] (0.18–151)	Freshwater and marine species exposed to micro- and nanoplastics via water and sediment/soil	Present review
0.14 (0.04–0.64)	71.6 (3.45–1991)	–	Freshwater and marine species exposed to micro- and nanoplastics	VKM (2019)
0.08 (0.04–0.11)	740 (610–1300)	–	Freshwater species exposed to microplastics. 25–75 percentile was used instead of confidence interval	Adam et al. (2019)
5.4 (0.93–31 mg/L)	5.97×10^{10} (1.6×10^{10}–22×10^{10})	–	Marine and freshwater species exposed to nanoplastics	Besseling et al. (2019)
1.67 (0.086–32.6)	1015 (101–10,223)	–	Marine and freshwater species exposed to microplastics	
–	64,000	–	Marine and freshwater species exposed to microplastics (10 to 5000 mm)	Burns and Boxall (2018)
–	33.3 (0.36–13,943)	–	Marine species exposed to microplastics	Everaert et al. (2018)
–	3214 (3.3900–84,261)	–	Marine species exposed via water and sediment to microplastics	Van Cauwenberghe (2016)

[a]Note that the HC_5 value for mass-based sediment exposure is derived from a minimum of necessary data and needs to be interpreted with caution

standardized testing, improved reporting of experimental designs, methods and results, as well as a higher focus on freshwater and terrestrial compartments, need to be prioritized in order to enable a sound risk assessment of plastic particles in the environment.

7.3.4 Direct and Indirect Effects at the Ecosystem/ Community Level

Cascading effects through different levels of biological organization is a central paradigm of ecotoxicology: contaminant-induced subcellular changes, such as enzymatic activity or gene expression, can impact higher levels of organization and affect organism's performance (e.g. locomotion, feeding, reproduction). These alterations might impact an entire population and could ultimately have consequences at the ecosystem level. With that said, directly linking effects at the lowest levels of biological organization to impacts on ecosystems is extremely challenging for any environmental contaminant (Galloway et al. 2017). The data currently available on nano- and microplastic ecotoxicity does not allow firm conclusions to be drawn about such links. However, certain endpoints observed at the individual level are indicators of potential indirect effects on other species and/or on the functioning of ecosystems. Such endpoints are therefore categorized as endpoints relevant at the ecosystem level. For example, behavioural changes at the individual level can affect prey-predator interactions (Fonte et al. 2016; Wen et al. 2018a) and impact entire trophic webs, or impaired burrowing activity of dwelling organisms can alter bioturbation and soil/sediment oxygenation (Green et al. 2016). Changes in microbial activity can also result in altered essential ecosystem processes, such as nutrient cycling (e.g. nitrogen and carbon cycles) (Green et al. 2017).

Among the studies reviewed in this chapter, endpoints relevant at the ecosystem level were most studied on three taxonomical groups: Annelida, Arthropoda and Chordata. The recorded endpoints were related to behaviour: feeding activity (Besseling et al. 2013, 2017; Browne et al. 2013; Cedervall et al. 2012; Green et al. 2016; Guven et al. 2018; Malinich et al. 2018; Mattsson et al. 2017; Wright et al. 2013), burial and burrow formation (Booth et al. 2016; Huerta Lwanga et al. 2016), cast production (Green et al. 2016; Prendergast-Miller et al. 2019), locomotion (Chae et al. 2018; Choi et al. 2018; Critchell and Hoogenboom 2018; de Felice et al. 2019; Frydkjær et al. 2017; Gambardella et al. 2017; Kim and An 2019; Lin et al. 2019b; Mattsson et al. 2017; Pitt et al. 2018; Skjolding et al. 2017; Tosetto et al. 2016, 2017; Ziajahromi et al. 2017), prey-predator interactions (de Sá et al. 2015; Ferreira et al. 2016; Fonte et al. 2016; Jacob et al. 2019; Luís et al. 2015; Mattsson et al. 2017; Wen et al. 2018a) and aggression (Critchell and Hoogenboom 2018). Studies focusing on such ecologically relevant endpoints are currently underrepresented (16% of the reviewed studies), although the available data shows that these endpoints can be impacted by plastic particles, especially locomotion (Cedervall et al. 2012; Choi et al. 2018; de Felice et al. 2019; Frydkjær et al. 2017; Kim and An 2019; Lin et al. 2019b; Mattsson et al. 2017), feeding activity (Besseling et al. 2013, 2017; Green et al. 2016; Guven et al. 2018; Mattsson et al. 2017; Wright et al. 2013) and prey-predator interactions (Fonte et al. 2016; Wen et al. 2018a).

Only a single study looked at the ecosystem-level effects on Cnidaria, more specifically on *P. damicornis* (Tang et al. 2018). The results obtained in this study suggest that acute exposure to PS particles can activate stress responses at the individual level, repressing detoxification and immune systems, which in turn can compromise the anti-stress capacity of exposed organisms. However, this study found a minimal impact in community shifts (symbiont density and chlorophyll content) in the short term. In a similar study, Reichert et al. (2018) suggested that species-specific effects might promote community shifts in coral reefs. For example, if growth, health and photosynthesis are affected, this might amplify the coral's susceptibility to other stressors such as increased seawater temperatures, contributing to shifts in coral reef assemblages. Like cnidarians, only one study considered the effects of nanoplastics at the ecosystem level in phytoplankton (González-Fernández et al. 2019). This study analysed the impact of PS-NH_2 (50 nm) on a diatom (*C. neogracile*), which led to changes of the concentration of associated bacterial communities. It is important to study effects following exposure to plastic particles in phytoplankton not only due to their susceptibility (as seen in the SSD) but also due to their importance in the ecosystem. As already stated, these organisms are at the base of the aquatic food web, and changes in their communities may disturb the productivity of an entire ecosystem (Prata et al. 2019). Moreover, particles may end up higher in the food web due to algae-particle interaction as the first step in the biomagnification (Nolte et al. 2017), as previously shown in other studies with suspension-feeding bivalves (Ward and Kach 2009). Finally, one study addressed the impacts of microplastics on the health and biological functioning of oysters (*O. edulis*) and on the structure of associated macrofaunal assemblages using an outdoor mesocosm experiment (Green 2016). The author found that exposure to high concentrations of microplastic resulted in alterations of assemblage structure, diversity, abundances and biomasses of several taxa in vegetated oyster habitats, whose cascade effects can lead to significant impacts in marine ecosystems.

Indirect, secondary effects are effects occurring on species not necessarily exposed to plastic particles but which are impacted by changes resulting from their direct exposure. In their mesocosm study, Green et al. (2016) exposed the lugworm *A. marina* to microplastics and observed a decrease in cast production, as well as decreased microbial biomass with increasing concentrations. One of the hypotheses discussed by the authors to explain the decreased microbial biomass was that reduced sand reworking by the worms would have resulted in less nutrients available in the sand to support primary productivity. No firm conclusion about indirect effects of microplastics could be drawn from this study, as microplastics could have directly affected microbial communities, but this scenario is one of the potential examples of indirect microplastic effects. In another recent study, reduced survival and reproduction were observed for the terrestrial invertebrate *Enchytraeus crypticus* following exposure to synthetic fibres (Selonen et al. 2020). However, fibre ingestion could not be confirmed, and the authors hypothesized that the observed effects could be due to changes in environmental conditions, such as microbial activity and physicochemical properties of the soil, resulting from microplastic exposure. In both cases, the authors (Green et al. 2016; Selonen et al. 2020) present indirect effects of microplastics as a hypothesis, but investigating microplastic indirect effects was not the main purpose of the study. Although highly ecologically relevant, studies on nano- and microplastic indirect effects are currently almost non-existent. Such studies are needed to help link effects at the organism level to impacts on the ecosystem level. Future studies should consider potential direct and indirect nano- and microplastic effects at the ecosystem level, to fill these major gaps in the field of plastic ecotoxicology.

7.3.5 *Interaction of Plastic Particles with Chemicals*

The challenge of assessing the impact of plastic particles in the environment is further complicated by the presence of chemicals, which can potentially pose additional hazards towards organisms. These chemicals comprise polymerization catalysts and additives, which are incorporated during production to endow plastics with specific characteristics (e.g. flame retardants, plasticizers, antioxidants, UV stabilizers and pigments) (Gallo et al. 2018) and non-intentionally added substances (NIAS). Furthermore, chemicals already present in the environment (e.g. polycyclic aromatic hydrocarbons (PAHs) and metals) may also be incorporated/adsorbed by plastic surfaces depending on the polymer physico-chemical properties (e.g. Teuten et al. 2009).

Few studies have identified nano- and microplastics as vectors for other contaminants (Trojan horse effect), and even fewer have focused on the presence and leaching of chemical additives. Of the 175 references reviewed, 48 addressed these combined effects, with a focus on chemicals present in the environment, such as PAHs (e.g. benzo(a)pyrene (BaP), phenanthrene, fluoranthene, pyrene), polychlorinated biphenyls (PCBs), organophosphates (e.g. chlorpyrifos), metals (e.g. gold, mercury, cadmium, chromium and copper), metal nanomaterials (gold and titanium nanoparticles) and pharmaceuticals (roxithromycin, cefalexin, carbamazepine, florfenicol, doxycycline and procainamide). Only a small percentage of these studies (12.5%) focused on chemicals known to be used as plastic additives (e.g. benzophenone, polybrominated diphenyl ethers (PBDEs), perfluorooctane sulfonates (PFOs), bisphenol A (BPA), triclosan), surfactants (e.g. nonylphenol), as well as chemical leachates extracted from plastic particles. In addition, the combined effects of plastic particles with natural acidic organic polymers (e.g. palmitic acid, humic acid and fulvic acid) were also considered in some of the reviewed publications.

Most of these studies were conducted in arthropods (28%), followed by fish (20%), molluscs (17%), phytoplankton (15% studies), annelids (9%), echinoderms (2%), nematodes (2%) and rotifers (2%). No studies on the combined effects of plastic particles and other contaminants were reported for cnidarians. Of the 57 studies reviewed for arthropods, 15 addressed the interaction between plastic particles and chemicals. These chemicals included benzophenone (Beiras et al. 2018), fluoranthene (Bergami et al. 2016, 2017; Horton et al. 2018; Vicentini et al. 2019), humic acid (Fadare et al. 2019; Wu et al. 2019a), PCBs (Gerdes et al. 2019; Lin et al. 2019a; Watts et al. 2015), phenanthrene (Ma et al. 2016), gold (Pacheco et al. 2018), BPA (Rehse et al. 2018), PAHs (Tosetto et al. 2016), palmitic acid (Vicentini et al. 2019) and roxithromycin (Zhang et al. 2019). Several effects at the subcellular, individual and population levels were seen in arthropods upon exposure to nano- or microplastics combined with these chemicals. The most reported effects where impacts on reproduction, mortality, development and growth. Eleven studies conducted on fish used microplastics sorbed with chemicals. In seven of those, the tested microplastics were purposely spiked with chemicals, such as BaP (Batel et al. 2018); antibiotics (Fonte et al. 2016); heavy metals such as mercury (Barboza et al. 2018), cadmium (Lu et al. 2018) and chromium (Luís et al. 2015); gold nanoparticles (Ferreira et al. 2016); and a cocktail of environmental contaminants comprising PCBs, PBDEs, PFOs and metals (Granby et al. 2018). Additionally, in four studies, the tested microplastics were deployed in environmental matrices (i.e. harbour, sewage effluent, urban bay), and further analyses confirmed the presence of environmental contaminants, such as surfactants and PAHs (Ašmonaitė et al. 2018a, b; Rochman et al. 2013; Tosetto et al. 2017). Interestingly, for every level of biological organization covered in these fish studies, the presence of chemicals sorbed on microplastics does not change the occurrence of adverse effects, indicating that microplastic-associated chemicals would play a minor role in microplastic effects. Studies on combined effects of micro- and nanoplastics and chemical exposure using molluscs included pyrene (Avio et al. 2015), carbamazepine (Brandts et al. 2018b), florfenicol (Guilhermino et al. 2018), mercury (Oliveira et al. 2018), fluoranthene (Paul-Pont et al. 2016; Rist et al. 2016), BaP (Pittura et al. 2018) and PCBs (Rochman et al. 2017). Effects at the cellular and subcellular levels were often reported for this taxonomical group, followed by impacts at the organ and individual level. Additionally, in one of the studies reviewed, no effects were reported for *M. galloprovincialis* exposed to benzophenone (Beiras et al. 2018). In the eight studies reported for phytoplankton, adverse effects of micro- and nanoplastics in combination with other contaminants were reported for metal mixtures (Baudrimont et al. 2020), copper (Bellingeri et al. 2019), titanium nanoparticles (Thiagarajan et al. 2019), fulvic and humic acid (Liu et al. 2019), chlorpyrifos (Garrido et al. 2019), doxycycline and procainamide (Prata et al. 2018), triclosan (Zhu et al. 2019) as well as leachate mixtures (Luo et al. 2019). Overall, the documented effects in these studies included reduction in growth, oxidative stress, membrane stability and reduction in protein content and natural pigments. From the 16 studies conducted with annelids, five included co-exposure with contaminants, namely, PCBs (Besseling et al. 2013, 2017), chlorpyrifos (sprayed to the surface of PE spheres (Rodríguez-Seijo et al. 2018b)), BaP (Gomiero et al. 2018), nonylphenol, phenanthrene, triclosan and PBDE-47 (sorbed to microplastics (Browne et al. 2013)). Of the effects found in annelids, alterations in behaviour (i.e. reduced feeding) were most commonly reported associated with exposure to PCBs (Besseling et al. 2013,

2017). Reduction in growth was also observed at lower concentrations when plastic particles were sprayed with chlorpyrifos (Rodríguez-Seijo et al. 2018b) or co-exposed with PCBs (Besseling et al. 2013). Of the reviewed studies for Echinodermata, only Beiras et al. (2018) utilized microplastics spiked with benzophenone-3, an organic, hydrophobic chemical found in cosmetic products, using *P. lividus* as a test organism. Even though ingestion of virgin and BP-3 spiked PE microplastics was observed at 1 and 10 mg/L, no acute toxicity was observed above concentrations considered environmentally relevant (low treatment = 20μg/L and high concentration treatment = 200 ng/L) (Beiras et al. 2018). When it comes to nematodes, in the study by Judy et al. (2019), microplastics were added to soil amended with municipal waste compost. The presence of trace metals was assessed in amended soils and in microplastics (PE, PET, PVC), and GC-MS analysis revealed the presence of phthalates in PVC, which could have accounted for the effects observed in exposed organisms. Only one study looked at combined effects of PE spheres and benzophenone using the rotifer *B. plicatilis*, for which no effects were reported (Beiras et al. 2018).

Overall, the studies reviewed on the joint toxicity of plastic particles and chemicals (either adsorbed to particles or additives) showed that their interaction can elicit a wide range of biological responses in exposed organisms. In addition, chemicals associated to plastic particles can also influence their bioavailability and potential transfer through food chains, possibly causing effects at the ecosystem level. Nonetheless, these findings need to be interpreted with caution as most of these studies differ in how they approach vectoral transfer kinetics and exposure mechanisms for chemicals under realistic natural conditions and thus overestimate the role of plastic particles as the delivery system of chemicals to organisms. The majority of these laboratory experiments use simplified exposure settings, in which clean organisms placed in clean media/sediment/soil are exposed to plastic particles pre-treated with chemicals. These controlled exposure settings create conditions that promote rapid dissolution of the chemicals from the plastic particles into the surrounding environmental compartment, which then become easily bioavailable to organisms through a more conventional exposure route (Diepens and Koelmans 2018; Booth and Sørensen 2020). Under more environmentally relevant exposure scenarios, currently available data suggests that chemicals accumulated in organisms are derived to a very small extent from ingested plastic particles, especially when compared to natural pathways of bioaccumulation as water, sediment and food (Koelmans et al. 2016; Besseling et al. 2017). For this reason, it is important to consider the relative importance of plastic particles as an exposure route for chemicals in the context of other uptake pathways that may be more relevant under realistic natural conditions (Lohmann 2017; Diepens and Koelmans 2018). To understand how plastic particles can act as vectors for other chemicals and what is the contribution that additives make to overall exposures, a thorough control of exposure mechanisms is therefore necessary. This will ensure that any observed biological effects are a consequence of exposure to the chemicals adsorbed and/or incorporated in the particles and not derived from their leaching, desorption and dissolution into environmental compartments (Booth and Sørensen 2020; Gallo et al. 2018;

Hermabessiere et al. 2017). In addition, there is a pressing need for studies addressing synergistic/antagonist effects following short- and long-term exposure to plastic particles in combination with contaminants of high concern, as well as studies on their cumulative effects in both terrestrial and aquatic species and potential biomagnification throughout food chains. For further information on the impacts of environmental contaminants and plastic additives in terrestrial and aquatic organisms, see reviews by Gallo et al. (2018) and Hermabessiere et al. (2017). For additional studies on the importance of exposure pathways for a range of chemicals present in plastic particles under natural conditions, the readers may refer to Koelmans et al. (2016), Lohmann (2017) and Diepens and Koelmans (2018).

7.4 Challenges and Future Directions

Exposure experiments focusing on the ecotoxicological effects of plastic particles in a wide range of organisms have increased exponentially over the past few years. A consensus from the reviewed literature is that plastic particles can impact organisms across successive levels of biological organization, covering effects from the subcellular level up to the ecosystem level (Galloway et al. 2017; VKM 2019). Nonetheless, our understanding on the mechanisms behind any toxic effects recorded is still minimal, partially due to a lack of attempt to link the physical and chemical properties of the particles being tested with the recorded toxic effects. Many of the reviewed studies relate to common chemical exposure endpoints rather than particle related endpoints, including how particles directly interact with the cellular environment and organisms, their uptake mechanisms, tissue distribution and subsequent impacts (e.g. tissue alterations due to inflammation or other physical impacts). Accordingly, understanding and distinguishing the potential physical and chemical effects of plastic particles across the whole spectrum of biological levels is needed to improve environmental risk assessment of plastic pollution, as a means to ensure a better protection and mitigation of its impacts in the different environmental compartments.

The comparability of existing ecotoxicological data is being hampered by numerous factors such as the use of wide array of experimental testing approaches, unrealistic environmental concentrations, lack of relevance in terms of particle characteristics (polymer type, shape or size), use of appropriate controls, incomplete/inadequate particle characterization (physico-chemical properties and chemical additives), variability in reporting units (e.g. in mass and/or particle number, % particles in food or sediment) and experimental conditions (e.g. exposure duration). Many of these limitations were found during the evaluation of data quality in the reviewed references, in which the use of appropriate controls, confirmation of exposure concentration and polymer type as well as presence of chemical leachates and particle size distributions were the most common issues. The ubiquitous nature of microplastic contamination, widespread geographical distribution, abundance and small size have also raised significant concerns regarding their interactive effects with chemicals, not only by increasing the bioavailability of contaminants in organisms but also by eliciting common toxic effects. This is especially true when considering the potential risk of chemical accumulation in higher trophic levels including humans, as well modifications in population structure and ecosystem dynamics (e.g. negative effects at lower trophic levels) that may potentially result in a reduced productivity of the whole ecosystem. However, the role of plastic particles as the delivery system of chemicals to organisms is currently overestimated and additional data is required to understand the relative importance of exposure to chemicals (either adsorbed or additives) from particles compared to other exposure pathways (e.g. water and natural diet).

This overview is consistent with the tendencies observed by other authors, calling into question the environmental relevance and proposed risks caused by nanoplastic and microplastic exposure (e.g. Burns and Boxall 2018; de Ruijter et al. 2020; Kögel et al. 2020; VKM 2019). To determine if these plastic particles are in fact posing significant risks to organisms, future work needs to focus on the development of reporting guidelines to improve the reproducibility and comparability of plastic-related research, as highlighted by Connors et al. (2017) and Cowger et al. (2020). Several research priorities are thus recommended to better understand the ecological risks of plastic particles in the terrestrial and aquatic environments:

1. **Standardization.** It is fundamental for ecotoxicological investigations to be comparable. A standardized approach from experimental design to reporting is required. To this end, quality assessments should be conducted throughout the whole duration of any laboratory studies (including concentrations and exposure conditions with quality assessment) to obtain reliable and comparable data.
2. **Environmental relevance.** Researchers should endeavour to conduct experiments which have relevance to current and future scenarios of plastic concentrations and characteristics in the different environmental compartments. These include partially degraded and irregularly shaped particles commonly found in the environment, with varying polymer types, sizes and surface properties. As fibres and fragments are prevalent in environmental samples, these should be prioritized in future studies.
3. **Particle vs. chemical effect.** The combination of particle and associated additives must be considered in ecotoxicological studies, such that it is possible to discriminate between effects derived from particles from those resulting from additive chemicals. Therefore, it is paramount that a thorough characterization of exposure materials is carried out, including the chemical profiles of organic and metal additives. To really understand whether plastic particles are relevant carriers for chemicals, environmentally realistic exposure settings also need to be taken into account when looking at particle-chemical interactions, more specifically leaching/desorption kinetics, chemical bioaccumulation from water/sediment/soil, natural diet and percentage of ingested particles.
4. **Ecosystem compartments.** As highlighted throughout this chapter, there is a disproportion between the number of studies conducted on marine, freshwater and terrestrial biota. Moving forward, it is important to direct attention towards

freshwater and terrestrial ecosystems, as these are considered the main sources and transport pathways of plastic particles to the marine environment.

5. **Test species.** Species utilized for ecotoxicological testing are generally focused on model organisms used for standard ecotoxicological testing. This originates a significant knowledge gap on the effects of plastic particles in other species that have critical roles in ecosystem balance. Species considered at highest risk of exposure due to their feeding strategies and position in the water column need to be prioritized in terms of ecotoxicity testing, e.g. planktonic species not included in ISO and OECD guidelines. Species ecology and time spent in various environmental compartments are also important considerations for choice of test species, with particular emphasis on early developmental stages that have been shown to be highly susceptible to the impacts of plastic particles. Moreover, given that soil/sediment is considered the ultimate sink for plastic particles and other conventional contaminants, increased testing with suspension and deposit feeders is also warranted.
6. **Physiological perspective.** Currently there is a lack of mechanistic understanding of the effects of microplastics and nanoplastics on biota. Additional efforts are needed to understand the differences in physical and chemical behaviour of plastic particles compared to conventional contaminants. The direct and indirect interaction of nano- and microplastics within the cellular environment and organisms, uptake mechanisms (size dependency), tissue distribution and impacts must therefore be comprehensibly assessed and linked to the physical and chemical properties of the particles being used. Modifications in experimental design and proper characterization of the particles (e.g. presence of additives) can also assist to explain the underlying mechanisms responsible for the observed responses and help distinguish physical from chemical toxicological effects.
7. **Integrated and multi-level approaches.** Long-term experiments with multiple species (e.g. model ecosystems) are required to examine effects with higher ecological relevance. Therefore, small- and large-scale mesocosm experiments mimicking environmentally relevant scenarios and covering links from primary producers (e.g. microalgae) to top predators (e.g. fish) are encouraged.

Acknowledgments This work was supported by the Research Council of Norway-funded projects MicroLEACH (Grant 295174) and REVEAL (Grant 656879). Handelens Miljøfond, the Norwegian Retailers Association, also supported this work through the project MicroOPT, in addition to the Norwegian Institute for Water Research (NIVA) strategic research programme.

References

Adam V, Yang T, Nowack B (2019) Toward an ecotoxicological risk assessment of microplastics: comparison of available hazard and exposure data in freshwaters: environmental risk assessment of microplastics. Environ Toxicol Chem 38(2):436–447. https://doi.org/10.1002/etc.4323

Anbumani S, Kakkar P (2018) Ecotoxicological effects of microplastics on biota: a review. Environ Sci Pollut Res 25(15):14373–14396. https://doi.org/10.1007/s11356-018-1999-x

Ašmonaitė G, Larsson K, Undeland I, Sturve J, Carney Almroth BM (2018a) Size matters: ingestion of relatively large microplastics contaminated with environmental pollutants posed little risk for fish health and fillet quality. Environ Sci Technol 52(24):14381–14391. https://doi.org/10.1021/acs.est.8b04849

Ašmonaitė G, Sundh H, Asker N, Carney Almroth BM (2018b) Rainbow trout maintain intestinal transport and barrier functions following exposure to polystyrene microplastics. Environ Sci Technol 52(24):14392–14401. https://doi.org/10.1021/acs.est.8b04848

Au SY, Bruce TF, Bridges WC, Klaine SJ (2015) Responses of *Hyalella azteca* to acute and chronic microplastic exposures. Environ Toxicol Chem 34(11):2564–2572. https://doi.org/10.1002/etc.3093

Avio CG, Gorbi S, Milan M, Benedetti M, Fattorini D, d'Errico G, Pauletto M, Bargelloni L, Regoli F (2015) Pollutants bioavailability and toxicological risk from microplastics to marine mussels. Environ Pollut 198:211–222. https://doi.org/10.1016/j.envpol.2014.12.021

Balbi T, Camisassi G, Montagna M, Fabbri R, Franzellitti S, Carbone C, Dawson K, Canesi L (2017) Impact of cationic polystyrene nanoparticles (PS-NH_2) on early embryo development of *Mytilus galloprovincialis*: effects on shell formation. Chemosphere 186:1–9. https://doi.org/10.1016/j.chemosphere.2017.07.120

Bank MS, Hansson SV (2019) The plastic cycle: a novel and holistic paradigm for the Anthropocene. Environ Sci Technol 53(13):7177–7179. https://doi.org/10.1021/acs.est.9b02942

Bank MS, Hansson SV (2021) The microplastic cycle: An introduction to a complex issue. In: Bank MS (ed) Microplastic in the environment: pattern and process. Springer, Amsterdam

Barboza LGA, Vieira LR, Branco V, Figueiredo N, Carvalho F, Carvalho C, Guilhermino L (2018) Microplastics cause neurotoxicity, oxidative damage and energy-related changes and interact with the bioaccumulation of mercury in the European seabass, Dicentrarchus labrax (Linnaeus, 1758). Aquat Toxicol 195:49–57. https://doi.org/10.1016/j.aquatox.2017.12.008

Batel A, Borchert F, Reinwald H, Erdinger L, Braunbeck T (2018) Microplastic accumulation patterns and transfer of benzo[a]pyrene to adult zebrafish (*Danio rerio*) gills and zebrafish embryos. Environ Pollut 235:918–930. https://doi.org/10.1016/j.envpol.2018.01.028

Baudrimont M, Arini A, Guégan C, Venel Z, Gigault J, Pedrono B, Prunier J, Maurice L, Ter Halle A, Feurtet-Mazel A (2020) Ecotoxicity of polyethylene nanoplastics from the North Atlantic oceanic gyre on freshwater and marine organisms (microalgae and filter-feeding bivalves). Environ Sci Pollut Res 27(4):3746–3755. https://doi.org/10.1007/s11356-019-04668-3

Beiras R, Bellas J, Cachot J, Cormier B, Cousin X, Engwall M, Gambardella C, Garaventa F, Keiter S, Le Bihanic F, López-Ibáñez S, Piazza V, Rial D, Tato T, Vidal-Liñán L (2018) Ingestion and contact with polyethylene microplastics does not cause acute toxicity on marine zooplankton. J Hazard Mater 360:452–460. https://doi.org/10.1016/j.jhazmat.2018.07.101

Bellingeri A, Bergami E, Grassi G, Faleri C, Redondo-Hasselerharm P, Koelmans AA, Corsi I (2019) Combined effects of nanoplastics and copper on the freshwater alga *Raphidocelis subcapitata*. Aquat Toxicol 210:179–187. https://doi.org/10.1016/j.aquatox.2019.02.022

Bergami E, Bocci E, Vannuccini ML, Monopoli M, Salvati A, Dawson KA, Corsi I (2016) Nano-sized polystyrene affects feeding, behavior and physiology of brine shrimp *Artemia franciscana* larvae. Ecotoxicol Environ Saf 123:18–25. https://doi.org/10.1016/j.ecoenv.2015.09.021

Bergami E, Pugnalini S, Vannuccini ML, Manfra L, Faleri C, Savorelli F, Dawson KA, Corsi I (2017) Long-term toxicity of surface-charged polystyrene nanoplastics to marine planktonic species *Dunaliella tertiolecta* and *Artemia franciscana*. Aquat Toxicol 189:159–169. https://doi.org/10.1016/j.aquatox.2017.06.008

Besseling E, Wegner A, Foekema EM, van den Heuvel-Greve MJ, Koelmans AA (2013) Effects of microplastic on fitness and PCB bioaccumulation by the lugworm *Arenicola marina* (L.). Environ Sci Technol 47(1):593–600. https://doi.org/10.1021/es302763x

Besseling E, Wang B, Lürling M, Koelmans AA (2014) Nanoplastic affects growth of *S. obliquus* and reproduction of *D. magna*. Environ Sci Technol 48(20):12336–12343. https://doi.org/10.1021/es503001d

Besseling E, Foekema EM, van den Heuvel-Greve MJ, Koelmans AA (2017) The effect of microplastic on the uptake of chemicals by the lugworm *Arenicola marina* (L.) under environmentally relevant exposure conditions. Environ Sci Technol 51(15):8795–8804. https://doi.org/10.1021/acs.est.7b02286

Besseling E, Redondo-Hasselerharm P, Foekema EM, Koelmans AA (2019) Quantifying ecological risks of aquatic micro- and nanoplastic. Crit Rev Environ Sci Technol 49(1):32–80. https://doi.org/10.1080/10643389.2018.1531688

Bhargava S, Chen Lee SS, Min Ying LS, Neo ML, Lay-Ming Teo S, Valiyaveettil S (2018) Fate of nanoplastics in marine larvae: a case study using barnacles, *Amphibalanus amphitrite*. ACS Sustain Chem Eng 6(5):6932–6940. https://doi.org/10.1021/acssuschemeng.8b00766

Blarer P, Burkhardt-Holm P (2016) Microplastics affect assimilation efficiency in the freshwater amphipod *Gammarus fossarum*. Environ Sci Pollut Res 23(23):23522–23532. https://doi.org/10.1007/s11356-016-7584-2

Booth AM, Sørensen L (2020) Microplastic fate and impacts in the environment. In: Rocha-Santos T, Costa M, Mouneyrac C (eds) Handbook of microplastics in the environment. Springer, pp 1–24. https://doi.org/10.1007/978-3-030-10618-8_29-1

Booth AM, Hansen BH, Frenzel M, Johnsen H, Altin D (2016) Uptake and toxicity of methylmethacrylate-based nanoplastic particles in aquatic organisms: ecotoxicity and uptake of nanoplastic particles. Environ Toxicol Chem 35(7):1641–1649. https://doi.org/10.1002/etc.3076

Bosker T, Olthof G, Vijver MG, Baas J, Barmentlo SH (2019) Significant decline of *Daphnia magna* population biomass due to microplastic exposure. Environ Pollut 250:669–675. https://doi.org/10.1016/j.envpol.2019.04.067

Botterell ZLR, Beaumont N, Dorrington T, Steinke M, Thompson RC, Lindeque PK (2018) Bioavailability and effects of microplastics on marine zooplankton: a review. Environ Pollut. https://doi.org/10.1016/j.envpol.2018.10.065

Bour A, Haarr A, Keiter S, Hylland K (2018) Environmentally relevant microplastic exposure affects sediment-dwelling bivalves. Environ Pollut 236:652–660. https://doi.org/10.1016/j.envpol.2018.02.006

Brandts I, Teles M, Tvarijonaviciute A, Pereira ML, Martins MA, Tort L, Oliveira M (2018a) Effects of polymethylmethacrylate nanoplastics on *Dicentrarchus labrax*. Genomics 110(6):435–441. https://doi.org/10.1016/j.ygeno.2018.10.006

Brandts I, Teles M, Gonçalves AP, Barreto A, Franco-Martinez L, Tvarijonaviciute A, Martins MA, Soares AMVM, Tort L, Oliveira M (2018b) Effects of nanoplastics on *Mytilus galloprovincialis* after individual and combined exposure with carbamazepine. Sci Total Environ 643:775–784. https://doi.org/10.1016/j.scitotenv.2018.06.257

Bråte ILN, Blázquez M, Brooks SJ, Thomas KV (2018) Weathering impacts the uptake of polyethylene microparticles from toothpaste in Mediterranean mussels (*M. galloprovincialis*). Sci Total Environ 626:1310–1318. https://doi.org/10.1016/j.scitotenv.2018.01.141

Browne MA, Niven SJ, Galloway TS, Rowland SJ, Thompson RC (2013) Microplastic moves pollutants and additives to worms, reducing functions linked to health and biodiversity. Curr Biol 23(23):2388–2392. https://doi.org/10.1016/j.cub.2013.10.012

Bruck S, Ford AT (2018) Chronic ingestion of polystyrene microparticles in low doses has no effect on food consumption and growth to the intertidal amphipod, Echinogammarus marinus? Environmental Pollution (Barking Essex: 1987) 233:1125–1130. https://doi.org/10.1016/j.envpol.2017.10.015

Bürkner P-C (2017) Brms: An R package for Bayesian multilevel models using Stan. J Stat Softw 80(1):1–28. https://doi.org/10.18637/jss.v080.i01

Bürkner P-C (2018) Advanced Bayesian multilevel modeling with the R package brms. R J 10(1):395–411

Burns EE, Boxall ABA (2018) Microplastics in the aquatic environment: evidence for or against adverse impacts and major knowledge gaps: microplastics in the environment. Environ Toxicol Chem 37(11):2776–2796. https://doi.org/10.1002/etc.4268

Canniff PM, Hoang TC (2018) Microplastic ingestion by *Daphnia magna* and its enhancement on algal growth. Sci Total Environ 633:500–507. https://doi.org/10.1016/j.scitotenv.2018.03.176

Cao D, Wang X, Luo X, Liu G, Zheng H (2017) Effects of polystyrene microplastics on the fitness of earthworms in an agricultural soil. IOP Conf Ser Earth Environ Sci 61:012148. https://doi.org/10.1088/1755-1315/61/1/012148

Capolupo M, Franzellitti S, Valbonesi P, Lanzas CS, Fabbri E (2018) Uptake and transcriptional effects of polystyrene microplastics in larval stages of the Mediterranean mussel *Mytilus galloprovincialis*. Environ Pollut 241:1038–1047. https://doi.org/10.1016/j.envpol.2018.06.035

Caruso G, Pedà C, Cappello S, Leonardi M, La Ferla R, Lo Giudice A, Maricchiolo G, Rizzo C, Maimone G, Rappazzo AC, Genovese L, Romeo T (2018) Effects of microplastics on trophic parameters, abundance and metabolic activities of seawater and fish gut bacteria in mesocosm conditions. Environ Sci Pollut Res 25(30):30067–30083. https://doi.org/10.1007/s11356-018-2926-x

Casado MP, Macken A, Byrne HJ (2013) Ecotoxicological assessment of silica and polystyrene nanoparticles assessed by a multitrophic test battery. Environ Int 51:97–105. https://doi.org/10.1016/j.envint.2012.11.001

Cedervall T, Hansson L-A, Lard M, Frohm B, Linse S (2012) Food chain transport of nanoparticles affects behaviour and fat metabolism in fish. PLoS One 7(2):e32254. https://doi.org/10.1371/journal.pone.0032254

Chae Y, Kim D, Kim SW, An Y-J (2018) Trophic transfer and individual impact of nano-sized polystyrene in a four-species freshwater food chain. Sci Rep 8(1):284. https://doi.org/10.1038/s41598-017-18849-y

Chen Q, Gundlach M, Yang S, Jiang J, Velki M, Yin D, Hollert H (2017) Quantitative investigation of the mechanisms of microplastics and nanoplastics toward zebrafish larvae locomotor activity. Sci Total Environ 584–585:1022–1031. https://doi.org/10.1016/j.scitotenv.2017.01.156

Choi JS, Jung Y-J, Hong N-H, Hong SH, Park J-W (2018) Toxicological effects of irregularly shaped and spherical microplastics in a marine teleost, the sheepshead minnow (*Cyprinodon variegatus*). Mar Pollut Bull 129(1):231–240. https://doi.org/10.1016/j.marpolbul.2018.02.039

Cole M, Galloway TS (2015) Ingestion of nanoplastics and microplastics by Pacific oyster larvae. Environ Sci Technol 49(24):14625–14632. https://doi.org/10.1021/acs.est.5b04099

Cole M, Lindeque P, Halsband C, Galloway TS (2011) Microplastics as contaminants in the marine environment: a review. Mar Pollut Bull 62(12):2588–2597. https://doi.org/10.1016/j.marpolbul.2011.09.025

Cole M, Lindeque P, Fileman E, Halsband C, Goodhead R, Moger J, Galloway TS (2013) Microplastic ingestion by zooplankton. Environ Sci Technol 47(12):6646–6655. https://doi.org/10.1021/es400663f

Cole M, Lindeque P, Fileman E, Halsband C, Galloway TS (2015) The impact of polystyrene microplastics on feeding, function and fecundity in the marine copepod *Calanus helgolandicus*. Environ Sci Technol 49(2):1130–1137. https://doi.org/10.1021/es504525u

Cole M, Coppock R, Lindeque PK, Altin D, Reed S, Pond DW, Sørensen L, Galloway TS, Booth AM (2019) Effects of nylon microplastic on feeding, lipid accumulation, and moulting in a coldwater copepod. Environ Sci Technol 53(12):7075–7082. https://doi.org/10.1021/acs.est.9b01853

Collard F, Gasperi J, Gabrielsen GW, Tassin B (2019) Plastic particle ingestion by wild freshwater fish: a critical review. Environ Sci Technol 53(22):12974–12988. https://doi.org/10.1021/acs.est.9b03083

Cong Y, Jin F, Tian M, Wang J, Shi H, Wang Y, Mu J (2019) Ingestion, egestion and post-exposure effects of polystyrene microspheres on marine medaka (*Oryzias melastigma*). Chemosphere 228:93–100. https://doi.org/10.1016/j.chemosphere.2019.04.098

Connors KA, Dyer SD, Belanger SE (2017) Advancing the quality of environmental microplastic research: advancing the quality of environmental microplastic research. Environ Toxicol Chem 36(7):1697–1703. https://doi.org/10.1002/etc.3829

Cowger W, Booth AM, Hamilton BM, Thaysen C, Primpke S, Munno K, Lusher AL, Dehaut A, Vaz VP, Liboiron M, Devriese LI, Hermabessiere L, Rochman C, Athey SN, Lynch JM, De Frond H, Gray A, Jones OAH, Brander S et al (2020) Reporting guidelines to increase the reproducibility and comparability of research on microplastics. Appl Spectrosc 74(9):1066–1077. https://doi.org/10.1177/0003702820930292

Critchell K, Hoogenboom MO (2018) Effects of microplastic exposure on the body condition and behaviour of planktivorous reef fish (Acanthochromis polyacanthus). PLoS One 13(3):e0193308. https://doi.org/10.1371/journal.pone.0193308

Cui R, Kim SW, An Y-J (2017) Polystyrene nanoplastics inhibit reproduction and induce abnormal embryonic development in the freshwater crustacean *Daphnia galeata*. Sci Rep 7(1):1–10. https://doi.org/10.1038/s41598-017-12299-2

de Felice B, Sabatini V, Antenucci S, Gattoni G, Santo N, Bacchetta R, Ortenzi MA, Parolini M (2019) Polystyrene microplastics ingestion induced behavioral effects to the cladoceran *Daphnia magna*. Chemosphere 231:423–431. https://doi.org/10.1016/j.chemosphere.2019.05.115

de Ruijter VN, Redondo-Hasselerharm PE, Gouin T, Koelmans AA (2020) Quality criteria for microplastic effect studies in the context of risk assessment: a critical review. Environ Sci Technol 54(19):11692–11705. https://doi.org/10.1021/acs.est.0c03057

de Sá LC, Luís LG, Guilhermino L (2015) Effects of microplastics on juveniles of the common goby (*Pomatoschistus microps*): confusion with prey, reduction of the predatory performance and efficiency, and possible influence of developmental conditions. Environ Pollut 196:359–362. https://doi.org/10.1016/j.envpol.2014.10.026

de Sá LC, Oliveira M, Ribeiro F, Rocha TL, Futter MN (2018) Studies of the effects of microplastics on aquatic organisms: what do we know and where should we focus our efforts in the future? Sci Total Environ 645:1029–1039. https://doi.org/10.1016/j.scitotenv.2018.07.207

Della Torre C, Bergami E, Salvati A, Faleri C, Cirino P, Dawson KA, Corsi I (2014) Accumulation and embryotoxicity of polystyrene nanoparticles at early stage of development of sea urchin embryos *Paracentrotus lividus*. Environ Sci Technol 48(20):12302–12311. https://doi.org/10.1021/es502569w

Détrée C, Gallardo-Escárate C (2017) Polyethylene microbeads induce transcriptional responses with tissue-dependent patterns in the mussel *Mytilus galloprovincialis*. J Molluscan Stud 83(2):220–225. https://doi.org/10.1093/mollus/eyx005

Détrée C, Gallardo-Escárate C (2018) Single and repetitive microplastics exposures induce immune system modulation and homeostasis alteration in the edible mussel *Mytilus galloprovincialis*. Fish Shellfish Immunol 83:52–60. https://doi.org/10.1016/j.fsi.2018.09.018

Diepens NJ, Koelmans AA (2018) Accumulation of plastic debris and associated contaminants in aquatic food webs. Environ Sci Technol 52(15):8510–8520. https://doi.org/10.1021/acs.est.8b02515

Ding J, Zhang S, Razanajatovo RM, Zou H, Zhu W (2018) Accumulation, tissue distribution, and biochemical effects of polystyrene microplastics in the freshwater fish red tilapia (*Oreochromis niloticus*). Environ Pollut 238:1–9. https://doi.org/10.1016/j.envpol.2018.03.001

EFSA CONTAM Panel (EFSA Panel on Contaminants in the Food Chain) (2016) Statement on the presence of microplastics and nanoplastics in food, with particular focus on seafood. EFSA J 14(6):4501. 30 pp

Everaert G, Van Cauwenberghe L, De Rijcke M, Koelmans AA, Mees J, Vandegehuchte M, Janssen CR (2018) Risk assessment of microplastics in the ocean: modelling approach and first conclusions. Environ Pollut 242:1930–1938. https://doi.org/10.1016/j.envpol.2018.07.069

Fadare OO, Wan B, Guo L-H, Xin Y, Qin W, Yang Y (2019) Humic acid alleviates the toxicity of polystyrene nanoplastic particles to *Daphnia magna*. Environ Sci Nano 6(5):1466–1477. https://doi.org/10.1039/C8EN01457D

Farkas J, Booth AM (2017) Are fluorescence-based chlorophyll quantification methods suitable for algae toxicity assessment of carbon nanomaterials? Nanotoxicology 11(4):569–577. https://doi.org/10.1080/17435390.2017.1329953

Ferreira P, Fonte E, Soares ME, Carvalho F, Guilhermino L (2016) Effects of multi-stressors on juveniles of the marine fish *Pomatoschistus microps*: gold nanoparticles, microplastics and temperature. Aquat Toxicol 170:89–103. https://doi.org/10.1016/j.aquatox.2015.11.011

Fonte E, Ferreira P, Guilhermino L (2016) Temperature rise and microplastics interact with the toxicity of the antibiotic cefalexin to juveniles of the common goby (*Pomatoschistus microps*): post-exposure predatory behaviour, acetylcholinesterase activity and lipid peroxidation. Aquat Toxicol 180:173–185. https://doi.org/10.1016/j.aquatox.2016.09.015

Franzellitti S, Capolupo M, Wathsala RHGR, Valbonesi P, Fabbri E (2019) The multixenobiotic resistance system as a possible protective response triggered by microplastic ingestion in Mediterranean mussels (*Mytilus galloprovincialis*): larvae and adult stages. Comp Biochem Physiol C Toxicol Pharmacol 219:50–58. https://doi.org/10.1016/j.cbpc.2019.02.005

Frydkjær CK, Iversen N, Roslev P (2017) Ingestion and egestion of microplastics by the cladoceran Daphnia magna: effects of regular and irregular shaped plastic and sorbed phenanthrene. Bull Environ Contam Toxicol 99(6):655–661. https://doi.org/10.1007/s00128-017-2186-3

Gallo F, Fossi C, Weber R, Santillo D, Sousa J, Ingram I, Nadal A, Romano D (2018) Marine litter plastics and microplastics and their toxic chemicals components: the need for urgent preventive measures. Environ Sci Eur 30(1):13. https://doi.org/10.1186/s12302-018-0139-z

Galloway TS, Cole M, Lewis C (2017) Interactions of microplastic debris throughout the marine ecosystem. Nat Ecol Evol 1(5). https://doi.org/10.1038/s41559-017-0116

Gambardella C, Morgana S, Ferrando S, Bramini M, Piazza V, Costa E, Garaventa F, Faimali M (2017) Effects of polystyrene microbeads in marine planktonic crustaceans. Ecotoxicol Environ Saf 145:250–257. https://doi.org/10.1016/j.ecoenv.2017.07.036

Gambardella C, Morgana S, Bramini M, Rotini A, Manfra L, Migliore L, Piazza V, Garaventa F, Faimali M (2018) Ecotoxicological effects of polystyrene microbeads in a battery of marine organisms belonging to different trophic levels. Mar Environ Res 141:313–321. https://doi.org/10.1016/j.marenvres.2018.09.023

Gardon T, Reisser C, Soyez C, Quillien V, Le Moullac G (2018) Microplastics affect energy balance and gametogenesis in the pearl oyster *Pinctada margaritifera*. Environ Sci Technol 52(9):5277–5286. https://doi.org/10.1021/acs.est.8b00168

Garrido S, Linares M, Campillo JA, Albentosa M (2019) Effect of microplastics on the toxicity of chlorpyrifos to the microalgae *Isochrysis galbana*, clone t-ISO. Ecotoxicol Environ Saf 173:103–109. https://doi.org/10.1016/j.ecoenv.2019.02.020

Gerdes Z, Ogonowski M, Nybom I, Ek C, Adolfsson-Erici M, Barth A, Gorokhova E (2019) Microplastic-mediated transport of PCBs? A depuration study with *Daphnia magna*. PLoS One 14(2). https://doi.org/10.1371/journal.pone.0205378

GESAMP (2019) Guidelines for the monitoring and assessment of plastic litter and microplastics in the ocean. In: Kershaw PJ, Turra A, Galgani F (eds) IMO/FAO/UNESCO-IOC/UNIDO/WMO/IAEA/UN/UNEP/UNDP/ISA Joint Group of Experts on the Scientific Aspects of Marine Environmental Protection, Rep. Stud. GESAMP No. 99, 130p

GESAMP (2020) Proceedings of the GESAMP International Workshop on assessing the risks associated with plastics and microplastics in the marine environment. In: Kershaw PJ, Carney Almroth B, Villarrubia-Gómez P, Koelmans AA, Gouin T (eds) IMO/FAO/UNESCO-IOC/UNIDO/WMO/IAEA/UN/UNEP/UNDP/ISA Joint Group of Experts on the Scientific Aspects of Marine Environmental Protection, Reports to GESAMP No. 103, 68 pp

Gomiero A, Strafella P, Pellini G, Salvalaggio V, Fabi G (2018) Comparative effects of ingested PVC micro particles with and without adsorbed benzo(a)pyrene vs. spiked sediments on the cellular and sub cellular processes of the benthic organism *Hediste diversicolor*. Front Mar Sci 5:99. https://doi.org/10.3389/fmars.2018.00099

Gonçalves C, Martins M, Sobral P, Costa PM, Costa MH (2019) An assessment of the ability to ingest and excrete microplastics by filter-feeders: a case study with the Mediterranean mussel. Environ Pollut 245:600–606. https://doi.org/10.1016/j.envpol.2018.11.038

González-Fernández C, Tallec K, Le Goïc N, Lambert C, Soudant P, Huvet A, Suquet M, Berchel M, Paul-Pont I (2018) Cellular responses of Pacific oyster (*Crassostrea gigas*) gametes exposed in vitro to polystyrene nanoparticles. Chemosphere 208:764–772. https://doi.org/10.1016/j.chemosphere.2018.06.039

González-Fernández C, Toullec J, Lambert C, Le Goïc N, Seoane M, Moriceau B, Huvet A, Berchel M, Vincent D, Courcot L, Soudant P, Paul-Pont I (2019) Do transparent exopolymeric particles (TEP) affect the toxicity of nanoplastics on *Chaetoceros neogracile*? Environ Pollut 250:873–882. https://doi.org/10.1016/j.envpol.2019.04.093

Granby K, Rainieri S, Rasmussen RR, Kotterman MJJ, Sloth JJ, Cederberg TL, Barranco A, Marques A, Larsen BK (2018) The influence of microplastics and halogenated contaminants in feed on toxicokinetics and gene expression in European seabass (*Dicentrarchus labrax*). Environ Res 164:430–443. https://doi.org/10.1016/j.envres.2018.02.035

Gray AD, Weinstein JE (2017) Size- and shape-dependent effects of microplastic particles on adult daggerblade grass shrimp (*Palaemonetes pugio*): uptake and retention of microplastics in grass shrimp. Environ Toxicol Chem. https://doi.org/10.1002/etc.3881

Green DS (2016) Effects of microplastics on European flat oysters, *Ostrea edulis* and their associated benthic communities. Environ Pollut (Barking, Essex: 1987) 216:95–103. https://doi.org/10.1016/j.envpol.2016.05.043

Green DS, Boots B, Sigwart J, Jiang S, Rocha C (2016) Effects of conventional and biodegradable microplastics on a marine ecosystem engineer (*Arenicola marina*) and sediment nutrient cycling. Environ Pollut 208(Part B):426–434. https://doi.org/10.1016/j.envpol.2015.10.010

Green DS, Boots B, O'Connor NE, Thompson R (2017) Microplastics affect the ecological functioning of an important biogenic habitat. Environ Sci Technol 51(1):68–77. https://doi.org/10.1021/acs.est.6b04496

Green DS, Colgan TJ, Thompson RC, Carolan JC (2019) Exposure to microplastics reduces attachment strength and alters the haemolymph proteome of blue mussels (*Mytilus edulis*). Environ Pollut 246:423–434. https://doi.org/10.1016/j.envpol.2018.12.017

Guilhermino L, Vieira LR, Ribeiro D, Tavares AS, Cardoso V, Alves A, Almeida JM (2018) Uptake and effects of the antimicrobial florfenicol, microplastics and their mixtures on freshwater exotic invasive bivalve *Corbicula fluminea*. Sci Total Environ 622–623:1131–1142. https://doi.org/10.1016/j.scitotenv.2017.12.020

Guven O, Bach L, Munk P, Dinh KV, Mariani P, Nielsen TG (2018) Microplastic does not magnify the acute effect of PAH pyrene on predatory performance of a tropical fish (*Lates calcarifer*). Aquat Toxicol 198:287–293. https://doi.org/10.1016/j.aquatox.2018.03.011

Haegerbaeumer A, Mueller M-T, Fueser H, Traunspurger W (2019) Impacts of micro- and nano-sized plastic particles on benthic invertebrates: a literature review and gap analysis. Front Environ Sci:7. https://doi.org/10.3389/fenvs.2019.00017

Hämer J, Gutow L, Köhler A, Saborowski R (2014) Fate of microplastics in the marine isopod *Idotea emarginata*. Environ Sci Technol 48(22):13451–13458. https://doi.org/10.1021/es501385y

Handy RD, van den Brink N, Chappell M, Mühling M, Behra R, Dušinská M, Simpson P, Ahtiainen J, Jha AN, Seiter J, Bednar A, Kennedy A, Fernandes TF, Riediker M (2012) Practical considerations for conducting ecotoxicity test methods with manufactured nanomaterials: what have we learnt so far? Ecotoxicology (London, England) 21(4):933–972. https://doi.org/10.1007/s10646-012-0862-y

Hankins C, Duffy A, Drisco K (2018) Scleractinian coral microplastic ingestion: potential calcification effects, size limits, and retention. Mar Pollut Bull 135:587–593. https://doi.org/10.1016/j.marpolbul.2018.07.067

Hartmann NB, Hüffer T, Thompson RC, Hassellöv M, Verschoor A, Daugaard AE, Rist S, Karlsson T, Brennholt N, Cole M, Herrling MP, Hess MC, Ivleva NP, Lusher AL, Wagner M (2019) Are we speaking the same language? Recommendations for a definition and categorization framework for plastic debris. Environ Sci Technol 53(3):1039–1047. https://doi.org/10.1021/acs.est.8b05297

Heindler FM, Alajmi F, Huerlimann R, Zeng C, Newman SJ, Vamvounis G, van Herwerden L (2017) Toxic effects of polyethylene terephthalate microparticles and Di(2-ethylhexyl)phthalate on the calanoid copepod, *Parvocalanus crassirostris*. Ecotoxicol Environ Saf 141:298–305. https://doi.org/10.1016/j.ecoenv.2017.03.029

Hermabessiere L, Dehaut A, Paul-Pont I, Lacroix C, Jezequel R, Soudant P, Duflos G (2017) Occurrence and effects of plastic additives on marine environments and organisms: a review. Chemosphere 182:781–793. https://doi.org/10.1016/j.chemosphere.2017.05.096

Hermsen E, Mintenig SM, Besseling E, Koelmans AA (2018) Quality criteria for the analysis of microplastic in biota samples: a critical review. Environ Sci Technol 52(18):10230–10240. https://doi.org/10.1021/acs.est.8b01611

Horton AA, Svendsen C, Williams RJ, Spurgeon DJ, Lahive E (2017) Large microplastic particles in sediments of tributaries of the River Thames, UK – abundance, sources and methods for effective quantification. Mar Pollut Bull 114(1):218–226. https://doi.org/10.1016/j.marpolbul.2016.09.004

Horton AA, Jürgens MD, Lahive E, van Bodegom PM, Vijver MG (2018) The influence of exposure and physiology on microplastic ingestion by the freshwater fish *Rutilus rutilus* (roach) in the river Thames, UK. Environ Pollut 236:188–194. https://doi.org/10.1016/j.envpol.2018.01.044

Huerta Lwanga E, Gertsen H, Gooren H, Peters P, Salánki T, van der Ploeg M, Besseling E, Koelmans AA, Geissen V (2016) Microplastics in the terrestrial ecosystem: implications for *Lumbricus terrestris* (Oligochaeta, Lumbricidae). Environ Sci Technol 50(5):2685–2691. https://doi.org/10.1021/acs.est.5b05478

Imhof HK, Laforsch C (2016) Hazardous or not – are adult and juvenile individuals of *Potamopyrgus antipodarum* affected by non-buoyant microplastic particles? Environ Pollut 218:383–391. https://doi.org/10.1016/j.envpol.2016.07.017

Imhof HK, Rusek J, Thiel M, Wolinska J, Laforsch C (2017) Do microplastic particles affect *Daphnia magna* at the morphological, life history and molecular level? PLoS One 12(11). https://doi.org/10.1371/journal.pone.0187590

Jacob H, Gilson A, Lanctôt C, Besson M, Metian M, Lecchini D (2019) No effect of polystyrene microplastics on foraging activity and survival in a post-larvae coral-reef fish, *Acanthurus triostegus*. Bull Environ Contam Toxicol 102(4):457–461. https://doi.org/10.1007/s00128-019-02587-0

Jemec A, Horvat P, Kunej U, Bele M, Kržan A (2016) Uptake and effects of microplastic textile fibers on freshwater crustacean *Daphnia magna*. Environ Pollut 219:201–209. https://doi.org/10.1016/j.envpol.2016.10.037

Jeong C-B, Won E-J, Kang H-M, Lee M-C, Hwang D-S, Hwang U-K, Zhou B, Souissi S, Lee S-J, Lee J-S (2016) Microplastic size-dependent toxicity, oxidative stress induction, and p-JNK and p-p38 activation in the monogonont rotifer (*Brachionus koreanus*). Environ Sci Technol 50(16):8849–8857. https://doi.org/10.1021/acs.est.6b01441

Jeong C-B, Kang H-M, Lee M-C, Kim D-H, Han J, Hwang D-S, Souissi S, Lee S-J, Shin K-H, Park HG, Lee J-S (2017) Adverse effects of microplastics and oxidative stress-induced MAPK/Nrf2 pathway-mediated defense mechanisms in the marine copepod *Paracyclopina nana*. Sci Rep 7. https://doi.org/10.1038/srep41323

Jin Y, Xia J, Pan Z, Yang J, Wang W, Fu Z (2018) Polystyrene microplastics induce microbiota dysbiosis and inflammation in the gut of adult zebrafish. Environ Pollut 235:322–329. https://doi.org/10.1016/j.envpol.2017.12.088

Jovanović B, Gökdağ K, Güven O, Emre Y, Whitley EM, Kideys AE (2018) Virgin microplastics are not causing imminent harm to fish after dietary exposure. Mar Pollut Bull 130:123–131. https://doi.org/10.1016/j.marpolbul.2018.03.016

Judy JD, Williams M, Gregg A, Oliver D, Kumar A, Kookana R, Kirby JK (2019) Microplastics in municipal mixed-waste organic outputs induce minimal short to long-term toxicity in key terrestrial biota. Environ Pollut 252:522–531. https://doi.org/10.1016/j.envpol.2019.05.027

Kallenbach EMF, Rødland ES, Buenaventura NT, Hurley R (2021) Microplastics in terrestrial and freshwater environments. In: Bank MS (ed) Microplastic in the environment: pattern and process. Springer, Amsterdam

Kaposi KL, Mos B, Kelaher BP, Dworjanyn SA (2014) Ingestion of microplastic has limited impact on a marine larva. Environ Sci Technol 48(3):1638–1645. https://doi.org/10.1021/es404295e

Karami A, Romano N, Galloway T, Hamzah H (2016) Virgin microplastics cause toxicity and modulate the impacts of phenanthrene on biomarker responses in African catfish (*Clarias gariepinus*). Environ Res 151:58–70. https://doi.org/10.1016/j.envres.2016.07.024

Karami A, Groman DB, Wilson SP, Ismail P, Neela VK (2017) Biomarker responses in zebrafish (*Danio rerio*) larvae exposed to pristine low-density polyethylene fragments. Environ Pollut 223:466–475. https://doi.org/10.1016/j.envpol.2017.01.047

Kim SW, An Y-J (2019) Soil microplastics inhibit the movement of springtail species. Environ Int 126:699–706. https://doi.org/10.1016/j.envint.2019.02.067

Kim D, Chae Y, An Y-J (2017) Mixture toxicity of nickel and microplastics with different functional groups on *Daphnia magna*. Environ Sci Technol 51(21):12852–12858. https://doi.org/10.1021/acs.est.7b03732

Koelmans AA, Bakir A, Burton GA, Janssen CR (2016) Microplastic as a vector for chemicals in the aquatic environment: critical review and model-supported reinterpretation of empirical studies. Environ Sci Technol 50(7):3315–3326. https://doi.org/10.1021/acs.est.5b06069

Koelmans AA, Mohamed Nor NH, Hermsen E, Kooi M, Mintenig SM, De France J (2019) Microplastics in freshwaters and drinking water: critical review and assessment of data quality. Water Res 155:410–422. https://doi.org/10.1016/j.watres.2019.02.054

Kögel T, Bjorøy Ø, Toto B, Bienfait AM, Sanden M (2020) Micro- and nanoplastic toxicity on aquatic life: determining factors. Sci Total Environ 709:136050. https://doi.org/10.1016/j.scitotenv.2019.136050

Kokalj AJ, Kunej U, Skalar T (2018) Screening study of four environmentally relevant microplastic pollutants: uptake and effects on *Daphnia magna* and *Artemia franciscana*. Chemosphere 208:522–529. https://doi.org/10.1016/j.chemosphere.2018.05.172

Kooi M, Koelmans AA (2019) Simplifying microplastic via continuous probability distributions for size, shape, and density. Environ Sci Technol Lett 6(9):551–557. https://doi.org/10.1021/acs.estlett.9b00379

Kühn S, Bravo Rebolledo EL, van Franeker JA (2015) Deleterious effects of litter on marine life. In: Bergmann M, Gutow L, Klages M (eds) Marine anthropogenic litter. Springer, Berlin, pp 75–116. https://doi.org/10.1007/978-3-319-16510-3_4

Lagarde F, Olivier O, Zanella M, Daniel P, Hiard S, Caruso A (2016) Microplastic interactions with freshwater microalgae: hetero-aggregation and changes in plastic density appear strongly dependent on polymer type. Environ Pollut 215:331–339. https://doi.org/10.1016/j.envpol.2016.05.006

Lambert S, Wagner M (2016) Characterisation of nanoplastics during the degradation of polystyrene. Chemosphere 145:265–268. https://doi.org/10.1016/j.chemosphere.2015.11.078

Law KL, Thompson RC (2014) Microplastics in the seas. Science 345(6193):144–145. https://doi.org/10.1126/science.1254065

Lee K-W, Shim WJ, Kwon OY, Kang J-H (2013) Size-dependent effects of micro polystyrene particles in the marine copepod *Tigriopus japonicus*. Environ Sci Technol 47(19):11278–11283. https://doi.org/10.1021/es401932b

Lei L, Liu M, Song Y, Lu S, Hu J, Cao C, Xie B, Shi H, He D (2018a) Polystyrene (nano)microplastics cause size-dependent neurotoxicity, oxidative damage and other adverse effects in *Caenorhabditis elegans*. Environ Sci Nano 5(8):2009–2020. https://doi.org/10.1039/C8EN00412A

Lei L, Wu S, Lu S, Liu M, Song Y, Fu Z, Shi H, Raley-Susman KM, He D (2018b) Microplastic particles cause intestinal damage and other adverse effects in zebrafish *Danio rerio* and nematode *Caenorhabditis elegans*. Sci Total Environ 619–620:1–8. https://doi.org/10.1016/j.scitotenv.2017.11.103

LeMoine CMR, Kelleher BM, Lagarde R, Northam C, Elebute OO, Cassone BJ (2018) Transcriptional effects of polyethylene microplastics ingestion in developing zebrafish (*Danio rerio*). Environ Pollut 243:591–600. https://doi.org/10.1016/j.envpol.2018.08.084

Leung J, Chan KYK (2018) Microplastics reduced posterior segment regeneration rate of the polychaete *Perinereis aibuhitensis*. Mar Pollut Bull 129(2):782–786. https://doi.org/10.1016/j.marpolbul.2017.10.072

Lin W, Jiang R, Xiong Y, Wu J, Xu J, Zheng J, Zhu F, Ouyang G (2019a) Quantification of the combined toxic effect of polychlorinated biphenyls and nano-sized polystyrene on *Daphnia magna*. J Hazard Mater 364:531–536. https://doi.org/10.1016/j.jhazmat.2018.10.056

Lin W, Jiang R, Hu S, Xiao X, Wu J, Wei S, Xiong Y, Ouyang G (2019b) Investigating the toxicities of different functionalized polystyrene nanoplastics on *Daphnia magna*. Ecotoxicol Environ Saf 180:509–516. https://doi.org/10.1016/j.ecoenv.2019.05.036

Liu Z, Cai M, Yu P, Chen M, Wu D, Zhang M, Zhao Y (2018) Age-dependent survival, stress defense, and AMPK in *Daphnia pulex* after short-term exposure to a polystyrene nanoplastic. Aquat Toxicol (Amsterdam, Netherlands) 204:1–8. https://doi.org/10.1016/j.aquatox.2018.08.017

Liu Z, Yu P, Cai M, Wu D, Zhang M, Huang Y, Zhao Y (2019) Polystyrene nanoplastic exposure induces immobilization, reproduction, and stress defense in the freshwater cladoceran *Daphnia pulex*. Chemosphere 215:74–81. https://doi.org/10.1016/j.chemosphere.2018.09.176

Lohmann R (2017) Microplastics are not important for the cycling and bioaccumulation of organic pollutants in the oceans-but should microplastics be considered POPs themselves? Integr Environ Assess Manag 13(3):460–465. https://doi.org/10.1002/ieam.1914

Long M, Paul-Pont I, Hégaret H, Moriceau B, Lambert C, Huvet A, Soudant P (2017) Interactions between polystyrene microplastics and marine phytoplankton lead to species-specific hetero-aggregation. Environ Pollut 228:454–463. https://doi.org/10.1016/j.envpol.2017.05.047

Lowry GV, Espinasse BP, Badireddy AR, Richardson CJ, Reinsch BC, Bryant LD, Bone AJ, Deonarine A, Chae S, Therezien M, Colman BP, Hsu-Kim H, Bernhardt ES, Matson CW, Wiesner MR (2012) Long-term transformation and fate of manufactured Ag nanoparticles in a simulated large scale freshwater emergent wetland. Environ Sci Technol 46(13):7027–7036. https://doi.org/10.1021/es204608d

Lu K, Qiao R, An H, Zhang Y (2018) Influence of microplastics on the accumulation and chronic toxic effects of cadmium in zebrafish (*Danio rerio*). Chemosphere 202:514–520. https://doi.org/10.1016/j.chemosphere.2018.03.145

Luan L, Wang X, Zheng H, Liu L, Luo X, Li F (2019) Differential toxicity of functionalized polystyrene microplastics to clams (*Meretrix meretrix*) at three key development stages of life history. Mar Pollut Bull 139:346–354. https://doi.org/10.1016/j.marpolbul.2019.01.003

Luís LG, Ferreira P, Fonte E, Oliveira M, Guilhermino L (2015) Does the presence of microplastics influence the acute toxicity of chromium(VI) to early juveniles of the common goby (*Pomatoschistus microps*)? A study with juveniles from two wild estuarine populations. Aquat Toxicol 164:163–174. https://doi.org/10.1016/j.aquatox.2015.04.018

Lundebye A-K, Lusher AL, Bank MS (2021) Microplastics in the ocean and seafood: implications for food security. In: Bank MS (ed) Microplastic in the environment: pattern and process. Springer, Amsterdam, pp xx–xx

Luo H, Xiang Y, He D, Li Y, Zhao Y, Wang S, Pan X (2019) Leaching behavior of fluorescent additives from microplastics and the toxicity of leachate to *Chlorella vulgaris*. Sci Total Environ 678:1–9. https://doi.org/10.1016/j.scitotenv.2019.04.401

Lusher A, Hollman P, Mendoza-Hill J (2017) Microplastics in fisheries and aquaculture: status of knowledge on their occurrence and implications for aquatic organisms and food safety. FAO

Ma Y, Huang A, Cao S, Sun F, Wang L, Guo H, Ji R (2016) Effects of nanoplastics and microplastics on toxicity, bioaccumulation, and environmental fate of phenanthrene in fresh water. Environ Pollut (Barking Essex: 1987) 219:166–173. https://doi.org/10.1016/j.envpol.2016.10.061

Magni S, Gagné F, André C, Della Torre C, Auclair J, Hanana H, Parenti CC, Bonasoro F, Binelli A (2018) Evaluation of uptake and chronic toxicity of virgin polystyrene microbeads in freshwater zebra mussel *Dreissena polymorpha* (Mollusca: Bivalvia). Sci Total Environ 631–632:778–788. https://doi.org/10.1016/j.scitotenv.2018.03.075

Mak CW, Ching-Fong Yeung K, Chan KM (2019) Acute toxic effects of polyethylene microplastic on adult zebrafish. Ecotoxicol Environ Saf 182:109442. https://doi.org/10.1016/j.ecoenv.2019.109442

Malinich TD, Chou N, Sepúlveda MS, Höök TO (2018) No evidence of microplastic impacts on consumption or growth of larval *Pimephales promelas*: no microplastic impacts on consumption or growth of larvae. Environ Toxicol Chem 37(11):2912–2918. https://doi.org/10.1002/etc.4257

Manfra L, Rotini A, Bergami E, Grassi G, Faleri C, Corsi I (2017) Comparative ecotoxicity of polystyrene nanoparticles in natural seawater and reconstituted seawater using the rotifer *Brachionus plicatilis*. Ecotoxicol Environ Saf 145:557–563. https://doi.org/10.1016/j.ecoenv.2017.07.068

Mao Y, Ai H, Chen Y, Zhang Z, Zeng P, Kang L, Li W, Gu W, He Q, Li H (2018) Phytoplankton response to polystyrene microplastics: perspective from an entire growth period. Chemosphere 208:59–68. https://doi.org/10.1016/j.chemosphere.2018.05.170

Mattsson K, Ekvall MT, Hansson L-A, Linse S, Malmendal A, Cedervall T (2015) Altered behavior, physiology, and metabolism in fish exposed to polystyrene nanoparticles. Environ Sci Technol 49(1):553–561. https://doi.org/10.1021/es5053655

Mattsson K, Johnson EV, Malmendal A, Linse S, Hansson L-A, Cedervall T (2017) Brain damage and behavioural disorders in fish induced by plastic nanoparticles delivered through the food chain. Sci Rep 7(1):11452. https://doi.org/10.1038/s41598-017-10813-0

Mazurais D, Ernande B, Quazuguel P, Severe A, Huelvan C, Madec L, Mouchel O, Soudant P, Robbens J, Huvet A, Zambonino-Infante J (2015) Evaluation of the impact of polyethylene microbeads ingestion in European sea bass (*Dicentrarchus labrax*) larvae. Mar Environ Res 112:78–85. https://doi.org/10.1016/j.marenvres.2015.09.009

Messinetti S, Mercurio S, Parolini M, Sugni M, Pennati R (2018) Effects of polystyrene microplastics on early stages of two marine invertebrates with different feeding strategies. Environ Pollut 237:1080–1087. https://doi.org/10.1016/j.envpol.2017.11.030

Mudunkotuwa A, Grassian H (2011) The devil is in the details (or the surface): impact of surface structure and surface energetics on understanding the behavior of nanomaterials in the environment. J Environ Monit 13(5):1135–1144. https://doi.org/10.1039/C1EM00002K

Nasser F, Lynch I (2016) Secreted protein eco-corona mediates uptake and impacts of polystyrene nanoparticles on *Daphnia magna*. J Proteome 137:45–51. https://doi.org/10.1016/j.jprot.2015.09.005

Nel AE, Mädler L, Velegol D, Xia T, Hoek EMV, Somasundaran P, Klaessig F, Castranova V, Thompson M (2009) Understanding biophysicochemical interactions at the nano–bio interface. Nat Mater 8(7):543–557. https://doi.org/10.1038/nmat2442

Nolte TM, Hartmann NB, Kleijn JM, Garnæs J, van de Meent D, Jan Hendriks A, Baun A (2017) The toxicity of plastic nanoparticles to green algae as influenced by surface modification, medium hardness and cellular adsorption. Aquat Toxicol 183:11–20. https://doi.org/10.1016/j.aquatox.2016.12.005

Ogonowski M, Schür C, Jarsén Å, Gorokhova E (2016) The effects of natural and anthropogenic microparticles on individual fitness in *Daphnia magna*. PLoS One 11(5):e0155063. https://doi.org/10.1371/journal.pone.0155063

Oliveira M, Ribeiro A, Hylland K, Guilhermino L (2013) Single and combined effects of microplastics and pyrene on juveniles (0+ group) of the common goby *Pomatoschistus microps* (Teleostei, Gobiidae). Ecol Indic 34:641–647. https://doi.org/10.1016/j.ecolind.2013.06.019

Oliveira P, Barboza LGA, Branco V, Figueiredo N, Carvalho C, Guilhermino L (2018) Effects of microplastics and mercury in the freshwater bivalve *Corbicula fluminea* (Müller, 1774): filtration rate, biochemical biomarkers and mercury bioconcentration. Ecotoxicol Environ Saf 164:155–163. https://doi.org/10.1016/j.ecoenv.2018.07.062

Ott WR (1990) A physical explanation of the lognormality of pollutant concentrations. J Air Waste Manage Assoc 40(10):1378–1383. https://doi.org/10.1080/10473289.1990.10466789

Pacheco A, Martins A, Guilhermino L (2018) Toxicological interactions induced by chronic exposure to gold nanoparticles and microplastics mixtures in *Daphnia magna*. Sci Total Environ 628–629:474–483. https://doi.org/10.1016/j.scitotenv.2018.02.081

Paul-Pont I, Lacroix C, González Fernández C, Hégaret H, Lambert C, Le Goïc N, Frère L, Cassone A-L, Sussarellu R, Fabioux C, Guyomarch J, Albentosa M, Huvet A, Soudant P (2016) Exposure of marine mussels Mytilus spp. to polystyrene microplastics: toxicity and influence on fluoranthene bioaccumulation. Environ Pollut 216:724–737. https://doi.org/10.1016/j.envpol.2016.06.039

Paul-Pont I, Tallec K, Gonzalez-Fernandez C, Lambert C, Vincent D, Mazurais D, Zambonino-Infante J-L, Brotons G, Lagarde F, Fabioux C, Soudant P, Huvet A (2018) Constraints and priorities for conducting experimental exposures of marine organisms to microplastics. Front Mar Sci 5. https://doi.org/10.3389/fmars.2018.00252

Phuong NN, Zalouk-Vergnoux A, Poirier L, Kamari A, Châtel A, Mouneyrac C, Lagarde F (2016) Is there any consistency between the microplastics found in the field and those used in laboratory experiments? Environ Pollut 211:111–123. https://doi.org/10.1016/j.envpol.2015.12.035

Pitt JA, Kozal JS, Jayasundara N, Massarsky A, Trevisan R, Geitner N, Wiesner M, Levin ED, Di Giulio RT (2018) Uptake, tissue distribution, and toxicity of polystyrene nanoparticles in developing zebrafish (*Danio rerio*). Aquat Toxicol 194:185–194. https://doi.org/10.1016/j.aquatox.2017.11.017

Pittura L, Avio CG, GiulianiME, d'Errico G, Keiter SH, Cormier B, Gorbi S, Regoli F (2018) Microplastics as vehicles of environmental PAHs to marine organisms: Combined chemical and physical hazards to the mediterranean mussels, *Mytilus galloprovincialis*. Front Marine Sci 5. https://doi.org/10.3389/fmars.2018.00103

PlasticsEurope (2019) PlasticsEurope—plastics—the facts 2019. An analysis of European plastics production, demand and waste data

Posthuma L, van Gils J, Zijp MC, van de Meent D, de Zwart D (2019) Species sensitivity distributions for use in environmental protection, assessment, and management of aquatic ecosystems for 12 386 chemicals. Environ Toxicol Chem 38(4):905–917. https://doi.org/10.1002/etc.4373

Prakash V, Dwivedi S, Gautam K, Seth M, Anbumani S (2020) Occurrence and ecotoxicological effects of microplastics on aquatic and terrestrial ecosystems. In: He D, Luo Y (eds) Microplastics in terrestrial environments: emerging contaminants and major challenges. Springer, Switzerland, pp 223–243. https://doi.org/10.1007/698_2020_456

Prata JC, Lavorante BRBO, Montenegro M d CBSM, Guilhermino L (2018) Influence of microplastics on the toxicity of the pharmaceuticals procainamide and doxycycline on the marine microalgae *Tetraselmis chuii*. Aquat Toxicol 197:143–152. https://doi.org/10.1016/j.aquatox.2018.02.015

Prata JC, da Costa JP, Lopes I, Duarte AC, Rocha-Santos T (2019) Effects of microplastics on microalgae populations: a critical review. Sci Total Environ 665:400–405. https://doi.org/10.1016/j.scitotenv.2019.02.132

Prata JC, Reis V, da Costa JP, Mouneyrac C, Duarte AC, Rocha-Santos T (2021) Contamination issues as a challenge in quality control and quality assurance in microplastics analytics. J Hazard Mater 403:123660. https://doi.org/10.1016/j.jhazmat.2020.123660

Prendergast-Miller MT, Katsiamides A, Abbass M, Sturzenbaum SR, Thorpe KL, Hodson ME (2019) Polyester-derived microfibre impacts on the soil-dwelling earthworm *Lumbricus terrestris*. Environ Pollut 251:453–459. https://doi.org/10.1016/j.envpol.2019.05.037

R Core Team (2020) R: A language and environment for statistical computing. R Foundation for Statistical Computing, Vienna, Austria. URL https://www.R-project.org/

Rainieri S, Conlledo N, Larsen BK, Granby K, Barranco A (2018) Combined effects of microplastics and chemical contaminants on the organ toxicity of zebrafish (*Danio rerio*). Environ Res 162:135–143. https://doi.org/10.1016/j.envres.2017.12.019

Redondo-Hasselerharm PE, Falahudin D, Peeters ETHM, Koelmans AA (2018) Microplastic effect thresholds for freshwater benthic macroinvertebrates. Environ Sci Technol 52(4):2278–2286. https://doi.org/10.1021/acs.est.7b05367

Rehse S, Kloas W, Zarfl C (2016) Short-term exposure with high concentrations of pristine microplastic particles leads to immobilisation of *Daphnia magna*. Chemosphere 153:91–99. https://doi.org/10.1016/j.chemosphere.2016.02.133

Rehse S, Kloas W, Zarfl C (2018) Microplastics reduce short-term effects of environmental contaminants. Part I: Effects of bisphenol A on freshwater zooplankton are lower in presence of polyamide particles. Int J Environ Res Public Health 15(2). https://doi.org/10.3390/ijerph15020280

Reichert J, Schellenberg J, Schubert P, Wilke T (2018) Responses of reef building corals to microplastic exposure. Environ Pollut 237:955–960. https://doi.org/10.1016/j.envpol.2017.11.006

Revel M, Lagarde F, Perrein-Ettajani H, Bruneau M, Akcha F, Sussarellu R, Rouxel J, Costil K, Decottignies P, Cognie B, Châtel A, Mouneyrac C (2019) Tissue-specific biomarker responses in the blue mussel mytilus spp. exposed to a mixture of microplastics at environmentally relevant concentrations. Front Environ Sci:7. https://doi.org/10.3389/fenvs.2019.00033

Ribeiro F, Garcia AR, Pereira BP, Fonseca M, Mestre NC, Fonseca TG, Ilharco LM, Bebianno MJ (2017) Microplastics effects in *Scrobicularia plana*. Mar Pollut Bull 122(1–2):379–391. https://doi.org/10.1016/j.marpolbul.2017.06.078

Rillig MC, Ziersch L, Hempel S (2017) Microplastic transport in soil by earthworms. Sci Rep 7(1):1–6. https://doi.org/10.1038/s41598-017-01594-7

Rist S, Assidqi K, Zamani NP, Appel D, Perschke M, Huhn M, Lenz M (2016) Suspended micro-sized PVC particles impair the performance and decrease survival in the Asian green mussel *Perna viridis*. Mar Pollut Bull 111(1):213–220. https://doi.org/10.1016/j.marpolbul.2016.07.006

Rist S, Baun A, Hartmann NB (2017) Ingestion of micro- and nanoplastics in *Daphnia magna* – quantification of body burdens and assessment of feeding rates and reproduction. Environ Pollut 228:398–407. https://doi.org/10.1016/j.envpol.2017.05.048

Rist S, Baun A, Almeda R, Hartmann NB (2019) Ingestion and effects of micro- and nanoplastics in blue mussel (Mytilus edulis) larvae. Mar Pollut Bull 140:423–430. https://doi.org/10.1016/j.marpolbul.2019.01.069

Rochman CM (2015) The complex mixture, fate and toxicity of chemicals associated with plastic debris in the marine environment. In: Bergmann M, Gutow L, Klages M (eds) Marine anthropogenic litter. Springer, Switzerland, pp 117–140. https://doi.org/10.1007/978-3-319-16510-3_5

Rochman CM, Hoh E, Kurobe T, Teh SJ (2013) Ingested plastic transfers hazardous chemicals to fish and induces hepatic stress. Sci Rep 3. https://doi.org/10.1038/srep03263

Rochman CM, Parnis JM, Browne MA, Serrato S, Reiner EJ, Robson M, Young T, Diamond ML, Teh SJ (2017) Direct and indirect effects of different types of microplastics on freshwater prey (*Corbicula fluminea*) and their predator (*Acipenser transmontanus*). PLoS One 12(11):e0187664. https://doi.org/10.1371/journal.pone.0187664

Rochman CM, Brookson C, Bikker J, Djuric N, Earn A, Bucci K, Athey S, Huntington A, McIlwraith H, Munno K, Frond HD, Kolomijeca A, Erdle L, Grbic J, Bayoumi M, Borrelle SB, Wu T, Santoro S, Werbowski LM et al (2019) Rethinking microplastics as a diverse contaminant suite. Environ Toxicol Chem 38(4):703–711. https://doi.org/10.1002/etc.4371

Rodríguez-Seijo A, Lourenço J, Rocha-Santos TAP, da Costa J, Duarte AC, Vala H, Pereira R (2017) Histopathological and molecular effects of microplastics in *Eisenia andrei* Bouché. Environ Pollut 220:495–503. https://doi.org/10.1016/j.envpol.2016.09.092

Rodríguez-Seijo A, da Costa JP, Rocha-Santos T, Duarte AC, Pereira R (2018a) Oxidative stress, energy metabolism and molecular responses of earthworms (*Eisenia fetida*) exposed to low-density polyethylene microplastics. Environ Sci Pollut Res Int 25(33):33599–33610. https://doi.org/10.1007/s11356-018-3317-z

Rodríguez-Seijo A, Santos B, da Silva EF, Cachada A, Pereira R (2018b) Low-density polyethylene microplastics as a source and carriers of agrochemicals to soil and earthworms. Environ Chem 16(1):8–17. https://doi.org/10.1071/EN18162

Santana MFM, Moreira FT, Pereira CDS, Abessa DMS, Turra A (2018) Continuous exposure to microplastics does not cause physiological effects in the cultivated mussel *Perna perna*. Arch Environ Contam Toxicol 74(4):594–604. https://doi.org/10.1007/s00244-018-0504-3

SAPEA (2019) A scientific perspective on microplastics in nature and society (Evidence Review Report No. 4). EU

Selonen S, Dolar A, Jemec Kokalj A, Skalar T, Parramon Dolcet L, Hurley R, van Gestel CAM (2020) Exploring the impacts of plastics in soil – the effects of polyester textile fibers on soil invertebrates. Sci Total Environ 700:134451. https://doi.org/10.1016/j.scitotenv.2019.134451

Sendra M, Staffieri E, Yeste MP, Moreno-Garrido I, Gatica JM, Corsi I, Blasco J (2019) Are the primary characteristics of polystyrene nanoplastics responsible for toxicity and ad/absorption in the marine diatom *Phaeodactylum tricornutum*? Environ Pollut (Barking, Essex: 1987) 249:610–619. https://doi.org/10.1016/j.envpol.2019.03.047

Seoane M, González-Fernández C, Soudant P, Huvet A, Esperanza M, Cid Á, Paul-Pont I (2019) Polystyrene microbeads modulate the energy metabolism of the marine diatom *Chaetoceros neogracile*. Environ Pollut (Barking, Essex: 1987) 251:363–371. https://doi.org/10.1016/j.envpol.2019.04.142

Skjolding LM, Ašmonaitė G, Jølck RI, Andresen TL, Selck H, Baun A, Sturve J (2017) An assessment of the importance of exposure routes to the uptake and internal localisation of fluorescent nanoparticles in zebrafish (*Danio rerio*), using light sheet microscopy. Nanotoxicology 11(3):351–359. https://doi.org/10.1080/17435390.2017.1306128

Snell TW, Hicks DG (2011) Assessing toxicity of nanoparticles using Brachionus manjavacas (Rotifera). Environ Toxicol 26(2):146–152. https://doi.org/10.1002/tox.20538

Song Y, Cao C, Qiu R, Hu J, Liu M, Lu S, Shi H, Raley-Susman KM, He D (2019) Uptake and adverse effects of polyethylene terephthalate microplastics fibers on terrestrial snails (*Achatina fulica*) after soil exposure. Environ Pollut 250:447–455. https://doi.org/10.1016/j.envpol.2019.04.066

Straub S, Hirsch PE, Burkhardt-Holm P (2017) Biodegradable and petroleum-based microplastics do not differ in their ingestion and excretion but in their biological effects in a freshwater invertebrate Gammarus fossarum. Int J Environ Res Public Health 14(7). https://doi.org/10.3390/ijerph14070774

Strungaru S-A, Jijie R, Nicoara M, Plavan G, Faggio C (2019) Micro- (nano) plastics in freshwater ecosystems: abundance, toxicological impact and quantification methodology. TrAC Trends Anal Chem 110:116–128. https://doi.org/10.1016/j.trac.2018.10.025

Sussarellu R, Suquet M, Thomas Y, Lambert C, Fabioux C, Pernet MEJ, Goïc NL, Quillien V, Mingant C, Epelboin Y, Corporeau C, Guyomarch J, Robbens J, Paul-Pont I, Soudant P, Huvet A (2016) Oyster reproduction is affected by exposure to polystyrene microplastics. Proc Natl Acad Sci 113(9):2430–2435. https://doi.org/10.1073/pnas.1519019113

Syakti AD, Jaya JV, Rahman A, Hidayati NV, Raza'i TS, Idris F, Trenggono M, Doumenq P, Chou LM (2019) Bleaching and necrosis of staghorn coral (*Acropora formosa*) in laboratory assays: immediate impact of LDPE microplastics. Chemosphere 228:528–535. https://doi.org/10.1016/j.chemosphere.2019.04.156

Tallec K, Huvet A, Di Poi C, González-Fernández C, Lambert C, Petton B, Le Goïc N, Berchel M, Soudant P, Paul-Pont I (2018) Nanoplastics impaired oyster free living stages, gametes and embryos. Environ Pollut 242:1226–1235. https://doi.org/10.1016/j.envpol.2018.08.020

Tang J, Ni X, Zhou Z, Wang L, Lin S (2018) Acute microplastic exposure raises stress response and suppresses detoxification and immune capacities in the scleractinian coral *Pocillopora damicornis*. Environ Pollut 243:66–74. https://doi.org/10.1016/j.envpol.2018.08.045

Tang J, Wang X, Yin J, Han Y, Yang J, Lu X, Xie T, Akbar S, Lyu K, Yang Z (2019) Molecular characterization of thioredoxin reductase in waterflea *Daphnia magna* and its expression regulation by polystyrene microplastics. Aquat Toxicol 208:90–97. https://doi.org/10.1016/j.aquatox.2019.01.001

Ter Halle A, Jeanneau L, Martignac M, Jardé E, Pedrono B, Brach L, Gigault J (2017) Nanoplastic in the North Atlantic subtropical gyre. Environ Sci Technol 51(23):13689–13697. https://doi.org/10.1021/acs.est.7b03667

Teuten EL, Saquing JM, Knappe DRU, Barlaz MA, Jonsson S, Björn A, Rowland SJ, Thompson RC, Galloway TS, Yamashita R, Ochi D, Watanuki Y, Moore C, Viet PH, Tana TS, Prudente M, Boonyatumanond R, Zakaria MP, Akkhavong K et al (2009) Transport and release of chemicals from plastics to the environment and to wildlife. Philos Trans R Soc Lond B Biol Sci 364(1526):2027–2045. https://doi.org/10.1098/rstb.2008.0284

Thiagarajan V, Natarajan L, Seenivasan R, Chandrasekaran N, Mukherjee A (2019) Influence of differently functionalized polystyrene microplastics on the toxic effects of P25 TiO2 NPs towards marine algae Chlorella sp. Aquat Toxicol 207:208–216. https://doi.org/10.1016/j.aquatox.2018.12.014

Thompson RC, Olsen Y, Mitchell RP, Davis A, Rowland SJ, John AWG, McGonigle D, Russell AE (2004) Lost at sea: where is all the plastic? Science (New York, NY) 304(5672):838. https://doi.org/10.1126/science.1094559

Tosetto L, Brown C, Williamson JE (2016) Microplastics on beaches: ingestion and behavioural consequences for beachhoppers. Mar Biol 163(10):199. https://doi.org/10.1007/s00227-016-2973-0

Tosetto L, Williamson JE, Brown C (2017) Trophic transfer of microplastics does not affect fish personality. Anim Behav 123:159–167. https://doi.org/10.1016/j.anbehav.2016.10.035

Ugolini A, Ungherese G, Ciofini M, Lapucci A, Camaiti M (2013) Microplastic debris in sandhoppers. Estuar Coast Shelf Sci 129:19–22. https://doi.org/10.1016/j.ecss.2013.05.026

Van Cauwenberghe L (2016) Occurrence, effects and risks of marine microplastics [Thesis submitted in fulfilment of the requirements for the degree of Doctor (PhD) in Applied Biological Sciences]. http://hdl.handle.net/1854/LU-7097647

Van Cauwenberghe L, Janssen CR (2014) Microplastics in bivalves cultured for human consumption. Environ Pollut (Barking, Essex: 1987) 193:65–70. https://doi.org/10.1016/j.envpol.2014.06.010

Van Cauwenberghe L, Claessens M, Vandegehuchte MB, Janssen CR (2015) Microplastics are taken up by mussels (*Mytilus edulis*) and lugworms (*Arenicola marina*) living in natural habitats. Environ Pollut 199:10–17. https://doi.org/10.1016/j.envpol.2015.01.008

Vicentini DS, Nogueira DJ, Melegari SP, Arl M, Köerich JS, Cruz L, Justino NM, Oscar BV, Puerari RC, da Silva MLN, Simioni C, Ouriques LC, Matias MS, de Castilhos Junior AB, Matias WG (2019) Toxicological evaluation and quantification of ingested metal-core nanoplastic by Daphnia magna through fluorescence and inductively coupled plasma-mass spectrometric methods. Environ Toxicol Chem 38(10):2101–2110. https://doi.org/10.1002/etc.4528

Vince J, Hardesty BD (2017) Plastic pollution challenges in marine and coastal environments: from local to global governance. Restor Ecol 25(1):123–128. https://doi.org/10.1111/rec.12388

VKM (2019) Microplastics; occurrence, levels and implications for environment and human health related to food.. Scientific Opinion of the Scientific Steering Committee of the Norwegian Scientific Committee for Food and Environment

von Moos N, Burkhardt-Holm P, Köhler A (2012) Uptake and effects of microplastics on cells and tissue of the blue mussel *Mytilus edulis L.* after an experimental exposure. Environ Sci Technol 46(20):11327–11335. https://doi.org/10.1021/es302332w

Wang J, Coffin S, Sun C, Schlenk D, Gan J (2019a) Negligible effects of microplastics on animal fitness and HOC bioaccumulation in earthworm *Eisenia fetida* in soil. Environ Pollut 249:776–784. https://doi.org/10.1016/j.envpol.2019.03.102

Wang W, Gao H, Jin S, Li R, Na G (2019b) The ecotoxicological effects of microplastics on aquatic food web, from primary producer to human: a review. Ecotoxicol Environ Saf 173:110–117. https://doi.org/10.1016/j.ecoenv.2019.01.113

Wang J, Li Y, Lu L, Zheng M, Zhang X, Tian H, Wang W, Ru S (2019c) Polystyrene microplastics cause tissue damages, sex-specific reproductive disruption and transgenerational effects in marine medaka (*Oryzias melastigma*). Environ Pollut 254:113024. https://doi.org/10.1016/j.envpol.2019.113024

Wang W, Ge J, Yu X (2020) Bioavailability and toxicity of microplastics to fish species: a review. Ecotoxicol Environ Saf 189:109913. https://doi.org/10.1016/j.ecoenv.2019.109913

Ward JE, Kach DJ (2009) Marine aggregates facilitate ingestion of nanoparticles by suspension-feeding bivalves. Mar Environ Res 68(3):137–142. https://doi.org/10.1016/j.marenvres.2009.05.002

Ward JE, Zhao S, Holohan BA, Mladinich KM, Griffin TW, Wozniak J, Shumway SE (2019) Selective ingestion and egestion of plastic particles by the blue mussel (*Mytilus edulis*) and eastern oyster (*Crassostrea virginica*): implications for using bivalves as bioindicators of microplastic pollution. Environ Sci Technol 53(15):8776–8784. https://doi.org/10.1021/acs.est.9b02073

Watts AJR, Urbina MA, Corr S, Lewis C, Galloway TS (2015) Ingestion of plastic microfibers by the crab *Carcinus maenas* and its effect on food consumption and energy balance. Environ Sci Technol 49(24):14597–14604. https://doi.org/10.1021/acs.est.5b04026

Weber A, Scherer C, Brennholt N, Reifferscheid G, Wagner M (2018) PET microplastics do not negatively affect the survival, development, metabolism and feeding activity of the freshwater invertebrate *Gammarus pulex*. Environ Pollut 234:181–189. https://doi.org/10.1016/j.envpol.2017.11.014

Wegner A, Besseling E, Foekema EM, Kamermans P, Koelmans AA (2012) Effects of nanopolystyrene on the feeding behavior of the blue mussel (*Mytilus edulis L.*). Environ Toxicol Chem/SETAC 31(11):2490–2497. https://doi.org/10.1002/etc.1984

Welden NAC, Cowie PR (2016) Long-term microplastic retention causes reduced body condition in the langoustine, *Nephrops norvegicus*. Environ Pollut 218:895–900. https://doi.org/10.1016/j.envpol.2016.08.020

Wen B, Zhang N, Jin S-R, Chen Z-Z, Gao J-Z, Liu Y, Liu H-P, Xu Z (2018a) Microplastics have a more profound impact than elevated temperatures on the predatory performance, digestion and energy metabolism of an Amazonian cichlid. Aquat Toxicol 195:67–76. https://doi.org/10.1016/j.aquatox.2017.12.010

Wen B, Jin S-R, Chen Z-Z, Gao J-Z, Liu Y-N, Liu J-H, Feng X-S (2018b) Single and combined effects of microplastics and cadmium on the cadmium accumulation, antioxidant defence and innate immunity of the discus fish (*Symphysodon aequifasciatus*). Environ Pollut 243:462–471. https://doi.org/10.1016/j.envpol.2018.09.029

Windsor FM, Durance I, Horton AA, Thompson RC, Tyler CR, Ormerod SJ (2019) A catchment-scale perspective of plastic pollution. Glob Chang Biol 25(4):1207–1221. https://doi.org/10.1111/gcb.14572

Woods MN, Stack ME, Fields DM, Shaw SD, Matrai PA (2018) Microplastic fiber uptake, ingestion, and egestion rates in the blue mussel (*Mytilus edulis*). Mar Pollut Bull 137:638–645. https://doi.org/10.1016/j.marpolbul.2018.10.061

Wright SL, Rowe D, Thompson RC, Galloway TS (2013) Microplastic ingestion decreases energy reserves in marine worms. Curr Biol 23(23):R1031–R1033. https://doi.org/10.1016/j.cub.2013.10.068

Wu J, Jiang R, Lin W, Ouyang G (2019a) Effect of salinity and humic acid on the aggregation and toxicity of polystyrene nanoplastics with different functional groups and charges. Environ Pollut 245:836–843. https://doi.org/10.1016/j.envpol.2018.11.055

Wu Y, Guo P, Zhang X, Zhang Y, Xie S, Deng J (2019b) Effect of microplastics exposure on the photosynthesis system of freshwater algae. J Hazard Mater 374:219–227. https://doi.org/10.1016/j.jhazmat.2019.04.039

Yu P, Liu Z, Wu D, Chen M, Lv W, Zhao Y (2018) Accumulation of polystyrene microplastics in juvenile *Eriocheir sinensis* and oxidative stress effects in the liver. Aquat Toxicol 200:28–36. https://doi.org/10.1016/j.aquatox.2018.04.015

Zhang C, Chen X, Wang J, Tan L (2017) Toxic effects of microplastic on marine microalgae *Skeletonema costatum*: interactions between microplastic and algae. Environ Pollut 220:1282–1288. https://doi.org/10.1016/j.envpol.2016.11.005

Zhang P, Yan Z, Lu G, Ji Y (2019) Single and combined effects of microplastics and roxithromycin on *Daphnia magna*. Environ Sci Pollut Res Int 26(17):17010–17020. https://doi.org/10.1007/s11356-019-05031-2

Zhao S, Zhu L, Li D (2015) Microplastic in three urban estuaries, China. Environ Pollut (Barking, Essex: 1987) 206:597–604. https://doi.org/10.1016/j.envpol.2015.08.027

Zhao T, Tan L, Huang W, Wang J (2019) The interactions between micro polyvinyl chloride (mPVC) and marine dinoflagellate *Karenia mikimotoi*: the inhibition of growth, chlorophyll and photosynthetic efficiency. Environ Pollut (Barking, Essex: 1987) 247:883–889. https://doi.org/10.1016/j.envpol.2019.01.114

Zhu D, Chen Q-L, An X-L, Yang X-R, Christie P, Ke X, Wu L-H, Zhu Y-G (2018) Exposure of soil collembolans to microplastics perturbs their gut microbiota and alters their isotopic composition. Soil Biol Biochem 116:302–310. https://doi.org/10.1016/j.soilbio.2017.10.027

Zhu Z, Wang S, Zhao F, Wang S, Liu F, Liu G (2019) Joint toxicity of microplastics with triclosan to marine microalgae *Skeletonema costatum*. Environ Pollut 246:509–517. https://doi.org/10.1016/j.envpol.2018.12.044

Ziajahromi S, Kumar A, Neale PA, Leusch FDL (2017) Impact of microplastic beads and fibers on waterflea (*Ceriodaphnia dubia*) survival, growth, and reproduction: implications of single and mixture exposures. Environ Sci Technol 10

Ziajahromi S, Kumar A, Neale PA, Leusch FDL (2018) Environmentally relevant concentrations of polyethylene microplastics negatively impact the survival, growth and emergence of sediment-dwelling invertebrates. Environ Pollut 236:425–431. https://doi.org/10.1016/j.envpol.2018.01.094

8

Additives and Associated Sorbed Contaminants: Impact on Food Safety

Esther Garrido Gamarro and Violetta Costanzo

Abstract Microplastics and nanoplastics may be found in the gastrointestinal tract of some aquatic animals and could potentially be ingested by humans if consumed whole. Information on the toxicity of plastic particles, as well as co-contaminants such as plastic additives, remains scarce. This represents a serious challenge to perform realistic risk assessments. An exposure assessment of selected plastic additives and co-contaminants of known toxicity associated with microplastics was carried out for shellfish in this study, which builds on an exposure assessment of microplastic additives and a limited number of associated contaminants in mussels conducted by the FAO in 2017. This study evaluates possible impacts to food safety by examining a diverse additives and associated sorbed contaminants. The results suggest that the levels of certain microplastic additives and sorbed co-contaminants in target animals (shrimp, prawns, clams, oysters, and mussels) do not pose a food safety threat to consumers. To get to further conclusions, an exposure assessment from the whole diet should be carried out and the toxicity of some of the most common polymers and plastic additives, as well as their mixtures, needs to be carefully evaluated.

8.1 Introduction

Plastic production has been increasing exponentially since the 1950s and was estimated to be 8300 million metric tons to date (Geyer et al. 2017). Since its first development in the 1800s, the production of plastic materials has changed to meet the needs of a variety of sectors and consumers and has enabled technological improvements and solutions. Due to their functional properties ("cheap and

E. Garrido Gamarro (✉) · V. Costanzo
Fisheries and Aquaculture Division, Food and Agriculture Organization of the United Nations (FAO), Rome, Italy
e-mail: esther.garridogamarro@fao.org

durable"), plastics have displaced many non-plastic materials, becoming the most utilized materials worldwide.

Plastics consist of a range of synthetic or semi-synthetic chemicals that are made of fossil resources and organic by-products. They are commonly divided into three categories: thermoplastics (polymers that can be re-melted), elastomers (elastic polymers that return to their original shape after being deformed), and thermosets (polymers that remain in a permanent solid state once hardened).

Depending on their specific use, polymers with different physical and chemical properties can be mixed and additives such as plasticizers, colourants, UV-stabilizers, flame retardants, and antioxidants can be added to improve the performance of the final product.

Recycling the complex mixtures of chemicals used for plastic production can be challenging, as well as the evaluation of their impact on the environment and human health. The extensive production of plastic requires efficient waste management systems, but most countries do not have the capacity to develop them.

Microplastic particles have been found in a variety of human food items, such as salt, beer, honey, and aquatic products, with seafood being the best-studied source of dietary intake of microplastics. Exposure to microplastic particles, their additives, and their sorbed co-contaminants depends on several factors, such as particle size, shape, chemical changes that occurred during processing and/or cooking of fisheries and aquaculture products, and consumption patterns.

A previous study on exposure to microplastics and associated co-contaminants suggests that exposure to this contaminant burden is typically less than 0.1% (FAO 2017). Microplastic contribution to the total dietary intake of additives and sorbed co-contaminants was estimated as very low, with maximum increases in BPA, PAHs, and PCBs load of less than 2%, 0.004%, and 0.006% respectively, after the ingestion of a portion of mussels (EFSA 2016). However, a recent study by Barboza et al. (2020) observed a clear correlation between microplastics intake in three species of wild-caught commercial fish and the levels of bisphenols in the muscle and liver. Higher microplastic concentrations in fish were correlated with higher levels of these compounds, whose concentration in the edible tissue exceeded the established limits for human safety set by the EFSA. Furthermore, a relation between the concentration of plastic-associated chemicals and ingested microplastics in marine organisms has already been hypothesized (Granby et al. 2018; Rochman et al. 2013; Teuten et al. 2009). These findings suggest that more investigations should be conducted on this subject to better identify the role of microplastics in the transfer of pollutants and which factors could influence the process. This chapter aims to provide an overview of the dietary exposure of microplastic particles, additives, and common microplastic co-contaminants through aquatic products using consumption data from the FAO/WHO database, while information on contamination levels of plastic pellets and microplastic ingestion by seafood are updated with the current literature. Moreover, four different groups of seafood were considered, to extend the exposure evaluation also to crustaceans and other bivalves. The final estimations are compared with the no observed effect levels (NOELs) and no observed adverse effect levels (NOAELs) and can be useful to provide a better understanding of the potential impacts on food safety.

8.2 Sorption of Environmental Contaminants by Microplastics

A potential threat to human health deriving from the exposure to microplastics is that these materials can scavenge and thus concentrate pollutants already present in production waters. The ingestion of contaminated plastic could lead to higher exposure to toxic chemicals, with possible endocrine disruption and carcinogenicity. The main process leading to the interaction between microplastics and hydrophobic organic chemicals (HOCs) in the water column does not involve the formation of covalent bonds; thus, its reversible nature preserves the likelihood of chemical desorption from the matrix (Endo and Koelmans 2016). Besides, plastic polymers are also recognized as possible vectors of heavy metals in the marine environment (Holmes et al. 2012, 2014), being experimentally able to accumulate concentrations even 800-fold higher than in seawater (Brennecke et al. 2016). Many field studies such as the International Pellet Watch have reported the concentration levels of persistent organic pollutants (POPs) and metals sorbed on beached and marine pellets, in addition to plastic additives (Tables 8.1 and 8.2).

Sorption processes can be classified into adsorption and absorption, depending on the mechanism of interaction between the polymer and the chemicals. Absorption mainly occurs when the molecules of pollutants diffuse into the bulk matrix of the polymer and interact with it through weak van der Waals forces or hydrophobic interactions. This process is mainly driven by the preferential partitioning of the chemical on plastic compared to water, which is usually linearly and positively correlated to its octanol-water partition coefficient (K_{ow}), a parameter that measures the level of hydrophobicity (Lee et al. 2014; Li et al. 2018c; O'Connor et al. 2016). Adsorption refers to a process that results in the sorption of molecules that are confined to the surface of microplastics (Endo and Koelmans 2016).

Absorption mainly occurs onto rubbery polymers (i.e. PE and PP), where external molecules pass through and associate within their matrix (Hüffer and Hofmann 2016; Müller et al. 2018; Teuten et al. 2009; Wang et al. 2018a). These polymers are generally recognized as the ones concentrating the highest amounts of HOCs and are then possibly more dangerous for marine life (Endo et al. 2005; Fisner et al. 2017; Hirai et al. 2011; O'Connor et al. 2016; Wang and Wang 2018; Wang et al. 2018b) and possibly human health.

Some polymers present several functional groups on their surface, conferring a certain degree of polarity. These are mostly PS and PVC, whose glassy nature also yields the formation of nanovoids and pores on the surface, which are the sites of sorption. In this case, the process is mainly led by adsorption, a mechanism through which the chemicals more efficiently bind to the plastic polymer through ionic,

Table 8.1 Maximum concentrations of persistent organic pollutants (POPs) and metals sorbed on plastic pellets from field studies

Location	Sampling	Extraction	PCB (ng/g)	PAH (ng/g)	DDT (ng/g)	DDE (ng/g)	PBDE (ng/g)	HCH (ng/g)	Cd (ng/g)	Pb (µg/g)	References
Portuguese Coast	Metal mesh sieving	n-Hexane, hexane-acetone	223	44,800	41		n.a.	n.a.	n.a.	n.a.	Antunes et al. (2013)
Greek beaches	N.S.	Hexane	290	500	42	15	n.a.	3.5	n.a.	n.a.	Karapanagioti et al. (2011)
South Devon, England	Plastic tweezers	Dilute aqua regia digestion	n.a.	n.a.	n.a.	n.a.	n.a.	n.a.	10	1.08	Ashton et al. (2010)
South West England	Plastic tweezers	Dilute aqua regia digestion	n.a.	n.a.	n.a.	n.a.	n.a.	n.a.	76.7	1.64	Holmes et al. (2012)
Portuguese coast	N.S.	Hexane	310	24,000	49	n.a.	n.a.	3.3	n.a.	n.a.	Mizukawa et al. (2013)
Japanese coast	Solvent rinsed stainless tweezers or by hand with PE gloves	Hexane	117	n.a.	n.a.	3.1	n.a.	n.a.	n.a.	n.a.	Mato et al. (2001)
Japanese beaches	Solvent rinsed stainless tweezers	n-Hexane	2300	n.a.	n.a.	n.a.	n.a.	n.a.	n.a.	n.a.	Endo et al. (2005)
Santos Bay, Brazil	Tweezers	DCM: n-hexane	n.a.	39,763	n.a.	n.a.	n.a.	n.a.	n.a.	n.a.	Fisner et al. (2017)
Portuguese coast	N.S.	Hexane, hexane: acetone	54	319.6	4.54	0.81	n.a.	n.a.	n.a.	n.a.	Frias et al. (2010)
Belgian coast	N.S.	Hexane: DCM	236	3007	n.a.	n.a.	n.a.	n.a.	n.a.	n.a.	Gauquie et al. (2015)
Central Pacific Gyre, Pacific Ocean	Neuston net and solvent rinsed stainless steel tweezers	DCM	78	868	4.8	2	9909	n.a.	n.a.	n.a.	Hirai et al. (2011)

Japanese beaches	Neuston net and solvent rinsed stainless steel tweezers	DCM	436	9297	198	198	230	n.a.	n.a.	n.a.	Hirai et al. (2011)
Tonkin Bay, Vietnam	Neuston net and solvent rinsed stainless steel tweezers	DCM	102	2024	108	42	412	n.a.	n.a.	n.a.	Hirai et al. (2011)
Seal Beach, California	Neuston net and solvent rinsed stainless steel tweezers	DCM	399	656	8	7.6	41	n.a.	n.a.	n.a.	Hirai et al. (2011)
Marbella Beach, Costa Rica	Neuston net and solvent rinsed stainless steel tweezers	DCM	61	284	124	10	180	n.a.	n.a.	n.a.	Hirai et al. (2011)
Hawaii	Squeeze, scoops, or directly in glass jars	DCM	980	500	22	n.a.	n.a.	n.a.	n.a.	n.a.	Rios et al. (2007)
Mexico	Squeeze, scoops, or directly in glass jars	DCM	n.a.	640	n.a.	n.a.	n.a.	n.a.	n.a.	n.a.	Rios et al. (2007)
California	Squeeze, scoops, or directly in glass jars	DCM	730	6200	1100	n.a.	n.a.	n.a.	n.a.	n.a.	Rios et al. (2007)
9 Japanese beaches	Solvent rinsed stainless tweezers	DCM	892	n.a.	n.a.	n.a.	n.a.	n.a.	n.a.	n.a.	Takada et al. (2006)
Japanese coast	Neuston net	N.S.	254	9370	n.a.	276	2.1	n.a.	n.a.	n.a.	Teuten et al. (2009)
Central Gyre	Neuston net	N.S.	23	959	n.a.	4.7	57	n.a.	n.a.	n.a.	Teuten et al. (2009)
Japan and South East Asia	Soap rinsed fingers or stainless steel tweezers	Hexane	453	n.a.	163	11.5	n.a.	1.23	n.a.	n.a.	Ogata et al. (2009)

(continued)

Table 8.1 (continued)

Location	Sampling	Extraction	PCB (ng/g)	PAH (ng/g)	DDT (ng/g)	DDE (ng/g)	PBDE (ng/g)	HCH (ng/g)	Cd (ng/g)	Pb (μg/g)	References
US. Coast	Soap rinsed fingers or stainless steel tweezers	Hexane	605	n.a.	267	128	n.a.	0.94	n.a.	n.a.	Ogata et al. (2009)
Africa	Soap rinsed fingers or stainless steel tweezers	Hexane	41	n.a.	4.49	0.91	n.a.	37.1	n.a.	n.a.	Ogata et al. (2009)
India	Soap rinsed fingers or stainless steel tweezers	Hexane	141	n.a.	29.8	15.8	n.a.	3.24	n.a.	n.a.	Ogata et al. (2009)
Australia	Soap rinsed fingers or stainless steel tweezers	Hexane	16	n.a.	6.69	0.57	n.a.	0.19	n.a.	n.a.	Ogata et al. (2009)
West Europe	Soap rinsed fingers or stainless steel tweezers	Hexane	169	n.a.	27.6	15.9	n.a.	1.12	n.a.	n.a.	Ogata et al. (2009)
South Atlantic Ocean	Manta trawl	DCM	590	n.a.	n.a.	n.a.	4.6	n.a.	n.a.	n.a.	Rochman et al. (2014)
Saint Helena British territory	Soap rinsed fingers or stainless steel tweezers	Hexane	7	n.a.	3.4	0.5	n.a.	19	n.a.	n.a.	Heskett et al. (2012)
Cocos Islands, Australia	Soap rinsed fingers or stainless steel tweezers	Hexane	6.5	n.a.	3.4	0.8	n.a.	1.7	n.a.	n.a.	Heskett et al. (2012)
Hawaii, USA	Soap rinsed fingers or stainless steel tweezers	Hexane	9.9	n.a.	3.4	0.5	n.a.	0.6	n.a.	n.a.	Heskett et al. (2012)
Island of Oahu, USA	Soap rinsed fingers or stainless steel tweezers	Hexane	1.5	n.a.	0.8	0.6	n.a.	<LOQ	n.a.	n.a.	Heskett et al. (2012)
Barbados, USA	Soap rinsed fingers or stainless steel tweezers	Hexane	1.7	n.a.	3.1	0.3	n.a.	<LOQ	n.a.	n.a.	Heskett et al. (2012)
Canary Islands, Spain	Soap-rinsed fingers or stainless steel tweezers	Hexane	9.0	n.a.	4.1	2.5	n.a.	0.6	n.a.	n.a.	Heskett et al. (2012)
North Pacific Central Gyre	Hand net, manta trawl	DCM	2856	14,459	336	126	n.a.	n.a.	n.a.	n.a.	Rios et al. (2010)

Table 8.2 Maximum concentrations of selected plastic additives in plastic pellets from field studies

Location	Sampling	Extraction	UV stabilizers (μg/g)	BPA (ng/g)	Antioxidants (μg/g)	NP (ng/g)	OP (ng/g)	References
Japanese coasts	Solvent rinsed stainless tweezers or by hand with PE gloves	Hexane	n.a.	n.a.	n.a.	16,000	n.a.	Mato et al. (2001)
Central Pacific Gyre, Pacific Ocean	Neuston net and solvent rinsed stainless steel tweezers	DCM	n.a.	283	n.a.	997	4	Hirai et al. (2011)
Japanese beaches	Neuston net and solvent rinsed stainless steel tweezers	DCM	n.a.	3.4	n.a.	1244	49	Hirai et al. (2011)
Tonkin Bay, Vietnam	Neuston net and solvent rinsed stainless steel tweezers	DCM	n.a.	263	n.a.	551	154	Hirai et al. (2011)
Seal Beach, California	Neuston net and solvent rinsed stainless steel tweezers	DCM	n.a.	26	n.a.	130	17	Hirai et al. (2011)
Marbella Beach, Costa Rica	Neuston net and solvent rinsed stainless steel tweezers	DCM	n.a.	730	n.a.	3936	14	Hirai et al. (2011)
9 Japanese beaches	Solvent rinsed stainless tweezers	DCM	n.a.	n.a.	n.a.	17,000	n.a.	Takada et al. (2006)
Japanese coast	Neuston net	N.S.	n.a.	<25	n.a.	n.s.	n.a.	Teuten et al. (2009)
Central Gyre	Neuston net	N.S.	n.a.	284	n.a.	2,660,000	n.a	Teuten et al. (2009)
Island of Kauai, Hawaii, USA	Stainless steel tweezers	n-Hexane	1133	n.a.	n.a.	n.a.	n.a.	Tanaka et al. (2020)
Geoje, South Korea	N.S.	DCM	82	n.a.	1620	n.a.	n.a.	Rani et al. (2017)
South Atlantic Ocean	Manta trawl	DCM	n.a.	4.9	n.a.	n.s	n.s	Rochman et al. (2014)

steric, non-covalent, or covalent bonds (i.e. π-π interactions) (Brennecke et al. 2016; Hüffer and Hofmann 2016; Velzeboer et al. 2014; Wang et al. 2019).

Sorption is also largely influenced by the surface area to volume ratio of a plastic particle, which increases as the size decreases (Brennecke et al. 2016; Li et al. 2019; Ma et al. 2016; Teuten et al. 2009; Zhan et al. 2016; Zhang et al. 2018). When exposed to weathering, the plastic surface can be subjected to embrittlement and fragmentation, steps that increase the surface/volume and provide more space, a larger contact area, and new sorption sites for external molecules (Napper et al. 2015; Wang et al. 2018b). UV rays-induced weathering, or photo-oxidation, can also lead to chemical alterations and loss of hydrophobicity through the creation of new oxygen-rich functional groups (i.e. carbonyl moieties). Salinity and pH can also play a role in sorption mechanisms. When pH is above the point of zero charge (PZC) of the plastic polymer, it assumes a negative charge that could result in electrostatic repulsion between its surface and other anionic chemicals (Holmes et al. 2014; Li et al. 2019; Wang et al. 2015). Salinity, on the other hand, can either increase the partitioning of nonpolar compounds (salting out) or decrease that of polar molecules due to competition in the adsorption sites (Karapanagioti and Klontza 2008; Llorca et al. 2018; Wang et al. 2015; Zhan et al. 2016; Zuo et al. 2019). This can result in a difference in sorption capacities between freshwater and seawater environments. Exposure time, chain length, and temperature have also been observed as influential factors in hydrophobic partitioning (Engler 2012; Llorca et al. 2018; Mato et al. 2001; Takada et al. 2006; Zhan et al. 2016; Zhang et al. 2018).

Finally, it must be noted that new kinds of biodegradable plastic polymers are being designed and are expected to be more easily and fully degraded over a short time, thus reducing their potential harm. Despite this assumption, Zuo et al. (2019) recently indicated that those highly rubbery MP could become even stronger vectors of organic chemicals. Evaluation of toxicity of the alternative materials and experimental studies are needed to clarify the possible harm.

8.2.1 Polycyclic Aromatic Hydrocarbons (PAHs)

The presence of polycyclic aromatic hydrocarbons (PAHs) in the environment can be the result of three different processes: the incomplete combustion of organic material (pyrolytic origin), spillage discharge of crude oil (petrogenic origin), or the post-depositional transformation of biogenic precursors (diagenetic origin). The contribution of petrogenic over pyrogenic sources (and vice versa) can be manifested through the calculation of the ratio between lighter (2–3 rings) and heavier congeners (4–6 rings), with higher values of this parameter indicating a major contribution of fossil sources (low molecular weight congeners). High molecular weight PAHs are generally the ones detected at higher concentrations on plastic pellets in the environment, implying combustion processes to be the main source of contamination (Gauquie et al. 2015; Rios et al. 2007). In a recent study, low molecular weight PAHs were mostly found on clearer materials, while high molecular weight PAHs were mostly detected on darker materials (Fisner et al. 2017). The colour of microplastics, along with their size and smell, is an important factor since some organisms may selectively feed on those pellets which resemble their prey (Chagnon et al. 2018; Hipfner et al. 2018; Ory et al. 2017; Savoca et al. 2016).

8.2.2 Polychlorinated Biphenyls (PCBs)

Polychlorinated biphenyls (PCBs) are a group of 209 lipophilic chemicals of a completely anthropogenic origin, which have been widely used from the 1930s to 1970s until they were banned because of their harmful nature.

The concentration of these pollutants in plastic pellets nowadays is mostly related to the presence of legacy PCBs in those industrialized countries that formerly used high amounts of them, which are still present in the environment due to their persistent nature (Ogata et al. 2009). When lower chlorinated congeners are detected inside an organism, their presence can be generally linked to the ingestion of contaminated MPs, since they would be more easily depleted along the trophic chain (Teuten et al. 2009). For this, only those highly substituted congeners would be more prone to be transferred and biomagnify in the trophic chain, and their exposure would then be mainly caused by prey ingestion (Yamashita et al. 2011). However, due to the high K_{ow} of highly chlorinated PCBs, they are more efficiently bound to the plastic and then less likely released from polymers (Colabuono et al. 2010). Their concentration is typically expressed as ICES-7 (ΣPCBs 28, 52, 101, 118, 138, 153, 180), which corresponds to the sum of seven indicator non-dioxin-like (NDL) congeners presenting the highest concentrations in technical mixtures and the environment (Webster et al. 2013). They are generally calculated as the sum of different congeners. The ones typically showing the highest concentrations are CB 118, 138, 153, 170, and 180 (Antunes et al. 2013; Colabuono et al. 2010; Gauquie et al. 2015; Mato et al. 2001; Yamashita et al. 2011). Congeners 138, 153, and 180 are the ones mostly detected in human serum and tissues (JECFA 2016).

Moreover, among PCBs, 12 congeners have a coplanar conformation (dioxin-like PCBs) that enables them to interact with xenobiotic receptors in the cell (i.e. aryl hydrocarbon receptor, AhR) and makes them able to interfere with the endocrine system, causing reproductive disorders among others (JECFA 2016; Pocar et al. 2005).

8.2.3 Dichlorodiphenyltrichloroethane (DDT)

Dichlorodiphenyltrichloroethane (DDT) is an organochlorine compound used as an insecticide. Total DDT concentration (tDDT) is usually expressed as the sum of DDT and its metabolites DDD and DDE (Antunes et al. 2013; Hirai et al. 2011; Ogata et al. 2009; Rios et al. 2007). When DDE/DDT ratio is very small, it indicates recent contamination. High levels of DDT in the environment and on plastics can be ascribed to their high production levels, especially by the USA, their intensive use as pesticides in agriculture, and as insecticides (JECFA 1961). While DDT production has been banned in some countries such as the USA since 1972, DDT and its derivatives are still in use as insecticides in some countries to prevent the spread of malaria.

8.2.4 Polybrominated Diphenyl Ethers (PBDEs)

Polybrominated diphenyl ethers (PBDEs) are a group of 209 anthropogenic chemicals produced and added to materials in order to improve their resistance to fire. They make up one of the major classes of brominated flame retardants (BFR). They are part of the group of lipophilic, persistent organic contaminants, can be accumulated through the food chain, and converted from one congener to another through metabolization. Their concentration in commercial products and, thus, their occurrence in the environment are not as high as those of PCBs. Industrial products are mostly made up of few congeners, so environmental data are available only for penta- hepta-, octa- and deca-BDEs, among which BDE-47 is the most abundant together with BDE-99, BDE-100, and BDE-209 (Darnerud et al. 2001). Their exposure is generally associated with hypothyroidism, since they can bind to thyroid hormone transporters, act as agonists/antagonists, or displace the bound hormones (Darnerud et al. 2001).

The congeners most commonly analysed to check contamination in feed and food are BDE-28, BDE-47, BDE-99, BDE-100, BDE-153, BDE-154, BDE-183, and BDE-209 (EFSA 2011). The log K_{ow} of these highly hydrophobic congeners ranges from 5.94 to more than 8 (Braekevelt et al. 2003). This class of POPs has among the greatest potential to cause harm to biological systems even at low concentrations (Abdelouahab et al. 2009; Carlson 1980; Fair et al. 2012).

8.2.5 Hexachlorocyclohexanes (HCHs)

Hexachlorocyclohexanes (HCHs) are long-range transported organochlorine pesticides that were mostly used either in agriculture or as insecticides. Several isomers can be present in the commercial mixtures, and when performing environmental analyses, their concentration is expressed as ΣHCH (α-HCH, β-HCH, γ-HCH, and δ-HCH). γ-HCH, the major component of pesticides, was found as the most abundant isomer in plastic samples from Mozambique and South Africa (Ogata et al. 2009). Log K_{ow} (γ-HCH = 3.8) for HCHs is lower than that for PCBs and DDTs, and they are thus supposed to partition the plastic pellets less than the other more hydrophobic chemicals (Mizukawa et al. 2013).

8.3 Desorption of Environmental Pollutants from Microplastics

New evidence implies a possible transfer of persistent, bioaccumulative, and toxic (PBT) pollutants into organisms from contaminated plastics (Avio et al. 2015; Browne et al. 2013; Chua et al. 2014; Engler 2012; Ryan et al. 1988; Tanaka et al. 2013; Wardrop et al. 2016; Yamashita et al. 2011). This would lead to an increased exposure to xenobiotic molecules once the plastic item is ingested.

Considering that the bond between a co-contaminant and the plastic substrate can be reversible, sorbed POPs will tend to desorb from the polymer into the water until equilibrium is finally attained (Andrady 2011). The desorption behaviour can be enhanced by the presence of some chemical surfactants, which are molecules able to increase the solubility of hydrophobic substances. These molecules possess an amphipathic nature, containing both hydrophilic and hydrophobic moieties, enabling them to interact with the two phases. Their mechanism of action consists of the lowering of aqueous phase polarity and the formation of micelles to include and contain the nonpolar chemicals. Among them, humic acids and linear alkyl benzene sulphonate, which can be found in the environment, have been found to significantly affect PCDD/DFs and PCBs leaching in some shredder residues and municipal solid waste (Sakai et al. 2000). In the gut, some digestive detergents are present and, through their surfactant properties, could induce the release of toxic substances from the ingested MPs, in a process that is negatively correlated with lipophilicity (Ahrens et al. 2001; Heinrich and Braunbeck 2019). This could suggest a higher degree of bioavailability.

8.3.1 Leaching of Additives from Microplastics

Many of the plastic additives (plasticizers, PBDEs flame retardants, antioxidants, and stabilizers) are not chemically bound to the polymer and can thus more easily migrate from the material. Only some reactive organic additives, such as some flame retardants, are polymerized with the plastic molecules becoming part of the polymer chain. All these chemicals are intentionally added during plastic manufacture in order to give plastics some specific features and improve their functional characteristics (e.g. UV and thermal resistance, flexibility). These compound are mixed within the bulk matrix, and their leaching behaviour can be influenced and enhanced by external factors, such as changes in temperature or pH, plastic ageing, and the presence of surfactants (Suhrhoff and Scholz-Böttcher 2016; US EPA 1992; Wei et al. 2019).

8.4 Microplastics and Nanoplastics Occurrence in Foods

Microplastic pollution has become a potential food safety threat that is especially relevant for fishery products as well as other seafood products including table salt. Although microplastics have been reported in products such as honey and sugar (Liebezeit and Liebezeit 2013) or beer (Liebezeit and Liebezeit 2014), aquatic products and water seem to be the best-studied source of dietary intake of microplastics.

8.4.1 Microplastics and Nanoplastics Occurrence in Fisheries and Aquaculture Products

Ingestion of microplastics by aquatic organisms have been reported in numerous studies; indeed, microplastics have been found in 12 out of the 25 most important species and genera that contribute to global marine fisheries (FAO 2017).

Small concentrations of microplastics of around one to two particles per fish have been observed in many important commercial species such as sardine, mackerel, anchovy, herring, and sprat from the Pacific, Atlantic, and Indian oceans, as well as the Mediterranean Sea (GESAMP 2016; Lusher et al. 2016). Other aquatic species of local and regional relevance from marine and freshwater environments have been found to contain microplastic particles too (FAO 2017).

Most microplastics have been observed in the gastrointestinal tract of fish, where most particles seem to concentrate after ingestion. However, several studies in marine organisms have shown that smaller microplastics and nanoplastics could be translocated in other organs such as the liver, although the translocation pathways are not well understood yet (Collard et al. 2017).

Bivalve molluscs have also been found to contain microplastic particles. One of the best-studied are mussels, where their occurrence has been reported in Europe, North America, Brazil, and China. The lowest level of microplastic concentration in mussels (less than 0.5 particle/g) was observed in Europe and the highest concentration of microplastics in mussels was reported in China, amounting to 4 particles/g (EFSA 2016).

Microplastics have also been observed in crustaceans, such as shrimps and in 83% of Norway lobsters in coastal waters of the North Sea and the Irish Sea, with average concentrations ranging from 0.03 to 1.92 particles/g in shrimp, mostly found in the digestive tract (Devriese et al. 2015; Murray and Cowie 2011).

An emerging threat to the consumption of fisheries and aquaculture products derives from the ability of microplastics to sorb environmental contaminants as persistent organic pollutants (POPs), metals, and pathogens, where its concentration can be several folds higher than in the water column. Humans can be exposed to these contaminants through the consumption of fisheries products especially bivalves and crustaceans, where the gastrointestinal tract is not removed. Another route of exposure might be the consumption of farmed species fed with contaminated fish or fishmeal (GESAMP 2016). Information on microplastics presence in different marine organisms (including commercial seafood species), sampling method/instrument, polymer type, and particle size are summarized in Table 8.3.

8.4.2 Microplastics and Nanoplastics Occurrence in Salt

Microplastics may contaminate table salt, as other sea products, and their occurrence has been reported in different countries such as Italy, Croatia, and Spain. The average microplastic content in Italy fluctuated from 1.57 (high-quality table salt) to 8.23 (low-quality table salt) particles/g, with a size range from 4 μm to 2100 μm. The results in Croatia fluctuated from and 27.13 (high-quality table salt) to 31.68 (low-quality table salt) particles/g with an average size range from 15 to 4628 μm (Renzi and Blašković 2018).

In Spain, microplastics were found in commercial table salt. The content went from 50 to 280 particles/kg, being PET the most frequently reported polymer, followed by PP and PE (Iñiguez et al. 2017).

8.4.3 Microplastics and Nanoplastics Occurrence in Water

The presence of microplastics has been reported in raw and treated drinking water. Depending on the water treatment, microplastic particles are found in different concentrations in water samples. Their average abundance was described from 1473 ± 34 to 3605 ± 497 particles/L in raw water and from 338 ± 76 to 628 ± 28 particles/L in treated water, depending on the treatment (Pivokonsky et al. 2018). A relevant finding is that some microplastic particles were reported to be down to the size of 1 μm and around 12 different microplastics compounds were identified, being PET, PP, and PE the most prominent polymers.

Table 8.3 Field studies on microplastic occurrence in fisheries and aquaculture products

Species	Particle amount	Sampling	Particle size	Polymers	Location	References
5 mesopelagic and 1 epipelagic fish species	1–83 items/fish (2.1 ± 5.78 items/fish)	Manta trawl	1–2.79 mm		North Pacific Central Gyre	Boerger et al. (2010)
10 species of fish (5 pelagic, 5 demersal)	1–15 items/fish (1.90 ± 0.10 items/fish)	Standard haul trawl	0.13 mm–14.3 mm	PA, LDPE, PS, RY, PEST	English Channel	Lusher et al. (2013)
Mytilus edulis *Crassostrea gigas*	0.36 ± 0.07 items/g ww 0.47 ± 0.16 items/g ww	Mussel farm, supermarket	5–10 µm 11–20 µm		North Sea, Atlantic Ocean	Van Cauwenberghe and Janssen (2014)
Mytilus edulis	Wild: 34 items/ind 106–126 items/ind Farmed: 75 items/ind, 178 items/ind	Collection at low tide, grocery store			Newfoundland, Nova Scotia, Canada	Mathalon and Hill (2014)
26 species of commercial fish	0–20 items/fish 1.40 ± 0.66 items/fish or 0.27 ± 0.63 items/fish	Stern trawlers	0.217–4.81 mm	Alkyd resin, PE, PP, RY, PA-6, PEST, acrylic,	Portuguese coast	Neves et al. (2015)
9 species of commercial bivalves	4.3–57.2 items/ind 2.1–10.5 items/g ww	Shanghai fishery market	5 µm–5 mm	PE, PET, PA	Coastal waters of China	Li et al. (2015)

Species	Particle amount	Sampling	Particle size	Polymers	Location	References
Indonesia: 11 fish species USA: 12 fish species and *Crassostrea gigas*	Indonesia: 0–21 items/fish (1.4 ± 3.7 items/fish) USA: 0–10/fish 0–2/ oyster (0.5 ± 1.4 items/ind)	Fish markets, fishermen	Indonesia: 3.5 ± 1.1 mm USA: 6.3 ± 6.7 mm (fish) 5.5 ± 5.8 mm (oysters)		Sulawesi (Indonesia) and California (USA)	Rochman et al. (2015)
Crangon crangon	0.68 ± 0.55 items/g ww Max 1.92 ± 0.61 items/g ww 1.23 ± 0.99 items/ind	Shrimp trawl and beam trawl net	200–1000 μm		Channel area and southern part of the North Sea (France, Belgium, Netherlands, UK)	Devriese et al. (2015)
Mytilus edulis *Arenicola marina*	0.2 ± 0.3 items/g ww Max 1.1 items/g ww 1.2 ± 2.8 items/g Max 11.3 items/g ww	Randomly on breakwaters Bait-pump or shovel	20–90 μm 15–100 μm	LDPE, HDPE, PS	French, Belgian, and Dutch North Sea coast	Van Cauwenberghe et al. (2015)
5 fish species (3 demersal and 2 pelagic)	0.03 ± 0.18 items/fish 0.19 ± 0.61 items/fish (pelagic)	Bottom/pelagic trawling	180 μm–50 cm	PE, PA, PP, PS, PET, PEST, PU	North and Baltic seas	Rummel et al. (2016)

(continued)

Table 8.3 (continued)

Species	Particle amount	Sampling	Particle size	Polymers	Location	References
Mytilus edulis	1.5–7.6 items/ind (4 items/ind) 0.9–4.6 items/g ww (2.2 items/g ww)	Tweezers at low tide, underwater with fishermen	< 250 μm Fibres: 33 μm–4.7 mm	CP, PET, PEST	Coastline of China	Li et al. (2016)
Perna perna	75% mussels contaminated with MPs	Byssus cut and removal	< 5000 μm		Santos estuary, São Paulo, Brazil	Santana et al. (2016)
Saccostrea cucullata	1.4–7.0 items/ind 1.5–7.2 items/g ww	Intertidal zone	20–5000 μm (mostly <100 μm)	PET, PP, PE, PS, CP, PVC, PA, EPS	Pearl River estuary, South China	Li et al. (2018a)
Penaeus monodon *Metapenaeus monoceros*	6.60 ± 2 items/ind 1.55–4.84 items/g ww (3.40 ± 1.23 items/g ww) 7.80 ± 2 items/ind 2.17–4.88 items/g ww (3.87 ± 1.05 items/g ww)	Offshore shrimp trawlers and bag net fishing	1–5 mm 250–500 μm	PA-6, RY	Northern Bay of Bengal, Bangladesh	Hossain et al. (2020)

Species	Particle amount	Sampling	Particle size	Polymers	Location	References
Venerupis philippinarum	0.07–5.47 items/g ww Wild: 8.4 ± 8.5 items/ind Max 12.7 ± 13.0 items/ind 0.9 ± 0.9 items/g Farmed: 11.3 ± 6.6 items/ind Max 15.4 ± 6.3 items/ind 1.7 ± 1.2 items/g Max 2.2 ± 0.8 items/g	0.5 m × 0.5 m quadrat, from farmed sites			Baynes sound, British Columbia	Davidson and Dudas (2016)
Emerita analoga	0.65 ± 1.64 items/ind Max 16 items/ind	Shovel, sand-coring tool		PP, polyacrylate, PEST, PE	Pacific coast of California	Horn et al. (2019)
Lithodes santolla	1–3 items/ind	Trap gears	3–>20 mm		Nassau Bay, Cape Horn, Chile	Andrade and Ovando (2017)

(continued)

Table 8.3 (continued)

Species	Particle amount	Sampling	Particle size	Polymers	Location	References
Saccostrea forskalii *Littoraria* sp. *Balanus amphitrite*	0.37 ± 0.03–0.57 ± 0.22 items/g ww 0.17 ± 0.08–0.23 ± 0.02 items/g ww 0.23 ± 0.10–0.43 ± 0.33 items/g ww	Intertidal zone at low tide		PS, PET, PA	Upper gulf of Thailand	Thushari et al. (2017)
Crassostrea gigas *Gammarus* sp. *Mytilus edulis* *Littorina littorea*	87 items/g dw 11 items/g dw 105 items/g dw 20 items/g dw	From littoral zones	1–5000 μm (Mostly 1–300 μm)		Dutch coast	Leslie et al. (2013)
Mytilus edulis	0.04–0.81 items/g ww (0.37 items/g ww)	Department stores, manual sampling at groyne and quayside	200 μm–1500 μm		Belgium, Netherlands	De Witte et al. (2014)
Chelon aurata *Rutilus kutum*	2.95 ± 1.98 items/fish 1.66 ± 1.23 items/fish		1.94 ± 0.71 mm 1.77 ± 0.53 mm	PA	Southern Caspian Sea	Zakeri et al. (2020)
Platichthys flesus *Diplodus vulgaris* *Dicentrarchus labrax*	0.18 ± 0.55 items/fish 3.14 ± 3.25 items/fish 0.30 ± 0.61 items/fish	Beam trawl		PE, PP, PEST, PA-6, RY, PAN	Mondego estuary, Portugal	Bessa et al. (2018)

Species	Particle amount	Sampling	Particle size	Polymers	Location	References
21 species of marine fish, 6 species of freshwater fish	1.1 ± 0.3–7.2 ± 2.8 items/fish 0.2 ± 0.1–17.2 ± 9.7 items/g	Fishery markets, fishermen	0.04–5 mm	CP, PET, PEST	Yangtze estuary, East China and South China Sea, Taihu lake	Jabeen et al. (2017)
Boops boops	2.47 ± 0.23–4.89 ± 0 0.45 items/fish	Bottom trawl nets and purse seine	1 nm–5 mm		Balearic Islands	Nadal et al. (2016)
6 marine fish species	1.62 ± 1.58–2.96 ± 5.21 items/fish (0–32 items/fish)	Hook and line	<5 mm		Texas Gulf Coast	Peters et al. (2017)
28 marine fish species	2.36 items/fish (1–35 items/fish)	Trawl net	656 ± 803 μm (range: 9.07–12074.11 μm)	Copolymers, alloys, PA, LDPE, PP, rubber	Mediterranean coast of Turkey	Güven et al. (2017)
26 marine fish species	0–3 items/fish	Tucker nets, fishermen	2.39 ± 0.28 mm (range: 1–3 mm)	PP, PE, PS, PVC, PAN	Saudi Arabian Red Sea coast	Baalkhuyur et al. (2018)
Hoplosternum littorale	3.6 items/fish (1–24 items/fish)	Fishermen	<1 mm–12 mm		Pajerù River, Northeast Brazil	Silva-Cavalcanti et al. (2017)
Scomberomorus cavalla *Rhizoprionodon lalandii*	2–6 items/fish 1–3 items/fish	Fishermen	2–5 mm 1–3 mm		Salvador, Northeastern Brazil	Miranda and de Carvalho-Souza (2016)
Rastrelliger kanagurta *Stolephorus waitei* *Chelon subviridis* *Johnius belangerii*	0–3 items /fish	Local markets		PP, PE, PS, PET, PA-6	Malaysia	Karami et al. (2017a)

(continued)

Table 8.3 (continued)

Species	Particle amount	Sampling	Particle size	Polymers	Location	References
Alepes djedaba *Epinephelus coioides* *Sphyraena jello* *Platycephalus indicus*	0.80 ± 0.12 items/g muscle 0.78 ± 0.22 items/g muscle 0.57 ± 0.17 items/g muscle 1.85 ± 0.46 items/g muscle	Fishmongers	<300 μm		Persian Gulf	Akhbarizadeh et al. (2018)
Mytilus galloprovincialis *Sardina pilchardus* *Pagellus erythrinus* *Mullus barbatus*	1.7 ± 0.2–2.0 ± 0.2 items/ind (1.9 ± 0.2 items/ind) 2.5 ± 0.3–5.3 ± 0.5 items/g ww 1.8 ± 0.2 items/fish 34.9 ± 7.9 items/g ww 1.9 ± 0.2 items/fish 27.8 ± 24.6 items/g ww 1.5 ± 0.3 items/fish 11.2 ± 2.8 items/g ww	Hand collection, trawling	0.1–0.5 mm	PE, PP, PET, PS, PTFE	Northern Ionian Sea	Digka et al. (2018)

Engraulis japonicus	2.3 ± 2.5 items/fish (0–15 items/fish)	Fishing	783 ± 1020 µm (150–6830 µm)	PE, PP, PS, ethylene/propylene copolymer, ethylene/propylene/diene terpolymer	Tokyo Bay	Tanaka and Takada (2016)
Mugil cephalus	Wild: 4.3 ± 14.5 items/fish Captive: 0.20 ± 0.48 items/fish	Fish markets, fish ponds	0.1–12 mm	PP, PE, PEST, PET, PA, PTFE	East coast of Hong Kong	Cheung et al. (2018)
Platycephalus indicus, Saurida tumbil, Sillago sihama, Cynoglossus abbreviatus, Penaeus semisulcatus	0.16–1.5 items/g	Trawl net	< 100 µm–> 1000 µm		Musa Estuary, Persian Gulf	Abbasi et al. (2018)
Crassostrea gigas *Mytilus edulis* *Tapes philippinarum* *Patinopecten yessoensis*	0.07 ± 0.06 items/g (0–0.19 items/g) 0.12 ± 0.11 items/g (0–0.35 items/g) 0.34 ± 0.31 items/g (0.03–1.08 items/g) 0.08 ± 0.08 items/g (0.01–0.17 items/g) Overall: 0–1.08 items/g ww 0–2.8 items/ind	Fishery markets	100–200 µm	PE, PP, PS PEST, silicon, PA, PEVA, PET, PU, acrylic, PTFE, PVC, PPS, PVA, PBT, polyethyl acrylate styrene, polyepoxides, styrene/acrylonitrile, and polystyrene ethylene butylene styrene	Seoul, Gwangju, Busan	Cho et al. (2019)

(continued)

Table 8.3 (continued)

Species	Particle amount	Sampling	Particle size	Polymers	Location	References
Cerithidea cingulata *Thais mutabilis* *Amiantis umbonella* *Amiantis purpuratus* *Pinctada radiata*	Clams and oyster: 0.2–2.2 items/g ww 3.9–6.9 items/ind Snails: 12.8–20.0 items/g ww 3.7–17.7 items/ind	Intertidal zone at low tide	10–> 5000 μm	PE, PET, PA	Persian Gulf	Naji et al. (2018)
Mytilus edulis	0.23 ± 0.09 items/g	Supermarket, collection from the wild	30–200 μm	PP, PE, PEST, ABS	Region Pays de la Loire, market in Nantes	Phuong et al. (2018)
Mytilus galloprovincialis	8.33 ± 3.58 items/g ww 3.6–12.4 items/ind 4.4–11.4 items/g ww	Local markets, collection from rocky bottom	0.75–6.00 mm (average: 1.15–2.29 mm)		Mediterranean Sea	Renzi et al. (2018)
Modiolus modiolus *Mytilus* spp.	0.086 ± 0.031 items/g ww 3.5 ± 1.29 items/ind 3.0 ± 0.9 items/g ww 3.2 ± 0.52 items/ind	Scuba diving, collection at low tide	0.2–> 2 mm	PET, PEST, poly (etherurethane)	Scottish coast	Catarino et al. (2018)

Species	Particle amount	Sampling	Particle size	Polymers	Location	References
Mytilus edulis	0.7–2.9 items/g ww 1.1–6.4 items/ind	Collection from the coast	8 μm–4.7 mm	PEST, PP, PE, RY	UK	Jiana Li et al. (2018b)
Mytilus galloprovincialis *Mytilus edulis*	0.04 ± 0.09–0.34 ± 0.33 items/g ww (0.13 ± 0.14 items/g ww)	Hand collection, department stores			Europe	Vandermeersch et al. (2015)
Pyropia spp.	Packaged: 0.9–3.0 items/g dw (1.8 ± 0.7 items/g dw) Processed: 1.0–2.8 items/g dw (1.8 ± 0.6 items/g dw)	Local markets	Packaged: 0.11–4.97 mm Processed: 0.07–4.74 mm	PEST, RY, PP, PA, CP	China, Yellow Sea	Li et al. (2020)
Mytilus edulis *Perna viridis*	1.52–5.36 items/g ww 0.77–8.22 items/ind	Tweezers at low tide	0.25–1 mm	PET, PE, RY, PVC, PP	Coastal waters of China	Qu et al. (2018)

(continued)

Table 8.3 (continued)

Species	Particle amount	Sampling	Particle size	Polymers	Location	References
Mytilus spp.	0–6.9 items/ind (1.5 ± 2.3 items/ind) 0–7.9 items/g ww (0.97 ± 2.61 items/g ww)	Hand collection, snorkelling, metal rake with net	70 µm–3870 µm	PET, PP, PE, PA, epoxy resin, PVC, PAN, styrene acrylonitrile, EVA, soloprene plastomer	Norway	Bråte et al. (2018)
Mytilus galloprovincialis *Chlamys farreri*	1.9–9.6 items/ind 2.0–12.8 items/g ww 5.2–19.4 items/ind 3.2–7.1 items/g ww	Fishery markets, coastal sampling	25 µm–5 mm	CP, PP, PTFE	China	DING et al. (2018)

8.5 Risk Profiling of Microplastics in Fisheries and Aquaculture Products

8.5.1 *Microplastics Dietary Intake*

The dietary intake of microplastics in foods depends on consumption habits, especially when dealing with fisheries and aquaculture products that mainly accumulate microplastics in the gastrointestinal tract, which minimizes the direct exposure to these particles when aquatic products are degutted. Previous studies and reviews have provided some estimates of human intake of microplastic particles through the consumption of different food commodities, and their results are illustrated in Table 8.4.

Exposure can be higher through consumption of small aquatic species such as crustaceans, echinoderms, bivalves, and small-sized fish that are commonly eaten whole. Microplastics have been also found in the muscle of canned, dried, and fresh commercial fish species, suggesting that evisceration may not be a completely efficient process, but the number of particles found in muscle is relatively low compared to the number of particles in the gastrointestinal tract (Abbasi et al. 2018; Akhbarizadeh et al. 2018; Karami et al. 2018, 2017a). In crustaceans such as shrimps, peeling practices that remove most of the digestive tract, the head, and the gills will reduce the exposure to plastics, as these parts are estimated to contain 90 percent of the microplastic particles (Devriese et al. 2015).

In Europe, human exposure to microplastics resulting from the consumption of bivalves may account for 1800 to 11,000 particles/year (Van Cauwenberghe and Janssen 2014). Mussels are among the most consumed bivalve molluscs, and the occurrence of microplastics in these products has been reported in several studies (EFSA 2016).

8.5.2 *Microplastics Uptake and Toxicity*

The physical characteristics of microplastic particles (size, shape), as well as the chemical characteristics, in addition to leaching of additives and pollutants or transport of pathogens, are the main factors for estimating the impact of these particles on food safety, but literature at this respect is indeed limited, and evidence of a possible transfer through the diet are lacking, as are the possible consequences.

Some studies have shown that small plastic particles are able to cross human placenta (Ragusa et al. 2021; Wick et al. 2010), which suggests a possible systemic increase in exposure as their size decreases (De Jong et al. 2008). The most frequent route of uptake would be absorption from the gut epithelium into the lymphatic system (Hussain et al. 2001). Absorption from the intestine is known to be very limited, and it should occur only for those microparticles <150 μm (EFSA 2016; FAO 2017). According to their size, particles are supposed to face different densities

Table 8.4 Previous estimates of exposure to microplastics through the consumption of seafood, water, and salt

Food	Reference intake	Derived MP intake	Country	References
Molluscs	72.1 g/day (top consumers) 11.8 g/day (minor consumers)	11,000 MP/year (top consumers) 1800 MP/year (minor consumers)	Europe	Van Cauwenberghe and Janssen (2014)
Fish muscle	300 g/week (adults) 50 g/week (children)	169–555 MP/week (adults) 28–92 MP/week (children)	Iran	Akhbarizadeh et al. (2018)
Bivalves	3.01 g/day	212 MP/year	South Korea	Cho et al. (2019)
Shellfish	4.03 g/day	283 MP/year	South Korea	Cho et al. (2019)
Mussels	82 g/year	123 MP/year	UK	Catarino et al. (2018)
Mussels	3.08 kg/year	4620 MP/year	Spain/ France/ Belgium	Catarino et al. (2018)
Mussels	225 g	7 µg 0.1 µg /kg bw/ day	Globally	EFSA (2016) and FAO (2017)
Fish Crustaceans Molluscs	15.21 kg/year 2.06 kg/year 2.65 kg/year	31–8323 MP/ year 206–17,716 MP/ year 0–27,825 MP/ year	Globally Globally Globally	Danopoulos et al. (2020), Danopoulos et al. (2020) and Danopoulos et al. (2020)
Water	2 l	85 µg/day 1.4 µg/kg bw/ day	Globally	WHO (2019)
Water	2.2 l/day (women) 3 l/day (men)	4400 MP/year (women) 5800 MP/year (men)	Globally	Kosuth et al. (2018)
Salt	5 g/day	1000 MP/year	China	Yang et al. (2015)
Salt	3.95 g/day	37 MP/year	Globally	Karami et al. (2017b)
Salt	5 g/day	40.6–1085.2 MP/ year	Italy	Renzi and Blašković (2018)
Salt	14.8–18.01 g/ day	64–302 MP/year	Turkey	Gündoğdu (2018)
Salt	2.3 g/day	40–680 MP/year	Globally	Kosuth et al. (2018)
Salt	10.06 g/day	0–42,600 MP/ year (average 3000)	Globally	Kim et al. (2018)
Salt	5 g/day	510 MP/year	Spain	Iñiguez et al. (2017)
Salt	5 g/day	117 µg/year	India	Seth and Shriwastav (2018)

inside the body, being either phagocyted by macrophage (> 0.5 μm), endocyted (<0.5 μm), cleared through splenic filtration (>0.2 μm), or eliminated via kidney filtration (<10 nm) (Monti et al. 2015; Yoo et al. 2011). Nevertheless, it has to be underlined that microplastics in the lymph will be excreted mostly through the spleen or with faeces through bile clearance in the liver if present in the blood (Yoo et al. 2011).

Among the possible adverse effects caused by microplastic exposure, oxidative stress and alteration of the immune function, possibly leading to immune depression, are the most likely to occur (Petit et al. 2002; Schirinzi et al. 2017). This is because some of the particles can be taken up in the lymphatic system by phagocytic cells, such as macrophages, as mentioned before.

Furthermore, it has to be underlined that PVC could cause additional toxicity because of the possible leaching of vinyl chloride, an extremely toxic monomer, classified as carcinogenic, that is used in the production process of this material and makes up 50 to 100% of the polymer by weight (Lithner et al. 2011). The toxicity of plastic is likely due to the unreacted residual monomers that constitute this material, as they may induce genotoxic effects. Lithner et al. (2011) have elaborated a hazard ranking of monomers and additives that are present in plastic polymers, indicating the hazard for human health. PVC is also the type of polymer that requires the highest amount of additives followed by PP, PE, and styrenics.

In addition to possible adverse health consequences of the polymers itself, the combined exposure to plastic additives and associated co-contaminants adsorbed by the surrounding environment must be taken into consideration. Many additives, such as BPA, phthalates, nonylphenols, and PBDE, are known to have an impact in the organism they enter in contact with, mostly through a mechanism of endocrine disruption. Besides, the main harm derives from the fact that these molecules, which give plastic some specific characteristics, are not strongly bound to the polymer matrix and tend to easily leach from it.

On the other hand, persistent organic pollutants (POPs) that interact and become associated with microplastic particles could not only impact and alter endocrine functions in the organism (i.e. PCBs, PBDEs) but also promote carcinogenicity (PAH). These outcomes mainly arise from the interaction with intracellular receptors and gene expression induction (JECFA 2016; Pocar et al. 2005). In consonance with the strength and type of interaction they have with the polymer's matrix, they could be released at different degrees.

The toxicity of dioxins and dioxin-like PCBs can be expressed through their corresponding toxic equivalency factor (TEF), implemented in the 1980s (Barnes 1991; Safe et al. 1985; Safe 1986) and later revised in 2005 (Van den Berg et al. 2006). This parameter expresses the relative toxicity of a compound in respect of a standard compound of known toxicity, the 2,3,7,8-tetrachlorodibenzo-p-dioxin (TCDD, TEF = 1). The toxic equivalency quotient (TEQ) is a value resulting from the sum of the weighted concentrations of each chemical multiplied by their TEF. Maximum uptake levels have been set for dioxin-like PCBs and dioxins in 2006 (European Commission 2006).

Allowable concentrations are estimated as provisional maximum tolerable daily intake (PTDI), provisional tolerable weekly intake (PTWI), and provisional tolerable monthly intake (PTMI) and can be expressed as WHO-TEQ. Their values are established during the hazard characterization step, by evaluating the no observed adverse effect level (NOAEL), and indicate the amount of a chemical that can be ingested daily, weekly, or monthly over a lifetime without adverse health consequences (JECFA 1995). In 2017, the FAO conducted an exposure exercise that considered mussels as the target species of greatest interest because of their high consumption and because it is consumed with the viscera, where microplastics tend to be located. The exposure assessment took into consideration the highest concentration of microplastics in mussels, which was reported at 4 particles/g in China (EFSA 2016). According to CIFOCOss, the highest reported consumption of mussels corresponded to the Belgian elderly (P95) and was estimated to be 250 g per day per person. Considering that the highest concentration of microplastics reported in mussels was 4 particles/g and the highest consumption was 250 g of mussels, it was estimated that a portion of mussels could contain up to 1000 microplastic particles, i.e. around 9 µg, depending on the volume and density of the particles.

The exposure assessment on microplastics in mussels confirmed that the intake of microplastics per day was 0.15 µg/kg for a person of 60 kg. These values were selected in order to describe the worst-case scenario and cover all the populations, including those at the highest risk.

Although the exposure assessment for mussels was carried out, the tolerable daily intake (TDI) for microplastic particles and most of its compounds has not yet been established; therefore, it is not possible to determine if this level of exposure is safe. However, the TDI for some plastic additives such as phthalates, BPA, alkylphenols, and brominated flame retardants, as well as associated sorbed contaminants such as PCBs, PAHs, and DDT, has been established, and the exposure assessment estimation shows that the level of the compounds present in microplastics from mussels was significantly lower than the TDI. Therefore, based on these assumptions, the intake of associated chemicals from ingested plastics via seafood consumption is minor compared to the total intake from the diet (FAO 2017).

As a complement, an exposure assessment exercise of microplastics in shellfish has been carried out with updated information from the current literature and also taking into account other contaminants that were not included in the previous estimates (e.g. HCHs). Procedures and outcomes are described in the following paragraph.

8.5.3 Case Study: Exposure Assessment of Microplastic Additives and Associated Sorbed Contaminants via Shellfish Consumption

Since plastic particles are more likely found in the intestinal tract, the consumption of all those seafood organisms whose GI is not removed can be the main route of exposure. Almost all shellfish are eaten whole, and in the present study, the attention is focused on mussels, shrimps, prawns, clams, and oysters. Data on daily seafood consumption were taken from the Chronic Individual Food Consumption Database (CIFOCOss) of the WHO. In order to perform an exposure assessment, countries reporting the highest consumption levels of shellfish were selected. These countries were China and Finland (mussels, clams, oysters) and the Netherlands (shrimps and prawns). Only the P95 consumers, meaning top consumers of these products from each country, were considered. The adults and elderly category presented the highest daily consumption levels (g/day).

Information about microplastic load in shellfish was taken from scientific papers, only considering the highest detected amount, described in Table 8.3. Plastic materials were assumed to be spherical with a diameter of 25 μm, which was among the most common plastic sizes found in a study by Van Cauwenberghe and Janssen (2014) in mussels. With this information, it was possible to calculate the volume of plastic particles. The density of polymers was taken from scientific papers and reviews, and only maximum values were considered (Andrady 2017; Avio et al. 2017). By knowing the volume and the density, it was then possible to determine the weight of each plastic polymer.

Data on microplastic contamination in bivalve molluscs were taken for clams (*Scapharca subcrenata*), mussels (*Mytilus galloprovincialis*), and oysters (*Saccostrea cucullata*) from three studies on commercial species (Table 8.3). For these species, the highest plastic load was estimated to be 10.5, 12.8, and 7.2 particles/g wet weight, respectively (DING et al. 2018; Li et al. 2015, 2018a). In addition to these, information on shrimp contamination were taken from a recent study, where up to 4.88 particles/g tissue (wet weight) were found in commercial brown shrimp (*Metapenaeus monoceros*) (Hossain et al. 2020).

Indirect exposure to microplastic-bound pollutants was estimated by using the highest reported contamination levels of plastic particles found in field studies (Tables 8.1 and 8.2). These data were used to provide an estimate of exposure to microplastic-bound contaminants through the consumption of shellfish.

The exposure assessment estimation consists of several steps. First, the maximum load of microplastics in shellfish (particles/g) was multiplied by the daily dietary intake of each commodity, or consumer P95 (g/day), thus measuring the total number of particles ingested every day. This result was then multiplied by the weight of each polymeric particle derived before, in order to obtain the estimated consumption of plastic per day (g/day). Then, this value was multiplied by the maximum concentration of each contaminant (ng/g) reported in field studies. With these estimations, it was possible to establish the daily intake of pollutants (ng/day) which, converted in picograms (pg) and divided by the average body weight of an adult (60 kg), would finally provide information on the daily dietary exposure to environmental contaminants mediated by microplastics (pg/ kg bw/day). This information is presented by commodity and polymer type in Table 8.5. The calculations were made based on the assumption that these chemicals were completely released from the microplastic particles.

The highest exposure to environmental contaminants associated with microplastics and plastic additives, as estimated in the present case study, could derive from the consumption of oysters, followed by mussels, clams, and finally shrimps and prawns (Table 8.5). This could cause concern because the three most contaminated shellfish groups are also the ones that are eaten whole, whereas shrimps are peeled, and most microplastics are removed.

Overall, nonylphenols are the group of xenobiotics that presented the highest microplastic-mediated exposure concentrations, which ranged from 0.25 ng/kg bw/day in PP and LDPE-contaminated shrimps to 2.33 ng/kg bw/day in PVC-contaminated mussels. These high levels can most likely be related to the use of this additive in the manufacture of plastics. After NPs, PAHs are the class of environmental pollutants that could bring out the highest harm through shellfish consumption, with a concentration ranging from 4.15 pg/kg bw/day in PP-contaminated shrimps to 39.26 pg/kg bw/day in PVC-contaminated mussels. Contrariwise, according to the present results, HCHs could be classified as the compounds whose sorbed concentration levels on microplastics could pose the least concern. In fact, the daily dietary exposure to HCHs varied from 0.00 pg/kg bw/day to 0.03 pg/kg bw/day, at maximum.

The results of the exercise demonstrate that polypropylene (PP) is the polymer that might raise the least concern when analysing microplastic-mediated exposure to xenobiotics, while PVC might be the most hazardous. Despite this, the dietary exposure to environmental contaminants on ingested plastics, as calculated in the present case study, can be considered negligible, compared to other sources. Also, there is still not enough clarity on tissue transfer dynamics and concentration of chemicals associated with microplastics, which can be influenced by factors as a fugacity gradient, so this can be considered as an estimate. In conclusion, when considering the outcomes of the present exercise, it is important to also keep into account the physical-chemical properties of each polymer type, as they could also influence the sorption/desorption of chemicals and then their concentration on the microplastics.

In addition to the previous analysis, China and Finland seafood consumption values were used to perform an estimate of the total dietary exposure to organic pollutants through shellfish consumption (Table 8.6). These two countries were chosen because they represent the ones with the highest shellfish consumption levels.

When only these two countries are considered in the analysis, the exposure levels to microplastic-associated contaminants seem to be higher. PVC is again the plastic polymer that can apparently cause the most significant exposure to environmental contaminants in organisms after consumption, followed by PET, PA, PS, HDPE, LDPE, and finally PP. Anyway, PVC is generally reported to sorb pollutants to a

Table 8.5 Overall maximum dietary exposure (pg/kg/day) of MP additives and associated sorbed pollutants resulting from the consumption of the four shellfish groups in the Netherlands, Finland, and China (age class: adults and elderly). Results are categorized by polymer type

Polymer	Seafood	MP-PAH (pg/kg bw/day)	MP-PCB (pg/kg bw/day)	MP-DDT (pg/kg bw/day)	MP-PBDE (pg/kg bw/day)	MP-HCH (pg/kg bw/day)	MP-BPA (pg/kg bw/day)	MP-NP (ng/kg bw/day)
HDPE	Shrimps and prawns	4.42	0.28	0.11	0.98	0.00	0.07	0.26
	Mussels	26.26	1.67	0.64	5.81	0.02	0.43	1.56
	Oysters	19.20	1.22	0.47	4.25	0.02	0.31	1.14
	Clams	6.94	0.44	0.17	1.53	0.01	0.11	0.41
LDPE	Shrimps and prawns	4.24	0.27	0.10	0.94	0.00	0.07	0.25
	Mussels	25.18	1.61	0.62	5.57	0.02	0.41	1.50
	Oysters	18.41	1.17	0.45	4.07	0.02	0.30	1.09
	Clams	6.65	0.42	0.16	1.47	0.01	0.11	0.39
PA	Shrimps and prawns	5.24	0.33	0.13	1.16	0.00	0.09	0.31
	Mussels	31.14	1.98	0.76	6.89	0.03	0.51	1.85
	Oysters	22.76	1.45	0.56	5.03	0.02	0.37	1.35
	Clams	8.23	0.52	0.20	1.82	0.01	0.13	0.49
PS	Shrimps and prawns	5.06	0.32	0.12	1.12	0.00	0.08	0.30
	Mussels	30.05	1.92	0.74	6.65	0.02	0.49	1.78
	Oysters	21.97	1.40	0.54	4.86	0.02	0.36	1.30
	Clams	7.94	0.51	0.19	1.76	0.01	0.13	0.47
PP	Shrimps and prawns	4.15	0.26	0.10	0.92	0.00	0.07	0.25
	Mussels	26.64	1.57	0.60	5.45	0.02	0.40	1.46
	Oysters	18.01	1.15	0.44	3.98	0.01	0.29	1.07
	Clams	6.51	0.41	0.16	1.44	0.01	0.11	0.39
PVC	Shrimps and prawns	6.61	0.42	0.16	1.46	0.01	0.11	0.39
	Mussels	39.26	2.50	0.96	8.68	0.03	0.64	2.33
	Oysters	28.70	1.83	0.70	6.35	0.02	0.47	1.70
	Clams	10.37	0.66	0.25	2.29	0.01	0.17	0.62

(continued)

Table 8.5 (continued)

Polymer	Seafood	MP-PAH (pg/kg bw/day)	MP-PCB (pg/kg bw/day)	MP-DDT (pg/kg bw/day)	MP-PBDE (pg/kg bw/ day)	MP-HCH (pg/kg bw/day)	MP-BPA (pg/kg bw/day)	MP-NP (ng/ kg bw/ day)
PET	Shrimps and prawns	6.38	0.41	0.16	1.41	0.01	0.10	0.38
	Mussels	37.90	2.42	0.93	8.38	0.03	0.62	2.25
	Oysters	27.71	1.77	0.68	6.13	0.02	0.45	1.65
	Clams	10.01	0.64	0.25	2.21	0.01	0.16	0.59

Table 8.6 Overall dietary exposure to MP additives and associated sorbed pollutants in Finland and China (age class: adults and elderly) resulting from shellfish consumption. Data on food consumption were taken from WHO CIFOCOss. Results are presented as overall exposure concentrations resulting from the combination of the four shellfish groups (mussels, oysters, clams, and shrimps and prawns), categorized by polymer type

Polymer	Country	MP-PAH (pg/kg bw/day)	MP-PCB (pg/kg bw/day)	MP-DDT (pg/kg bw/day)	MP-PBDE (pg/kg bw/ day)	MP-HCH (pg/kg bw/day)	MP-BPA (pg/kg bw/day)	MP-NP (ng/ kg bw/ day)
HDPE	China	41.52	2.65	1.02	9.18	0.03	0.68	2.47
	Finland	22.43	1.43	0.55	4.96	0.02	0.37	1.33
LDPE	China	39.81	2.54	0.98	8.80	0.03	0.65	2.36
	Finland	21.51	1.37	0.53	4.76	0.02	0.35	1.28
PA	China	49.22	3.14	1.21	10.89	0.04	0.80	2.92
	Finland	26.60	1.70	0.65	5.88	0.02	0.43	1.58
PS	China	47.51	3.03	1.17	10.51	0.04	0.77	2.82
	Finland	25.67	1.64	0.63	5.68	0.02	0.42	1.52
PP	China	38.95	2.48	0.96	8.62	0.03	0.63	2.31
	Finland	21.05	1.34	0.52	4.65	0.02	0.34	1.25
PVC	China	62.06	3.96	1.52	13.73	0.05	1.01	3.68
	Finland	33.53	2.14	0.82	7.42	0.03	0.55	1.99
PET	China	59.92	3.82	1.47	13.25	0.05	0.98	3.56
	Finland	32.38	2.06	0.79	7.16	0.03	0.53	1.92

lower amount compared to rubbery plastics and especially to PE. Our calculations, in fact, analyse the sorptive capacities of plastic materials by only taking into consideration their densities and not their physico-chemical properties. With respect to this, it should be noted that cooking and food processing can sometimes lead to physicochemical changes in the plastic particles. A recent study has measured a reduction of approximately 14% of microplastics in cooked mussels, with also a possible size reduction of the particles (Renzi et al. 2018). Moreover, high temperatures could also enhance the release of chemical compounds from microplastics (Bach et al. 2013).

Finally, it is now possible to compare the results of the exposure assessment of associated chemicals from ingested plastics via seafood consumption (Table 8.5; Table 8.6) with the no observed effect levels (NOELs) and no observed adverse effect levels (NOAELs) established by international expert committees such as the Joint FAO/WHO Expert Committee on Food Additives (JECFA) or EFSA (Table 8.7). The purpose of this last step is to check whether the highest load of microplastic-bound pollutants could lead to a significant threat to humans after the consumption of shellfish or not.

The NOEL values are not indicative of the tolerable daily intake (TDI) of contaminants but can provide useful information on the threshold of toxicological concern. In the present case study, the dietary exposures to MP-bound pollutant concentrations were in the order of nanograms (ng/kg bw/day), well below the NOELs set by the literature.

PAHs can be metabolized, resulting in genotoxic effects and carcinogenicity, especially the ones presenting a higher number of aromatic rings (Scientific Committee on Food 2002b). For this reason, no TDI can be estimated. Mean and high daily intake in adults have been estimated to be 4 and 10 ng/kg bw, with children exposure being more than twofolds higher (JECFA 2006a). Benzo(a)pyrene can be used as a marker of PAH contamination in food, but the sum of benzo[a]pyrene and chrysene (PAH2); the sum of benzo[a]pyrene, chrysene, benz[a]anthracene, and benzo[b]fluoranthene (PAH4); and the sum of benzo[a]pyrene, benz[a]anthracene, benzo[b]fluoranthene, benzo[k]fluoranthene, benzo[ghi]perylene, chrysene, dibenz[a,h]anthracene, and indeno[1,2,3-cd]pyrene (PAH8) have been suggested as

Table 8.7 No observed effect levels (NOELs), benchmark dose lower confidence limit (BMDL), no observed adverse effect levels (NOAELs), allowable daily intake (ADI), and tolerable daily intake (TDI) values established by international authorities

Chemical	NOEL /NOAEL/BMDL	ADI/TDI	Source
PCBs	0.04 mg/kg bw/day	20 ng/kg bw	Faroon et al. (2003) and JECFA (1990)
PAHs	0.17 (PAH2), 0.34 (PAH4),0.49 (PAH8) mg/kg bw	Not established, carcinogen	EFSA (2008)
Benzo(a) pyrene	100 μg/kg bw	Not established, carcinogen	JECFA (2006b)
γ-HCH	0.47 mg/kg bw/day	0–0.005 mg/kg bw	JECFA (2002)
DDT	1 mg/kg bw/day	0.01 mg/kg bw	JECFA (2001)
PBDEs	Not established	Not established	JECFA (2006c)
BPA	5 mg/kg bw/day	4 μg/kg bw	EFSA (2015) and Scientific Committee on Food (2002a)
NP	15 mg/kg/day	5 μg/kg bw	Bontje et al. (2004) and Nielsen et al. (2000)

more suitable alternatives, with the last two parameters being the most appropriate markers for genotoxic and carcinogenic PAHs in food commodities (EFSA 2008).

For DDT, the no observed effect level (NOEL) and provisional tolerable daily intake (PTDI) are 1 mg/kg bw/day and 0–0.01 mg/kg bw/day, respectively, with the highest observed average intake of 0.68 mg/man/day (JECFA 1961).

The tolerable daily intake (TDI) temporarily established for BPA in foodstuff was set as 4 µg/kg bw/day (EFSA 2015), and the daily intake in adults in Europe has been estimated up to 1.5 µg/kg bw (Scientific Committee on Food 2002a). As it concerns intake for nonylphenol, the TDI was estimated at 5 µg/kg bw/day by the Danish Institute of Safety and Toxicology (Nielsen et al. 2000). It can be concluded that the levels of microplastic-bound contaminants found in the selected commodities (shrimp, prawns, clams, oyster, and mussels) are below the ones reported in Table 8.7. Based on these estimation values, it could be assumed that the transfer of environmental pollutants and additives mediated by microplastic particles in shellfish is negligible. In fact, the contribution made by other sources such as the ingestion of contaminated prey items and subsequent transfer of POPs from food to organism supplies most of the contaminant burden (FAO 2017). However, emerging additives from MP should be further explored, and plasticizer additives such as phthalates and organophosphate flame retardants have not yet been investigated.

8.5.4 Limitations for Food Safety Risk Assessment

The fate of plastic in the human body and its possible food safety impact are unknown. Although it is thought that the particles below 1.5 µm can penetrate into the capillaries of the organs, while larger particles will be excreted (Yoo et al. 2011), there are many knowledge gaps such as toxicological data of commonly ingested plastics and its compounds.

The best-studied dietary sources of microplastics are fisheries and aquaculture products, which are important food commodities in certain areas. However, the toxicity of most plastic monomers, polymers, and additives present in microplastics has never been evaluated by relevant international expert scientific committees such as the JECFA. International expert committees such as the JECFA are key to evaluate the potential toxicity, considering newly generated scientific data and establishing the basis for the risk analysis exercises (risk assessment, risk management, and risk communication (Fig. 8.1)).

In order to perform a proper exposure and then risk assessment of plastic particles, and plastic as a vector of additives and associated sorbed pollutants, researchers should develop new techniques to better understand toxicity and transfer mechanisms. In addition to that, improvement should be made to detect smaller plastic fragments, especially those in the range of the nanoparticles, which are not much studied. These are, in fact, the ones that could mostly enhance negative consequences on the organism both because of their ability to cross biological barriers and their higher sorption capacity.

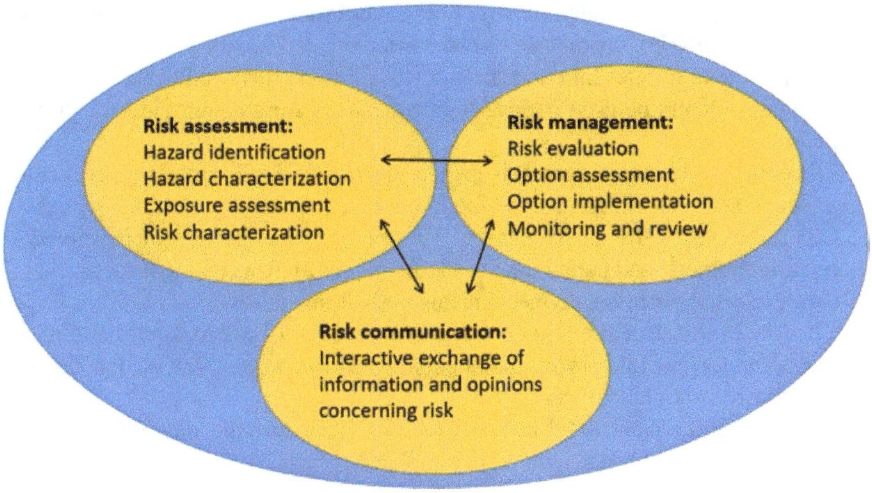

Fig. 8.1 FAO/WHO risk analysis framework. (Adapted from FAO 2017)

8.6 Research Gaps

Although most microplastics are found in the gastrointestinal tract of aquatic animals, there is a limited understanding of the presence of microplastics and specially nanoplastics in edible parts of fisheries and aquaculture products. This mainly occurs because there are no standard methodologies to analyse plastic particles in foods. While many researchers are developing analytical methods, there are still research gaps:

- Standard tissue digestion and polymer identification and quantification protocols should be determined.
- Future research should focus more on nanoplastics, which may also be the ones eliciting the highest exposure due to their size.
- As some particles have been found in the edible tissues of fish, future studies should also include muscle analysis.
- More studies should focus their attention on other food commodities that could be contaminated by microplastics, such as table salt or seaweed.
- More investigations on additive leaching and contaminant desorption processes under gut conditions should be carried out, to understand the exposure to xenobiotic chemicals through plastic ingestion.

Data that would allow a food safety risk assessment is limited and the consequences of microplastic exposure through diet are poorly understood. Most studies have investigated the effects of the inhalation of plastic particles, and very few of them analysed the consequences of a dietary uptake. As it concerns the risk assessment of microplastics, the following research gaps were identified:

- Chronic exposure to microplastics and nanoplastics in humans via ingestion, inhalation and, skin contact should be carried out.
- Environmentally relevant concentrations of plastics, their additives, and sorbed pollutants should be used in dose-response assays for a better understanding of their toxicity.
- The exposure to microorganisms present on plastic debris should be evaluated, as some pathogens might also be found on them.
- Consequences of the dietary intake of different concentrations of plastic additives, monomers, and contaminants should be clarified, as many of these chemicals might trigger response mechanisms even at low doses.
- Changes in the structure and chemical composition of microplastics and nanoplastics in foods while processing or cooking, as well as their interaction with the food matrix, should be further studied.
- Once the consequences of microplastic exposure through diets are clarified, mechanisms of control should be put in place.

8.7 Conclusions

Micro- and nanoscale plastic particles are widely distributed in the aquatic environment, and aquatic animals are exposed to them, which results in the presence of microplastics and nanoplastics in fisheries and aquaculture commodities.

Most microplastics and nanoplastics are found in the gastrointestinal tract of aquatic species, and evisceration could lead to a substantial decrease in the exposure to plastics in the final consumer. Nevertheless, there are fisheries and aquaculture commodities such as small fish, crustaceans, and bivalve mollusc that are commonly consumed whole, so the exposure through these products is higher.

There is evidence of a sorptive behaviour of microplastics and nanoplastics, which has been observed both in the laboratory and in the field. Because of their hydrophobicity and high surface/volume, plastic polymers can concentrate organic contaminants by several folds compared to the water column and might also host biofilm-forming microorganisms on their surface.

Microplastic contaminants and additives added during the manufacturing process to add specific features of the final product have been found to have adverse effects on humans and animals, mostly related to carcinogenicity and reproductive toxicity.

Data on human dietary intake of microplastics are very scarce. An exposure assessment was carried out to evaluate human exposure to environmental contaminants and plastic additives through microplastic-contaminated seafood. The worst-case estimate of the exposure to contaminants through seafood indicates the contribution of microplastic contaminants, and additives through fisheries and aquaculture products are negligible compared to other sources. Besides, the exposure scenario assumed the complete release of contaminants and additives once in the gut, which is unlikely. Moreover, even in this case, the levels of xenobiotics were several orders of magnitude lower than threshold values (NOAEL) indicated by international authorities.

Microplastic dietary intake for other food commodities such as salt or water should be carried out to understand the overall exposure through food and open the possibility for food safety risk assessment.

Literature on the actual effect of microplastics and nanoplastics in humans is extremely poor, and more research is needed to understand the toxicity of the most common polymers and plastic additives, as well as their mixture.

8.8 Glossary

8.8.1 Microplastics and Nanoplastics Definition

Recently, many studies have focused their attention on the possible harm caused by micro- and nano-sized plastic particles on human and animal health. The size of the particles seems to be one of the main aspects to possibly pose a food safety threat.

There is an ongoing debate about how to define micro- and nanoplastic particles. One of the most accepted definition describes them as plastic particles consisting of a heterogeneous mixture of different shaped materials in the range from 0.1 μm to 5000 μm in their longest dimensions (EFSA 2016; Lusher et al. 2017), while nanoplastics are defined as plastic particles whose size ranges from 0.001 μm to 0.1 μm (Klaine et al. 2012).

Plastics can also be classified according to the process that generated them. Primary microplastics are intentionally produced (e.g. plastic and/or cosmetics manufacture), while secondary microplastics are the result of fragmentation of larger materials are discharged into the environment (GESAMP 2015).

8.8.2 Microplastics and Nanoplastics Composition

Depending on the intended use, polymers with different physical and chemical properties can be mixed. Additionally, additives such as plasticizers, flame retardants, colourants, or antioxidants are normally included in various percentages to improve their performance. When plastics reach the environment, they can also sorb and accumulate many hydrophobic environmental contaminants, being a potential vector for additives and sorbed contaminants to the organisms. Among the sorbed hydrophobic pollutants are polychlorinated biphenyls (PCBs), polycyclic aromatic hydrocarbons (PAHs), and chlorinated pesticides, all of them belonging to the group of persistent organic pollutants (POPs), as they are persistent, bioaccumulative, and toxic (PBT) substances. Trace metals and microorganisms such as pathogenic bacteria or viruses might also sorb on microplastics (FAO 2017; GESAMP 2016).

8.8.2.1 Monomers and Polymers

Monomers such as ethylene, propylene, and styrene are the building blocks of polymers that lead to the production of a variety of materials. The most common polymers are acrylonitrile butadiene styrene (ABS), acrylic, epoxy resin, expanded polystyrene (EPS), polyethylene high density (HDPE), polyethylene low density (LDPE), polycarbonate, polycaprolactone, polyethylene (PE), polyethylene terephthalate (PET), poly (glycolic) acid, poly(lactide), poly(methyl methacrylate), polypropylene (PP), polystyrene (PS), polyurethane, polyethylene linear low density, polyamide (Nylon) 4, 6, 11, 66 (PA), polyvinyl alcohol, polyvinyl chloride (PVC), styrene-butadiene rubber, and thermoplastic polyurethane (FAO 2017). All these monomers and polymers can be expected to be part of microplastic particles present in the environment and therefore enter different food value chains.

8.8.2.2 Flame Retardants

Today, there are more than 175 chemicals classified as flame retardants (FRs) (Alaee et al. 2003). Some of these compounds are commonly added to polymers to reduce their flammability. Polybrominated diphenyl ethers (PBDEs) and hexabromocyclododecanes (HBCDs) are the most utilized brominated FRs in plastic manufacture and are commonly added to polystyrene, polyesters, polyolefins, polyamides, epoxies, and ABS. Some HBCDs and PBDEs are simply blended with the polymers and therefore are more likely to leach out of the products, while others can be incorporated into the polymers (Hutzinger and Thoma 1987), a consequence that poses an environmental and food safety concern. PBDEs and HBCD are listed by the Stockholm Convention as Persistent Organic Pollutants (POPs) and are associated with hepatotoxicity, kidney toxicity, endocrine-disrupting effects, and teratogenicity (Muirhead et al. 2006; Yogui and Sericano 2009). Nowadays, organophosphorus flame retardants (OPFRs) are extensively used as additives in order to replace the brominated ones. These compounds have been seen to induce toxicity in vitro and in vivo up to a certain degree (e.g. oxidative stress, cytotoxicity, endocrine disruption), but information on their accumulation and biomagnification in the food chain is still scarce (reviewed by Du et al. 2019).

8.8.2.3 Plasticizers

Substances such as phthalates and BP are used to enhance flexibility and softness and to reduce brittleness. These chemicals are usually added to synthetic polymers used in food packaging such as polycarbonate and epoxy resins but also PE, PP, and/

or PVC (EFSA 2007; FAO/WHO 2009). Both phthalates and BPA have been found to act as endocrine disruptors, causing fertility problems, cardiovascular diseases, development disorders, and reproductive cancers, while the toxicity of other BPs either remains unknown or information is not sufficient (Chen et al. 2016; Ma et al. 2019; Rochester 2013).

8.8.2.4 Antioxidants and Stabilizers

Nonylphenols (NPs) are a group of organic compounds belonging to the family of alkylphenols and are extensively used as a stabilizer in food packaging and as antioxidants in polymers such as rubber, vinyl, polyolefins, polystyrenes, and PVC (GESAMP 2016; USEPA 2010). NPs are known to be endocrine disruptors and have been reported to exert synergistic effects following their co-occurrence with other compounds (Soares et al. 2008; Vethaak et al. 2005).

References

Abbasi S, Soltani N, Keshavarzi B, Moore F, Turner A, Hassanaghaei M (2018) Microplastics in different tissues of fish and prawn from the Musa Estuary, Persian Gulf. Chemosphere. https://doi.org/10.1016/j.chemosphere.2018.04.076

Abdelouahab N, Suvorov A, Pasquier JC, Langlois MF, Praud JP, Takser L (2009) Thyroid disruption by low-dose BDE-47 in prenatally exposed lambs. Neonatology. https://doi.org/10.1159/000209316

Ahrens MJ, Hertz J, Lamoureux EM, Lopez GR, McElroy AE, Brownawell BJ (2001) The role of digestive surfactants in determining bioavailability of sediment-bound hydrophobic organic contaminants to 2 deposit-feeding polychaetes. Mar Ecol Prog Ser 212:145–157. https://doi.org/10.3354/meps212145

Akhbarizadeh R, Moore F, Keshavarzi B (2018) Investigating a probable relationship between microplastics and potentially toxic elements in fish muscles from northeast of Persian Gulf. Environ Pollut 232:154–163. https://doi.org/10.1016/j.envpol.2017.09.028

Alaee M, Arias P, Sjödin A, Bergman Å (2003) An overview of commercially used brominated flame retardants, their applications, their use patterns in different countries/regions and possible modes of release. Environ Int. https://doi.org/10.1016/S0160-4120(03)00121-1

Andrade C, Ovando F (2017) First record of microplastics in stomach content of the southern king crab Lithodes santolla (Anomura: Lithodidae), Nassau bay, Cape Horn, Chile. An del Inst la Patagon 45:59–65. https://doi.org/10.4067/s0718-686x2017000300059

Andrady AL (2011) Microplastics in the marine environment. Mar Pollut Bull 62:1596–1605. https://doi.org/10.1016/j.marpolbul.2011.05.030

Andrady AL (2017) The plastic in microplastics: a review. Mar Pollut Bull 119:12–22. https://doi.org/10.1016/j.marpolbul.2017.01.082

Antunes JC, Frias JGL, Micaelo AC, Sobral P (2013) Resin pellets from beaches of the Portuguese coast and adsorbed persistent organic pollutants. Estuar Coast Shelf Sci 130:62–69. https://doi.org/10.1016/j.ecss.2013.06.016

Ashton K, Holmes L, Turner A (2010) Association of metals with plastic production pellets in the marine environment. Mar Pollut Bull 60:2050–2055. https://doi.org/10.1016/j.marpolbul.2010.07.014

Avio CG, Gorbi S, Milan M, Benedetti M, Fattorini D, D'Errico G, Pauletto M, Bargelloni L, Regoli F (2015) Pollutants bioavailability and toxicological risk from microplastics to marine mussels. Environ Pollut 198:211–222. https://doi.org/10.1016/j.envpol.2014.12.021

Avio CG, Gorbi S, Regoli F (2017) Plastics and microplastics in the oceans: from emerging pollutants to emerged threat. Mar Environ Res 128:2–11. https://doi.org/10.1016/j.marenvres.2016.05.012

Baalkhuyur FM, Bin Dohaish E-JA, Elhalwagy MEA, Alikunhi NM, AlSuwailem AM, Røstad A, Coker DJ, Berumen ML, Duarte CM (2018) Microplastic in the gastrointestinal tract of fishes along the Saudi Arabian Red Sea coast. Mar Pollut Bull 131:407–415. https://doi.org/10.1016/j.marpolbul.2018.04.040

Bach C, Dauchy X, Severin I, Munoz J-F, Etienne S, Chagnon M-C (2013) Effect of temperature on the release of intentionally and non-intentionally added substances from polyethylene terephthalate (PET) bottles into water: chemical analysis and potential toxicity. Food Chem 139:672–680. https://doi.org/10.1016/j.foodchem.2013.01.046

Barboza LGA, Cunha SC, Monteiro C, Fernandes JO, Guilhermino L (2020) Bisphenol A and its analogs in muscle and liver of fish from the North East Atlantic Ocean in relation to microplastic contamination. Exposure and risk to human consumers. J Hazard Mater. https://doi.org/10.1016/j.jhazmat.2020.122419

Barnes DG (1991) Toxicity equivalents and EPA's risk assessment of 2,3,7,8-TCDD. Sci Total Environ 104:73–86. https://doi.org/10.1016/0048-9697(91)90008-3

Bessa F, Barría P, Neto JM, Frias JPGL, Otero V, Sobral P, Marques JC (2018) Occurrence of microplastics in commercial fish from a natural estuarine environment. Mar Pollut Bull 128:575–584. https://doi.org/10.1016/j.marpolbul.2018.01.044

Boerger CM, Lattin GL, Moore SL, Moore CJ (2010) Plastic ingestion by planktivorous fishes in the North Pacific Central Gyre. Mar Pollut Bull 60:2275–2278. https://doi.org/10.1016/j.marpolbul.2010.08.007

Bontje D, Hermens J, Vermeire T, Damstra T (2004) Integrated risk assessment: nonylphenol case study. Report prepared for the WHO/UNEP/ILO International Programme on Chemical Safety 4:64–75

Braekevelt E, Tittlemier SA, Tomy GT (2003) Direct measurement of octanol-water partition coefficients of some environmentally relevant brominated diphenyl ether congeners. Chemosphere. https://doi.org/10.1016/S0045-6535(02)00841-X

Bråte ILN, Hurley R, Iversen K, Beyer J, Thomas KV, Steindal CC, Green NW, Olsen M, Lusher A (2018) Mytilus spp. as sentinels for monitoring microplastic pollution in Norwegian coastal waters: a qualitative and quantitative study. Environ Pollut 243:383–393. https://doi.org/10.1016/j.envpol.2018.08.077

Brennecke D, Duarte B, Paiva F, Caçador I, Canning-Clode J (2016) Microplastics as vector for heavy metal contamination from the marine environment. Estuar Coast Shelf Sci. https://doi.org/10.1016/j.ecss.2015.12.003

Browne MA, Niven SJ, Galloway TS, Rowland SJ, Thompson RC (2013) Microplastic moves pollutants and additives to worms, reducing functions linked to health and biodiversity. Curr Biol 23:2388–2392. https://doi.org/10.1016/j.cub.2013.10.012

Carlson GP (1980) Induction of xenobiotic metabolism in rats by short-term administration of brominated diphenyl ethers. Toxicol Lett. https://doi.org/10.1016/0378-4274(80)90143-5

Catarino AI, Macchia V, Sanderson WG, Thompson RC, Henry TB (2018) Low levels of microplastics (MP) in wild mussels indicate that MP ingestion by humans is minimal compared to exposure via household fibres fallout during a meal. Environ Pollut. https://doi.org/10.1016/j.envpol.2018.02.069

Chagnon C, Thiel M, Antunes J, Ferreira JL, Sobral P, Ory NC (2018) Plastic ingestion and trophic transfer between Easter Island flying fish (Cheilopogon rapanouiensis) and yellowfin tuna (Thunnus albacares) from Rapa Nui (Easter Island). Environ Pollut. https://doi.org/10.1016/j.envpol.2018.08.042

Chen D, Kannan K, Tan H, Zheng Z, Feng YL, Wu Y, Widelka M (2016) Bisphenol analogues other than. BPA: environmental occurrence, human exposure, and toxicity – a review. Environ Sci Technol. https://doi.org/10.1021/acs.est.5b05387

Cheung L, Lui C, Fok L (2018) Microplastic contamination of wild and captive flathead grey mullet (Mugil cephalus). Int J Environ Res Public Health 15:597. https://doi.org/10.3390/ijerph15040597

Cho Y, Shim WJ, Jang M, Han GM, Hong SH (2019) Abundance and characteristics of microplastics in market bivalves from South Korea. Environ Pollut. https://doi.org/10.1016/j.envpol.2018.11.091

Chua EM, Shimeta J, Nugegoda D, Morrison PD, Clarke BO (2014) Assimilation of polybrominated diphenyl ethers from microplastics by the marine amphipod, Allorchestes compressa. Environ Sci Technol 48:8127–8134. https://doi.org/10.1021/es405717z

Colabuono FI, Taniguchi S, Montone RC (2010) Polychlorinated biphenyls and organochlorine pesticides in plastics ingested by seabirds. Mar Pollut Bull 60:630–634. https://doi.org/10.1016/j.marpolbul.2010.01.018

Collard F, Gilbert B, Compère P, Eppe G, Das K, Jauniaux T, Parmentier E (2017) Microplastics in livers of European anchovies (Engraulis encrasicolus, L.). Environ Pollut. https://doi.org/10.1016/j.envpol.2017.07.089

Danopoulos E, Jenner LC, Twiddy M, Rotchell JM (2020) Microplastic contamination of seafood intended for human consumption: a systematic review and meta-analysis. Environ Health Perspect. https://doi.org/10.1289/EHP7171

Darnerud PO, Eriksen GS, Jóhannesson T, Larsen PB, Viluksela M (2001) Polybrominated diphenyl ethers: occurrence, dietary exposure, and toxicology. Environ Health Perspect 109:49–68. https://doi.org/10.1289/ehp.01109s149

Davidson K, Dudas SE (2016) Microplastic ingestion by wild and cultured Manila clams (Venerupis philippinarum) from Baynes Sound, British Columbia. Arch Environ Contam Toxicol 71:147–156. https://doi.org/10.1007/s00244-016-0286-4

De Jong WH, Hagens WI, Krystek P, Burger MC, Sips AJAM, Geertsma RE (2008) Particle size-dependent organ distribution of gold nanoparticles after intravenous administration. Biomaterials 29:1912–1919. https://doi.org/10.1016/j.biomaterials.2007.12.037

De Witte B, Devriese L, Bekaert K, Hoffman S, Vandermeersch G, Cooreman K, Robbens J (2014) Quality assessment of the blue mussel (Mytilus edulis): comparison between commercial and wild types. Mar Pollut Bull 85:146–155. https://doi.org/10.1016/j.marpolbul.2014.06.006

Devriese LI, van der Meulen MD, Maes T, Bekaert K, Paul-Pont I, Frère L, Robbens J, Vethaak AD (2015) Microplastic contamination in brown shrimp (Crangon crangon, Linnaeus 1758) from coastal waters of the Southern North Sea and Channel area. Mar Pollut Bull 98:179–187. https://doi.org/10.1016/j.marpolbul.2015.06.051

Digka N, Tsangaris C, Torre M, Anastasopoulou A, Zeri C (2018) Microplastics in mussels and fish from the northern Ionian Sea. Mar Pollut Bull. https://doi.org/10.1016/j.marpolbul.2018.06.063

Ding J-F, Li J-X, Sun C-J, He C-F, Jiang F-H, Gao F-L, Zheng L (2018) Separation and identification of microplastics in digestive system of bivalves. Chinese J Anal Chem 46:690–697. https://doi.org/10.1016/S1872-2040(18)61086-2

Du J, Li H, Xu S, Zhou Q, Jin M, Tang J (2019) A review of organophosphorus flame retardants (OPFRs): occurrence, bioaccumulation, toxicity, and organism exposure. Environ Sci Pollut Res 26:22126–22136. https://doi.org/10.1007/s11356-019-05669-y

EFSA (2007) Opinion of the Scientific Panel on food additives, flavourings, processing aids and materials in contact with food (AFC) related to 2,2-BIS(4-HYDROXYPHENYL) PROPANE. EFSA J 5:428. https://doi.org/10.2903/j.efsa.2007.428

EFSA (2008) Polycyclic aromatic hydrocarbons in food – scientific opinion of the panel on contaminants in the food chain. EFSA J 724:1–114. https://doi.org/10.2903/j.efsa.2008.724

EFSA (2011) Scientific opinion on Polybrominated diphenyl ethers (PBDEs) in food. EFSA J 9:1–274. https://doi.org/10.2903/j.efsa.2011.2156

EFSA (2015) Scientific Opinion on the risks to public health related to the presence of bisphenol A (BPA) in foodstuffs. EFSA J 13:3978. https://doi.org/10.2903/j.efsa.2015.3978

EFSA (2016) Presence of microplastics and nanoplastics in food, with particular focus on seafood. EFSA J 14. https://doi.org/10.2903/j.efsa.2016.4501

Endo S, Koelmans AA (2016) Sorption of hydrophobic organic compounds to plastics in the marine environment: equilibrium. In: Hazardous chemicals associated with plastics in the marine environment, Handbook of environmental chemistry. Springer, Cham

Endo S, Takizawa R, Okuda K, Takada H, Chiba K, Kanehiro H, Ogi H, Yamashita R, Date T (2005) Concentration of polychlorinated biphenyls (PCBs) in beached resin pellets: variability among individual particles and regional differences. Mar Pollut Bull 50:1103–1114. https://doi.org/10.1016/j.marpolbul.2005.04.030

Engler RE (2012) The complex interaction between marine debris and toxic chemicals in the ocean. Environ Sci Technol 46:12302–12315. https://doi.org/10.1021/es3027105

European Commission (2006) Commission Regulation (EC) No 1881/2006 of 19 December 2006 setting maximum levels for certain contaminants in foodstuffs (Text with EEA relevance) 2006, 5–24

Fair PA, Stavros HC, Mollenhauer MAM, Dewitt JC, Henry N, Kannan K, Yun SH, Bossart GD, Keil DE, Peden-Adams MM (2012) Immune function in female B6C3F1 mice is modulated by DE-71, a commercial polybrominated diphenyl ether mixture. J Immunotoxicol. https://doi.org/10.3109/1547691X.2011.643418

FAO (2017) 2017. Microplastics in fisheries and aquaculture: status of knowledge on their occurrence and implications for aquatic organisms and food safety, FAO Fisheries and Aquaculture Technical Paper 615

FAO/WHO (2009) BISPHENOL A (BPA) – current state of knowledge and future actions by WHO and FAO. WHO, pp 1–6

Faroon OM, Keith LS, Smith-Simon C, De Rosa CT (2003) Concise international chemical assessment document 55: polychlorinated biphenyls: human health aspects. IPCS Concise Int Chem Assess Doc

Fisner M, Majer A, Taniguchi S, Bícego M, Turra A, Gorman D (2017) Colour spectrum and resin-type determine the concentration and composition of Polycyclic Aromatic Hydrocarbons (PAHs) in plastic pellets. Mar Pollut Bull. https://doi.org/10.1016/j.marpolbul.2017.06.072

Frias JPGL, Sobral P, Ferreira AM (2010) Organic pollutants in microplastics from two beaches of the Portuguese coast. Mar Pollut Bull 60:1988–1992. https://doi.org/10.1016/j.marpolbul.2010.07.030

Gauquie J, Devriese L, Robbens J, De Witte B (2015) A qualitative screening and quantitative measurement of organic contaminants on different types of marine plastic debris. Chemosphere 138:348–356. https://doi.org/10.1016/j.chemosphere.2015.06.029

GESAMP (2015) Sources, fate and effects of microplastics in the marine environment: a global assessment. Reports Stud. GESAMP

GESAMP (2016) Sources, fate and effects of microplastics in the marine environment: part two of a global assessment. Reports Stud. GESAMP

Geyer R, Jambeck JR, Law KL (2017) Production, use, and fate of all plastics ever made. Sci Adv 3:e1700782. https://doi.org/10.1126/sciadv.1700782

Granby K, Rainieri S, Rasmussen RR, Kotterman MJJ, Sloth JJ, Cederberg TL, Barranco A, Marques A, Larsen BK (2018) The influence of microplastics and halogenated contaminants in feed on toxicokinetics and gene expression in European seabass (Dicentrarchus labrax). Environ Res. https://doi.org/10.1016/j.envres.2018.02.035

Gündoğdu S (2018) Contamination of table salts from Turkey with microplastics. Food Addit Contam – Part A Chem Anal Control Expo Risk Assess. https://doi.org/10.1080/19440049.2018.1447694

Güven O, Gökdağ K, Jovanović B, Kıdeyş AE (2017) Microplastic litter composition of the Turkish territorial waters of the Mediterranean Sea, and its occurrence in the gastrointestinal tract of fish. Environ Pollut. https://doi.org/10.1016/j.envpol.2017.01.025

Heinrich P, Braunbeck T (2019) Bioavailability of microplastic-bound pollutants in vitro: the role of adsorbate lipophilicity and surfactants. Comp Biochem Physiol Part – C Toxicol Pharmacol 221:59–67. https://doi.org/10.1016/j.cbpc.2019.03.012

Heskett M, Takada H, Yamashita R, Yuyama M, Ito M, Geok YB, Ogata Y, Kwan C, Heckhausen A, Taylor H, Powell T, Morishige C, Young D, Patterson H, Robertson B, Bailey E, Mermoz J (2012) Measurement of persistent organic pollutants (POPs) in plastic resin pellets from remote islands: toward establishment of background concentrations for International Pellet Watch. Mar Pollut Bull. https://doi.org/10.1016/j.marpolbul.2011.11.004

Hipfner JM, Galbraith M, Tucker S, Studholme KR, Domalik AD, Pearson SF, Good TP, Ross PS, Hodum P (2018) Two forage fishes as potential conduits for the vertical transfer of microfibres in Northeastern Pacific Ocean food webs. Environ Pollut. https://doi.org/10.1016/j.envpol.2018.04.009

Hirai H, Takada H, Ogata Y, Yamashita R, Mizukawa K, Saha M, Kwan C, Moore C, Gray H, Laursen D, Zettler ER, Farrington JW, Reddy CM, Peacock EE, Ward MW (2011) Organic micropollutants in marine plastics debris from the open ocean and remote and urban beaches. Mar Pollut Bull 62:1683–1692. https://doi.org/10.1016/j.marpolbul.2011.06.004

Holmes LA, Turner A, Thompson RC (2012) Adsorption of trace metals to plastic resin pellets in the marine environment. Environ Pollut 160:42–48. https://doi.org/10.1016/j.envpol.2011.08.052

Holmes LA, Turner A, Thompson RC (2014) Interactions between trace metals and plastic production pellets under estuarine conditions. Mar Chem 167:25–32. https://doi.org/10.1016/j.marchem.2014.06.001

Horn D, Miller M, Anderson S, Steele C (2019) Microplastics are ubiquitous on California beaches and enter the coastal food web through consumption by Pacific mole crabs. Mar Pollut Bull 139:231–237. https://doi.org/10.1016/j.marpolbul.2018.12.039

Hossain MS, Rahman MS, Uddin MN, Sharifuzzaman SM, Chowdhury SR, Sarker S, Nawaz Chowdhury MS (2020) Microplastic contamination in Penaeid shrimp from the Northern Bay of Bengal. Chemosphere 238:124688. https://doi.org/10.1016/j.chemosphere.2019.124688

Hüffer T, Hofmann T (2016) Sorption of non-polar organic compounds by micro-sized plastic particles in aqueous solution. Environ Pollut 214:194–201. https://doi.org/10.1016/j.envpol.2016.04.018

Hussain N, Jaitley V, Florence AT (2001) Recent advances in the understanding of uptake of microparticulates across the gastrointestinal lymphatics. Adv Drug Deliv Rev 50:107–142. https://doi.org/10.1016/S0169-409X(01)00152-1

Hutzinger O, Thoma H (1987) Polybrominated dibenzo-p-dioxins and dibenzofurans: the flame retardant issue. Chemosphere. https://doi.org/10.1016/0045-6535(87)90181-0

Iñiguez ME, Conesa JA, Fullana A (2017) Microplastics in Spanish table salt. Sci Rep. https://doi.org/10.1038/s41598-017-09128-x

Jabeen K, Su L, Li J, Yang D, Tong C, Mu J, Shi H (2017) Microplastics and mesoplastics in fish from coastal and fresh waters of China. Environ Pollut 221:141–149. https://doi.org/10.1016/j.envpol.2016.11.055

JECFA (1961) Evaluation of the carcinogenic hazards of food additives. Fifth report of the Joint FAO/WHO Expert Committee on Food Additives, Geneva. WHO Technical Report Series – 220

JECFA (1990) Evaluation of certain food additives and contaminants. Thirty-fifth report of the joint FAO/WHO Expert Committee on Food Additives World Health Organization Technical Report Series 789

JECFA (1995) Application of risk analysis to food standards issues. Report of the Joint FAO/WHO Expert Consultation, Geneva. 13-17 March 1995

JECFA (2001) Pesticide residues in food – 2000. Report of the Joint Meeting of the FAO Panel of Experts on Pesticide Residues in Food and the Environment and the WHO Core Assessment Group. FAO Plant Production and Protection Paper 163, 2001. Pestic. residues food 1–249.

JECFA (2002) Pesticide residues in food – 2002. Report of the Joint Meeting of the FAO Panel of Experts on Pesticide Residues in Food and the Environment and the WHO Core Assessment

Group on Pesticide Residues. FAO Plant Production and Protection Paper 172, 2002., Pesticide residues in food

JECFA (2006a) Joint FAO/WHO expert committee on food additives. Sixty-fourth meeting Rome, 8–17 February 2005

JECFA (2006b) Evaluation of certain food contaminants. Sixty-fourth report of the Joint FAO/ WHO Expert Committee on Food Additives. WHO Technical Report Series, 930. Available at: http://apps.who.int/food-additives-contaminants-jecfadatabase/chemical.aspx?chemID=4306, World Health Organization – Technical Report Series

JECFA (2006c) Safety evaluation of certain contaminants in food. Prepared by the Sixty-fourth meeting of the Joint FAO/WHO Expert Committee on Food Additives (JECFA)

JECFA (2016) Evaluation of certain food additives and contaminants. Eightieth report of the Joint FAO/WHO Expert Committee on Food Additives, Geneva. WHO technical report series – 995. World Heal. Organ. Rep. Ser

Karami A, Golieskardi A, Ho YB, Larat V, Salamatinia B (2017a) Microplastics in eviscerated flesh and excised organs of dried fish. Sci Rep. https://doi.org/10.1038/s41598-017-05828-6

Karami A, Golieskardi A, Keong Choo C, Larat V, Galloway TS, Salamatinia B (2017b) The presence of microplastics in commercial salts from different countries. Sci Rep. https://doi.org/10.1038/srep46173

Karami A, Golieskardi A, Choo CK, Larat V, Karbalaei S, Salamatinia B (2018) Microplastic and mesoplastic contamination in canned sardines and sprats. Sci Total Environ. https://doi.org/10.1016/j.scitotenv.2017.09.005

Karapanagioti HK, Klontza I (2008) Testing phenanthrene distribution properties of virgin plastic pellets and plastic eroded pellets found on Lesvos island beaches (Greece). Mar Environ Res. https://doi.org/10.1016/j.marenvres.2007.11.005

Karapanagioti HK, Endo S, Ogata Y, Takada H (2011) Diffuse pollution by persistent organic pollutants as measured in plastic pellets sampled from various beaches in Greece. Mar Pollut Bull. https://doi.org/10.1016/j.marpolbul.2010.10.009

Kim JS, Lee HJ, Kim SK, Kim HJ (2018) Global pattern of microplastics (MPs) in commercial food-grade salts: sea salt as an indicator of seawater MP pollution. Environ Sci Technol. https://doi.org/10.1021/acs.est.8b04180

Klaine SJ, Koelmans AA, Horne N, Carley S, Handy RD, Kapustka L, Nowack B, von der Kammer F (2012) Paradigms to assess the environmental impact of manufactured nanomaterials. Environ Toxicol Chem. https://doi.org/10.1002/etc.733

Kosuth M, Mason SA, Wattenberg EV (2018) Anthropogenic contamination of tap water, beer, and sea salt. PLoS One. https://doi.org/10.1371/journal.pone.0194970

Lee H, Shim WJ, Kwon JH (2014) Sorption capacity of plastic debris for hydrophobic organic chemicals. Sci Total Environ 470–471:1545–1552. https://doi.org/10.1016/j.scitotenv.2013.08.023

Leslie HA, van Velzen MJM, Vethaak AD (2013) Microplastic survey of the Dutch environment: novel data set of microplastics in North Sea sediments, treated wastewater effluents and marine biota. IVM Inst Environ Stud 476:1–30

Li J, Yang D, Li L, Jabeen K, Shi H (2015) Microplastics in commercial bivalves from China. Environ Pollut 207:190–195. https://doi.org/10.1016/j.envpol.2015.09.018

Li J, Qu X, Su L, Zhang W, Yang D, Kolandhasamy P, Li D, Shi H (2016) Microplastics in mussels along the coastal waters of China. Environ Pollut 214:177–184. https://doi.org/10.1016/j.envpol.2016.04.012

Li HX, Ma LS, Lin L, Ni ZX, Xu XR, Shi HH, Yan Y, Zheng GM, Rittschof D (2018a) Microplastics in oysters Saccostrea cucullata along the Pearl River Estuary, China. Environ Pollut 236:619–625. https://doi.org/10.1016/j.envpol.2018.01.083

Li J, Green C, Reynolds A, Shi H, Rotchell JM (2018b) Microplastics in mussels sampled from coastal waters and supermarkets in the United Kingdom. Environ Pollut. https://doi.org/10.1016/j.envpol.2018.05.038

Li J, Zhang K, Zhang H (2018c) Adsorption of antibiotics on microplastics. Environ Pollut. https://doi.org/10.1016/j.envpol.2018.02.050

Li Y, Li M, Li Z, Yang L, Liu X (2019) Effects of particle size and solution chemistry on Triclosan sorption on polystyrene microplastic. Chemosphere 231:308–314. https://doi.org/10.1016/j.chemosphere.2019.05.116

Li Q, Feng Z, Zhang T, Ma C, Shi H (2020) Microplastics in the commercial seaweed nori. J Hazard Mater. https://doi.org/10.1016/j.jhazmat.2020.122060

Liebezeit G, Liebezeit E (2013) Non-pollen particulates in honey and sugar. Food Addit Contam – Part A Chem Anal Control Expo Risk Assess. https://doi.org/10.1080/19440049.2013.843025

Liebezeit G, Liebezeit E (2014) Synthetic particles as contaminants in German beers. Food Addit Contam – Part A Chem Anal Control Expo Risk Assess. https://doi.org/10.1080/19440049.2014.945099

Lithner D, Larsson A, Dave G (2011) Environmental and health hazard ranking and assessment of plastic polymers based on chemical composition. Sci Total Environ 409:3309–3324. https://doi.org/10.1016/j.scitotenv.2011.04.038

Llorca M, Schirinzi G, Martínez M, Barceló D, Farré M (2018) Adsorption of perfluoroalkyl substances on microplastics under environmental conditions. Environ Pollut 235:680–691. https://doi.org/10.1016/j.envpol.2017.12.075

Lusher AL, McHugh M, Thompson RC (2013) Occurrence of microplastics in the gastrointestinal tract of pelagic and demersal fish from the English Channel. Mar Pollut Bull 67:94–99. https://doi.org/10.1016/j.marpolbul.2012.11.028

Lusher AL, O'Donnell C, Officer R, O'Connor I (2016) Microplastic interactions with North Atlantic mesopelagic fish. ICES J Mar Sci. https://doi.org/10.1093/icesjms/fsv241

Lusher AL, Welden NA, Sobral P, Cole M (2017) Sampling, isolating and identifying microplastics ingested by fish and invertebrates. Anal Methods 9:1346–1360. https://doi.org/10.1039/C6AY02415G

Ma Y, Huang A, Cao S, Sun F, Wang L, Guo H, Ji R (2016) Effects of nanoplastics and microplastics on toxicity, bioaccumulation, and environmental fate of phenanthrene in fresh water. Environ Pollut 219:166–173. https://doi.org/10.1016/j.envpol.2016.10.061

Ma Y, Liu H, Wu J, Yuan L, Wang Y, Du X, Wang R, Marwa PW, Petlulu P, Chen X, Zhang H (2019) The adverse health effects of bisphenol A and related toxicity mechanisms. Environ Res. https://doi.org/10.1016/j.envres.2019.108575

Mathalon A, Hill P (2014) Microplastic fibers in the intertidal ecosystem surrounding Halifax Harbor, Nova Scotia. Mar Pollut Bull 81:69–79. https://doi.org/10.1016/j.marpolbul.2014.02.018

Mato Y, Isobe T, Takada H, Kanehiro H, Ohtake C, Kaminuma T (2001) Plastic resin pellets as a transport medium for toxic chemicals in the marine environment. Environ Sci Technol 35:318–324. https://doi.org/10.1021/es0010498

Miranda de, D. A., & de Carvalho-Souza, G. F (2016) Are we eating plastic-ingesting fish? Mar Pollut Bull. https://doi.org/10.1016/j.marpolbul.2015.12.035

Mizukawa K, Takada H, Ito M, Geok YB, Hosoda J, Yamashita R, Saha M, Suzuki S, Miguez C, Frias J, Antunes JC, Sobral P, Santos I, Micaelo C, Ferreira AM (2013) Monitoring of a wide range of organic micropollutants on the Portuguese coast using plastic resin pellets. Mar Pollut Bull 70:296–302. https://doi.org/10.1016/j.marpolbul.2013.02.008

Monti DM, Guarnieri D, Napolitano G, Piccoli R, Netti P, Fusco S, Arciello A (2015) Biocompatibility, uptake and endocytosis pathways of polystyrene nanoparticles in primary human renal epithelial cells. J Biotechnol. https://doi.org/10.1016/j.jbiotec.2014.11.004

Muirhead EK, Skillman AD, Hook SE, Schultz IR (2006) Oral Exposure of PBDE-47 in Fish: Toxicokinetics and Reproductive Effects in Japanese Medaka (Oryzias latipes) and Fathead Minnows (Pimephales promelas). Environ Sci Technol 40:523–528. https://doi.org/10.1021/es0513178

Müller A, Becker R, Dorgerloh U, Simon F-G, Braun U (2018) The effect of polymer aging on the uptake of fuel aromatics and ethers by microplastics. Environ Pollut 240:639–646. https://doi.org/10.1016/j.envpol.2018.04.127

Murray F, Cowie PR (2011) Plastic contamination in the decapod crustacean Nephrops norvegicus (Linnaeus, 1758). Mar Pollut Bull 62:1207–1217. https://doi.org/10.1016/j.marpolbul.2011.03.032

Nadal MA, Alomar C, Deudero S (2016) High levels of microplastic ingestion by the semipelagic fish bogue Boops boops (L.) around the Balearic Islands. Environ Pollut. https://doi.org/10.1016/j.envpol.2016.04.054

Naji A, Nuri M, Vethaak AD (2018) Microplastics contamination in molluscs from the northern part of the Persian Gulf. Environ Pollut. https://doi.org/10.1016/j.envpol.2017.12.046

Napper IE, Bakir A, Rowland SJ, Thompson RC (2015) Characterisation, quantity and sorptive properties of microplastics extracted from cosmetics. Mar Pollut Bull 99:178–185. https://doi.org/10.1016/j.marpolbul.2015.07.029

Neves D, Sobral P, Ferreira JL, Pereira T (2015) Ingestion of microplastics by commercial fish off the Portuguese coast. Mar Pollut Bull 101:119–126. https://doi.org/10.1016/j.marpolbul.2015.11.008

Nielsen E, Østergaard G, Inger Thorup O, Ladefoged O, Jelnes JE (2000) Toxicological evaluation and limit values for nonylphenol, nonylphenol ethoxylates, tricresyl phosphates and benzoic acid. Inst Food Saf Toxicol – Danish Vet Food Adm 43

O'Connor IA, Golsteijn L, Hendriks AJ (2016) Review of the partitioning of chemicals into different plastics: consequences for the risk assessment of marine plastic debris. Mar Pollut Bull 113:17–24. https://doi.org/10.1016/j.marpolbul.2016.07.021

Ogata Y, Takada H, Mizukawa K, Hirai H, Iwasa S, Endo S, Mato Y, Saha M, Okuda K, Nakashima A, Murakami M, Zurcher N, Booyatumanondo R, Zakaria MP, Dung LQ, Gordon M, Miguez C, Suzuki S, Moore C, Karapanagioti HK, Weerts S, McClurg T, Burres E, Smith W, Van Velkenburg M, Lang JS, Lang RC, Laursen D, Danner B, Stewardson N, Thompson RC (2009) International Pellet Watch: Global monitoring of persistent organic pollutants (POPs) in coastal waters. 1. Initial phase data on PCBs, DDTs, and HCHs. Mar Pollut Bull 58:1437–1446. https://doi.org/10.1016/j.marpolbul.2009.06.014

Ory NC, Sobral P, Ferreira JL, Thiel M (2017) Amberstripe scad Decapterus muroadsi (Carangidae) fish ingest blue microplastics resembling their copepod prey along the coast of Rapa Nui (Easter Island) in the South Pacific subtropical gyre. Sci Total Environ. https://doi.org/10.1016/j.scitotenv.2017.01.175

Peters CA, Thomas PA, Rieper KB, Bratton SP (2017) Foraging preferences influence microplastic ingestion by six marine fish species from the Texas Gulf Coast. Mar Pollut Bull. https://doi.org/10.1016/j.marpolbul.2017.06.080

Petit A, Catelas I, Antoniou J, Zukor DJ, Huk OL (2002) Differential apoptotic response of J774 macrophages to alumina and ultra-high-molecular-weight polyethylene particles. J Orthop Res 20:9–15. https://doi.org/10.1016/S0736-0266(01)00077-8

Phuong NN, Zalouk-Vergnoux A, Kamari A, Mouneyrac C, Amiard F, Poirier L, Lagarde F (2018) Quantification and characterization of microplastics in blue mussels (Mytilus edulis): protocol setup and preliminary data on the contamination of the French Atlantic coast. Environ Sci Pollut Res 25:6135–6144. https://doi.org/10.1007/s11356-017-8862-3

Pivokonsky M, Cermakova L, Novotna K, Peer P, Cajthaml T, Janda V (2018) Occurrence of microplastics in raw and treated drinking water. Sci Total Environ. https://doi.org/10.1016/j.scitotenv.2018.08.102

Pocar P, Fischer B, Klonisch T, Hombach-Klonisch S (2005) Molecular interactions of the aryl hydrocarbon receptor and its biological and toxicological relevance for reproduction. Reproduction 129:379–389. https://doi.org/10.1530/rep.1.00294

Qu X, Su L, Li H, Liang M, Shi H (2018) Assessing the relationship between the abundance and properties of microplastics in water and in mussels. Sci Total Environ 621:679–686. https://doi.org/10.1016/j.scitotenv.2017.11.284

Ragusa A, Svelato A, Santacroce C, Catalano P, Notarstefano V, Carnevali O, Papa F, Rongioletti MCA, Baiocco F, Draghi S, D'Amore E, Rinaldo D, Matta M, Giorgini E (2021) Plasticenta:

first evidence of microplastics in human placenta. Environ Int. https://doi.org/10.1016/j.envint.2020.106274

Rani M, Shim WJ, Han GM, Jang M, Song YK, Hong SH (2017) Benzotriazole-type ultraviolet stabilizers and antioxidants in plastic marine debris and their new products. Sci Total Environ. https://doi.org/10.1016/j.scitotenv.2016.11.033

Renzi M, Blašković A (2018) Litter & microplastics features in table salts from marine origin: Italian versus Croatian brands. Mar Pollut Bull. https://doi.org/10.1016/j.marpolbul.2018.06.065

Renzi M, Guerranti C, Blašković A (2018) Microplastic contents from maricultured and natural mussels. Mar Pollut Bull. https://doi.org/10.1016/j.marpolbul.2018.04.035

Rios LM, Moore C, Jones PR (2007) Persistent organic pollutants carried by synthetic polymers in the ocean environment. Mar Pollut Bull 54:1230–1237. https://doi.org/10.1016/j.marpolbul.2007.03.022

Rios LM, Jones PR, Moore C, Narayan UV (2010) Quantitation of persistent organic pollutants adsorbed on plastic debris from the Northern Pacific Gyre's "eastern garbage patch". J Environ Monit. https://doi.org/10.1039/c0em00239a

Rochester JR (2013) Bisphenol A and human health: a review of the literature. Reprod Toxicol. https://doi.org/10.1016/j.reprotox.2013.08.008

Rochman CM, Hoh E, Kurobe T, Teh SJ (2013) Ingested plastic transfers hazardous chemicals to fish and induces hepatic stress. Sci Rep 3:1–7. https://doi.org/10.1038/srep03263

Rochman CM, Lewison RL, Eriksen M, Allen H, Cook AM, Teh SJ (2014) Polybrominated diphenyl ethers (PBDEs) in fish tissue may be an indicator of plastic contamination in marine habitats. Sci Total Environ. https://doi.org/10.1016/j.scitotenv.2014.01.058

Rochman CM, Tahir A, Williams SL, Baxa DV, Lam R, Miller JT, Teh FC, Werorilangi S, Teh SJ (2015) Anthropogenic debris in seafood: plastic debris and fibers from textiles in fish and bivalves sold for human consumption. Sci Rep. https://doi.org/10.1038/srep14340

Rummel CD, Löder MGJ, Fricke NF, Lang T, Griebeler EM, Janke M, Gerdts G (2016) Plastic ingestion by pelagic and demersal fish from the North Sea and Baltic Sea. Mar Pollut Bull 102:134–141. https://doi.org/10.1016/j.marpolbul.2015.11.043

Ryan PG, Connell AD, Gardner BD (1988) Marine pollution bulletin plastic ingestion and PCBs in seabirds: is there a relationship? Mar Pollut Bull 19:174–176

Safe SH (1986) Comparative toxicology and mechanism of action of polychlorinated dibenzo-p-dioxins and dibenzofurans. Annu Rev Pharmacol Toxicol 26:371–399. https://doi.org/10.1146/annurev.pharmtox.26.1.371

Safe S, Bandiera S, Sawyer T (1985) PCBs: structure-function relationships and mechanism of action. Environ Health Perspect 60:47–56. https://doi.org/10.1289/ehp.856047

Sakai S, Urano S, Takatsuki H (2000) Leaching behavior of PCBs and PCDDs/DFs from some waste materials. Waste Manag. https://doi.org/10.1016/S0956-053X(99)00316-5

Santana MFM, Ascer LG, Custódio MR, Moreira FT, Turra A (2016) Microplastic contamination in natural mussel beds from a Brazilian urbanized coastal region: rapid evaluation through bioassessment. Mar Pollut Bull 106:183–189. https://doi.org/10.1016/j.marpolbul.2016.02.074

Savoca MS, Wohlfeil ME, Ebeler SE, Nevitt GA (2016) Marine plastic debris emits a keystone infochemical for olfactory foraging seabirds. Sci Adv 2:1–9. https://doi.org/10.1126/sciadv.1600395

Schirinzi GF, Pérez-Pomeda I, Sanchís J, Rossini C, Farré M, Barceló D (2017) Cytotoxic effects of commonly used nanomaterials and microplastics on cerebral and epithelial human cells. Environ Res 159:579–587. https://doi.org/10.1016/j.envres.2017.08.043

Scientific Committee on Food (2002a) Opinion of the Scientific Committee on Food on Bisphenol A. SCF/CS/PM/3936 Final. Eur Comm Heal Consum Prot Dir

Scientific Committee on Food (2002b) Opinion of the Scientific Committee on Food on the risks to human health of Polycyclic Aromatic Hydrocarbons in Food SCF/CS/CNTM/PAH/29 ADD1. Annex

Seth CK, Shriwastav A (2018) Contamination of Indian sea salts with microplastics and a potential prevention strategy. Environ Sci Pollut Res. https://doi.org/10.1007/s11356-018-3028-5

Silva-Cavalcanti JS, Silva JDB, de França EJ, de Araújo MCB, Gusmão F (2017) Microplastics ingestion by a common tropical freshwater fishing resource. Environ Pollut 221:218–226. https://doi.org/10.1016/j.envpol.2016.11.068

Soares A, Guieysse B, Jefferson B, Cartmell E, Lester JN (2008) Nonylphenol in the environment: a critical review on occurrence, fate, toxicity and treatment in wastewaters. Environ Int. https://doi.org/10.1016/j.envint.2008.01.004

Suhrhoff TJ, Scholz-Böttcher BM (2016) Qualitative impact of salinity, UV radiation and turbulence on leaching of organic plastic additives from four common plastics – A lab experiment. Mar Pollut Bull. https://doi.org/10.1016/j.marpolbul.2015.11.054

Takada H, Mato Y, Endo S, Yamashita R, Zakaria MP (2006) Pellet watch : global monitoring of persistent organic pollutants (POPs) using beached plastic resin pellets. English

Tanaka K, Takada H (2016) Microplastic fragments and microbeads in digestive tracts of planktivorous fish from urban coastal waters. Sci Rep. https://doi.org/10.1038/srep34351

Tanaka K, Takada H, Yamashita R, Mizukawa K, Fukuwaka MA, Watanuki Y (2013) Accumulation of plastic-derived chemicals in tissues of seabirds ingesting marine plastics. Mar Pollut Bull 69:219–222. https://doi.org/10.1016/j.marpolbul.2012.12.010

Tanaka K, Takada H, Ikenaka Y, Nakayama SMM, Ishizuka M (2020) Occurrence and concentrations of chemical additives in plastic fragments on a beach on the island of Kauai. Hawaii Mar Pollut Bull. https://doi.org/10.1016/j.marpolbul.2019.110732

Teuten EL, Saquing JM, Knappe DRU, Barlaz MA, Jonsson S, Björn A, Rowland SJ, Thompson RC, Galloway TS, Yamashita R, Ochi D, Watanuki Y, Moore C, Viet PH, Tana TS, Prudente M, Boonyatumanond R, Zakaria MP, Akkhavong K, Ogata Y, Hirai H, Iwasa S, Mizukawa K, Hagino Y, Imamura A, Saha M, Takada H (2009) Transport and release of chemicals from plastics to the environment and to wildlife. Philos Trans R Soc B Biol Sci 364:2027–2045. https://doi.org/10.1098/rstb.2008.0284

Thushari GGN, Senevirathna JDM, Yakupitiyage A, Chavanich S (2017) Effects of microplastics on sessile invertebrates in the eastern coast of Thailand: an approach to coastal zone conservation. Mar Pollut Bull 124:349–355. https://doi.org/10.1016/j.marpolbul.2017.06.010

US EPA (1992) Plastic pellets in the aquatic environment: sources and recommendations. United States Environmental Protection Agency

USEPA (2010) Nonylphenol (NP) and Nonylphenol Ethoxylates (NPEs) Action Plan. U.S. Environmental Protection Agency

Van Cauwenberghe L, Janssen CR (2014) Microplastics in bivalves cultured for human consumption. Environ Pollut. https://doi.org/10.1016/j.envpol.2014.06.010

Van Cauwenberghe L, Claessens M, Vandegehuchte MB, Janssen CR (2015) Microplastics are taken up by mussels (Mytilus edulis) and lugworms (Arenicola marina) living in natural habitats. Environ Pollut 199:10–17. https://doi.org/10.1016/j.envpol.2015.01.008

Van den Berg M, Birnbaum LS, Denison M, De Vito M, Farland W, Feeley M, Fiedler H, Hakansson H, Hanberg A, Haws L, Rose M, Safe S, Schrenk D, Tohyama C, Tritscher A, Tuomisto J, Tysklind M, Walker N, Peterson RE (2006) The 2005 World Health Organization reevaluation of human and mammalian toxic equivalency factors for dioxins and dioxin-like compounds. Toxicol Sci 93:223–241. https://doi.org/10.1093/toxsci/kfl055

Vandermeersch G, Van Cauwenberghe L, Janssen CR, Marques A, Granby K, Fait G, Kotterman MJJ, Diogène J, Bekaert K, Robbens J, Devriese L (2015) A critical view on microplastic quantification in aquatic organisms. Environ Res 143:46–55. https://doi.org/10.1016/j.envres.2015.07.016

Velzeboer I, Kwadijk CJAF, Koelmans AA (2014) Strong sorption of PCBs to nanoplastics, microplastics, carbon nanotubes, and fullerenes. Environ Sci Technol. https://doi.org/10.1021/es405721v

Vethaak AD, Lahr J, Schrap SM, Belfroid AC, Rijs GBJ, Gerritsen A, de Boer J, Bulder AS, Grinwis GCM, Kuiper RV, Legler J, Murk TAJ, Peijnenburg W, Verhaar HJM, de Voogt P (2005) An integrated assessment of estrogenic contamination and biological effects in the

aquatic environment of The Netherlands. Chemosphere 59:511–524. https://doi.org/10.1016/j.chemosphere.2004.12.053
Wang W, Wang J (2018) Comparative evaluation of sorption kinetics and isotherms of pyrene onto microplastics. Chemosphere. https://doi.org/10.1016/j.chemosphere.2017.11.078
Wang F, Shih KM, Li XY (2015) The partition behavior of perfluorooctanesulfonate (PFOS) and perfluorooctanesulfonamide (FOSA) on microplastics. Chemosphere 119:841–847. https://doi.org/10.1016/j.chemosphere.2014.08.047
Wang F, Wang F, Zeng EY (2018a) Sorption of toxic chemicals on microplastics. In: Microplastic contamination in aquatic environments. Elsevier, pp 225–247. https://doi.org/10.1016/B978-0-12-813747-5.00007-2
Wang Z, Chen M, Zhang L, Wang K, Yu X, Zheng Z, Zheng R (2018b) Sorption behaviors of phenanthrene on the microplastics identified in a mariculture farm in Xiangshan Bay, southeastern China. Sci Total Environ 628–629:1617–1626. https://doi.org/10.1016/j.scitotenv.2018.02.146
Wang J, Liu X, Liu G, Zhang Z, Wu H, Cui B, Bai J, Zhang W (2019) Size effect of polystyrene microplastics on sorption of phenanthrene and nitrobenzene. Ecotoxicol Environ Saf 173:331–338. https://doi.org/10.1016/j.ecoenv.2019.02.037
Wardrop P, Shimeta J, Nugegoda D, Morrison PD, Miranda A, Tang M, Clarke BO (2016) Chemical pollutants Sorbed to ingested microbeads from personal care products accumulate in fish. Environ Sci Technol 50:4037–4044. https://doi.org/10.1021/acs.est.5b06280
Webster L, Roose P, Bersuder P, Kotterman MJJ, Haarich M, Vorkamp K (2013) Determination of polychlorinated biphenyls (PCBs) in sediment and biota. ICES Tech Mar Environ Sci 53
Wei X-F, Linde E, Hedenqvist MS (2019) Plasticiser loss from plastic or rubber products through diffusion and evaporation. npj Mater Degrad. https://doi.org/10.1038/s41529-019-0080-7
WHO (2019) Microplastics in Drinking-Water; Licence: CC BY-NC-SA 3.0 IGO: World Health Organization: Geneva
Wick P, Malek A, Manser P, Meili D, Maeder-Althaus X, Diener L, Diener PA, Zisch A, Krug HF, Von Mandach U (2010) Barrier capacity of human placenta for nanosized materials. Environ Health Perspect 118:432–436. https://doi.org/10.1289/ehp.0901200
Yamashita R, Takada H, Fukuwaka MA, Watanuki Y (2011) Physical and chemical effects of ingested plastic debris on short-tailed shearwaters, Puffinus tenuirostris, in the North Pacific Ocean. Mar Pollut Bull 62:2845–2849. https://doi.org/10.1016/j.marpolbul.2011.10.008
Yang D, Shi H, Li L, Li J, Jabeen K, Kolandhasamy P (2015) Microplastic pollution in table salts from China. Environ Sci Technol. https://doi.org/10.1021/acs.est.5b03163
Yogui GT, Sericano JL (2009) Polybrominated diphenyl ether flame retardants in the U.S. marine environment: a review. Environ Int 35:655–666. https://doi.org/10.1016/j.envint.2008.11.001
Yoo JW, Doshi N, Mitragotri S (2011) Adaptive micro and nanoparticles: temporal control over carrier properties to facilitate drug delivery. Adv Drug Deliv Rev 63:1247–1256. https://doi.org/10.1016/j.addr.2011.05.004
Zakeri M, Naji A, Akbarzadeh A, Uddin S (2020) Microplastic ingestion in important commercial fish in the southern Caspian Sea. Mar Pollut Bull. https://doi.org/10.1016/j.marpolbul.2020.111598
Zhan Z, Wang J, Peng J, Xie Q, Huang Y, Gao Y (2016) Sorption of 3,3′,4,4′-tetrachlorobiphenyl by microplastics: a case study of polypropylene. Mar Pollut Bull 110:559–563. https://doi.org/10.1016/j.marpolbul.2016.05.036
Zhang X, Zheng M, Wang L, Lou Y, Shi L, Jiang S (2018) Sorption of three synthetic musks by microplastics. Mar Pollut Bull 126:606–609. https://doi.org/10.1016/j.marpolbul.2017.09.025
Zuo LZ, Li HX, Lin L, Sun YX, Diao ZH, Liu S, Zhang ZY, Xu XR (2019) Sorption and desorption of phenanthrene on biodegradable poly(butylene adipate co-terephthalate) microplastics. Chemosphere. https://doi.org/10.1016/j.chemosphere.2018.09.173

9

Relationship between Microplastic Pollution and Antibiotic Resistance

Nachiket P. Marathe and Michael S. Bank

Abstract Microplastic pollution is a big and rapidly growing environmental problem. Although the direct effects of microplastic pollution are increasingly studied, the indirect effects are hardly investigated, especially in the context of spreading of disease and antibiotic resistance genes, posing an apparent hazard for human health. Microplastic particles provide a hydrophobic surface that provides substrate for attachment of microorganisms and readily supports formation of microbial biofilms. Pathogenic bacteria such as fish pathogens *Aeromonas* spp., *Vibrio* spp., and opportunistic human pathogens like *Escherichia coli* are present in these biofilms. Moreover, some of these pathogens are shown to be multidrug resistant. The presence of microplastics is known to enhance horizontal gene transfer in bacteria and thus, may contribute to dissemination of antibiotic resistance. Microplastics can also adsorb toxic chemicals like antibiotics and heavy metals, which are known to select for antibiotic resistance. Microplastics may, thus, serve as vectors for transport of pathogens and antibiotic resistance genes in the aquatic environment. In this book chapter, we provide background information on microplastic biofouling ("plastisphere concept"), discuss the relationship between microplastic and antibiotic resistance, and identify knowledge gaps and directions for future research.

9.1 Introduction

Microplastic (<5 mm, GESAMP 2019) pollution is a widespread and global environmental problem that is projected to increase in upcoming decades creating significant challenges for its management and prevention (Borrelle et al. 2020; Jambeck

N. P. Marathe (✉)
Institute of Marine Research, Bergen, Norway
e-mail: Nachiket.Marathe@hi.no

M. S. Bank
Institute of Marine Research, Bergen, Norway

University of Massachusetts Amherst, Amherst, MA, USA
e-mail: Michael.Bank@hi.no

et al. 2015). Transport of microplastic from land via headwater streams and large rivers to the ocean (Hurley et al. 2018; Jambeck et al. 2015; van Wijnen et al. 2019) is an important component of the microplastic pollution cycle, and plastic particles can now be found globally throughout all ecosystem components including the atmosphere, terrestrial landscapes, aquatic freshwater and marine environments, and all types of biota including seafood species commonly consumed by humans (Bank and Hansson 2019).

Microplastics represent a novel substrate for marine bacteria including both fish and human pathogens (Dang and Lovell 2016; McCormick et al. 2014; Zettler et al. 2013) and are also a reservoir for metal resistance and antibiotic resistance genes. The role of microplastics in the spread of antibiotic resistance is a relatively new research topic that has garnered significant interest by scientists (Bank et al. 2020; Bowley et al. 2021; Guo et al. 2020; Parthasarathy et al. 2019; Radisic et al. 2020). The indirect effects of microplastics have not been well studied especially in the context of seafood safety and global food security, and these effects may pose a significant hazard for human health regarding the spread of disease (Bank et al. 2020; Guo et al. 2020). The specific objectives of this chapter were to (1) provide background information on microplastic biofouling and describe the concept of the "plastisphere" (Zettler et al. 2013), (2) discuss the relationship of microplastic and antibiotic resistance, and (3) identify knowledge gaps and directions for future research.

9.2 The Plastisphere Concept

One of the critical mechanisms of the microplastic antibiotic resistance connection is the "plastisphere" concept. This concept was originally presented in the seminal paper by Zettler et al. (2013) who reported that microbial communities attached to plastic debris were diverse and composed of heterotrophs, autotrophs, predators, and symbionts and were distinct from the surrounding marine waters. These plastic particle surfaces represented a novel substrate and/or ecological habitat within the water column and on the surface of the open ocean (Amaral-Zettler et al. 2015, 2020; Bowley et al. 2021; Oberbeckmann et al. 2018; Wright et al. 2020; Zettler et al. 2013). Microplastic particles have hydrophobic surfaces, with no net charge, upon entering the ocean as virgin artificial materials; however, they can quickly become colonized by microbial biofilms (Bowley et al. 2021; Wright et al. 2020; Zettler et al. 2013). The development of this concept was important for forming scientific questions regarding the overall direct and indirect impacts of microplastic pollution primarily because of the long residence time of microplastic in the environment and the potential for long-range transport and the associated risks of transfer of pathogens and disease (Bowley et al. 2021). Pathogenic microbes such as *Vibrio* spp. have been reported in high abundance within the plastisphere (Amaral-Zettler et al. 2020; Bowley et al. 2021; Zettler et al. 2013; Zhang et al. 2020) and although not all *vibrios* are pathogenic, they often prefer lower salinity found in coastal and estuarine areas, thus highlighting the importance of the plastisphere regarding its distribution, abundance, fate, and transport (Bowley et al. 2021; Thompson et al. 2004). These identified risks and the processes related to microplastic and microbe interactions are complex and are influenced by ocean currents (Hale et al. 2020), sources, fate and transport dynamics, trophic transfer and food web complexity, horizontal gene transfer and attachment properties (Arias-Andres et al. 2019), buoyancy and sinking properties of microplastics, variation, and uptake by farmed (Sun et al. 2020a) and wild seafood taxa, leading to subsequent human exposures (Bowley et al. 2021; Zhou et al. 2020).

9.3 Antibiotic Resistance

The introduction of antibiotics for the treatment of infectious disease is one of the most important advances in healthcare. The global spread of resistance in bacteria, particularly in human pathogens, presents major challenges for treatment and preventing the spread of infections (Ventola 2015). Annually, in the European Union/European Economic Area, an estimated more than 33,000 deaths and more than 800,000 cases of "impacted life-years" are attributable to infections caused by antibiotic-resistant pathogens, with direct and indirect estimated costs of more than 1.5 B€ (Cassini et al. 2019). The World Health Organization (WHO) has predicted the advent of infectious diseases for which no antibiotic treatment will be available (WHO 2019).

Antibiotic resistance is a natural phenomenon. Misuse and over use of antibiotics has led to the development, selection, and global spread of antibiotic resistance (Roberts and Zembower 2020). Selection pressure from the presence of antibiotics or other antimicrobial compounds like heavy metals and biocides leads to the enrichment of antibiotic-resistant bacteria and antibiotic resistance genes (ARGs) in bacteria from humans, animals, and the environment (Francino 2016; Gullberg et al. 2014; Marathe et al. 2013; Seiler and Berendonk 2012). Horizontal gene transfer (HGT) is a fundamental force driving bacterial evolution and contributes to the dissimilation of ARGs in both clinical and environmental bacteria (Boto 2010; Emamalipour et al. 2020; Jain et al. 2003). Antimicrobial compounds like antibiotics, biocides, and heavy metals can drive the development of antibiotic resistance and stimulate horizontal transfer of antibiotic resistance genes (Andersson and Hughes 2014; Jutkina et al. 2018; Zhang et al. 2018), thus aiding selection and dissemination of antibiotic resistance.

9.4 Microplastics and Antibiotic Resistance

Microorganisms attach themselves to surfaces forming a complex matrix of biopolymers and microbial cells known as biofilm (Dang and Lovell 2016). Formation of biofilms protect bacteria from unfavorable conditions in the environment (Donlan 2002). Biofilms are ubiquitous in aquatic environments and play an important role in various biological and ecological processes (Guo et al. 2018). Aquatic biofilms serve as a sink for various contaminants, like heavy metals, and antibiotics that are known to select for antibiotic resistance and stimulate horizontal transfer of antibiotic resistance genes (Gullberg et al. 2014; Guo et al. 2018; Jutkina et al. 2018; Richard et al. 2019). Accordingly, antibiotic resistance genes have been detected in natural aquatic biofilms (Balcázar et al. 2015; Guo et al. 2018).

Microplastic particles provide a hydrophobic surface that readily supports formation of microbial biofilms, where environmental conditions are the main drivers of biofilm formation (Oberbeckmann et al. 2018; Rummel et al. 2017). Pathogenic bacteria such as fish pathogens *Aeromonas* spp., *Vibrio* spp., and opportunistic human pathogens like *E. coli* can invariably be present in these biofilms (Kirstein et al. 2016; Silva et al. 2019; Viršek et al. 2017). Microplastics can selectively enrich both antibiotics and antibiotic-resistant bacteria on their surfaces in landfill leachates, freshwater, as well as in sea water (Su et al. 2020; Sun et al. 2020b; Wang et al. 2020; Wu et al. 2019). Thus, microplastics may serve as a vector for transport of pathogens in the aquatic environment.

Several methods have been used for detecting and quantifying ARGs associated with marine plastics including selective isolation of resistant bacteria and phenotypic antibiotic sensitivity testing, whole genome sequencing, shotgun metagenomics, and quantitative polymerase chain reaction (qPCR). Culture-based methods involving isolation of bacteria on a culture media followed by antibiotic sensitivity testing is a traditional approach used for studying antibiotic resistance (Khan et al. 2019). Zhang et al. (2020) carried out isolation and characterization of antibiotic-resistant marine bacteria from microplastic particles collected from marine aquaculture sites using a combination of seven antibiotics and a non-selective media. They showed presence of several multidrug-resistant marine bacteria including pathogenic *Vibrio* species on microplastics (Zhang et al. 2020). In contrast, other studies carried out selective isolation of pathogens like *Vibrio* spp. (Laverty et al. 2020) and *E. coli* (Song et al. 2020) showing presence of multidrug-resistant pathogens on marine microplastics. Recently, a study reported whole genome sequences (WGS) of antibiotic-resistant fish pathogens isolated from marine plastics (Radisic et al. 2020). With the advent of next-generation sequencing technology, WGS analysis of pathogens has become common and affordable tool for genotyping and epidemiology in clinics (Quainoo et al. 2017). WGS analyses are effective in elucidating the total metabolic potential of microorganisms and understanding the genetic basis of antibiotic resistance (Grevskott et al. 2020; Hendriksen et al. 2019). Although this is true, WGS data on microplastic-associated bacteria is largely lacking.

Only a small proportion of bacteria present in an environment can be cultivated in the lab. This limits detection and quantification of antibiotic resistance genes present in uncultivable bacteria using traditional methods (Lloyd et al. 2018; Stewart 2012). Methods like qPCR analysis or shotgun metagenomics, that use the total genomic DNA extracted from a given sample, partly overcome this limitation. Using qPCR, Wang et al. (2020) showed enrichment of ARGs like *sul1*, *tet*A, *tet*C, *tet*X, and *ermE* on plastic particles in both freshwater and sea water (Wang et al. 2020), while another study showed selective enrichment of *strB*, bla_{TEM}, *ermB*, *tet*M, and *tet*Q on microplastic particles in landfill leachates (Shi et al. 2020). These studies selected a limited number of ARGs for their analysis. In contrast, using recently developed high-throughput qPCR screening that can analyze more than 200 ARGs, Lu et al. (2020) showed presence of between 34 and 43 different ARGs on the surface of microplastic particles collected from vegetable soil (Lu et al. 2020).

Shotgun metagenomics gives an overview of the total bacteria and associated genes present in a given sample (Simon and Daniel 2011). Using this method Yang et al. (2019) found a total of 64 ARG subtypes that provide resistance against 13 different classes of antibiotics on macroplastics and microplastics collected from the North Pacific Gyre. Along with enrichment of ARGs, the study also found enrichment of metal resistance genes on microplastics (Yang et al. 2019). This study and several of the earlier described studies show presence of clinically important ARGs, like *sul1*, *tet*A, *tet*C, *tet*X, *ermE*, *aac(3)*, *macB*, and bla_{TEM}, that are invariably found in human pathogens, on microplastic particles (Alcock et al. 2020), thus suggesting that microplastics in the environment act as reservoirs for clinically important antibiotic resistance genes.

Microplastics originate from a variety of processes and invariably ends up in the marine environment via streams and large rivers (Hurley et al. 2018; Jambeck et al. 2015). High levels of microplastics reach the wastewater treatment plants (WWTP) (Dris et al. 2015). Although most of the microplastics are removed during primary and secondary waste treatment, smaller microplastics may still be present in the treated effluents (Talvitie et al. 2017). Treated effluents have low concentrations of microplastic particles but the high volume of effluents released may leads to considerable contamination of the aquatic ecosystem (Murphy et al. 2016; Talvitie et al. 2017). WWTPs receive municipal and/or hospital waste which invariably contains both human pathogens and clinically important antibiotic resistance genes (Le et al. 2016; Marathe et al. 2017, 2018, 2019; Rizzo et al. 2013). Treated effluents from waste water treatment plants are recognized as one of the major sources of environmental pollution with antibiotic resistance genes and resistant pathogens (Alexander et al. 2020; Czekalski et al. 2014; Karkman et al. 2019). The presence of microplastic particles in waste water effluents, thus, presents opportunities for antibiotic-resistant pathogens to colonize and form biofilms on plastic particles. This may lead to further dissemination of resistance in the marine environment via microplastics. Although this is true, there is limited knowledge on the impact of microplastics from treated effluents from WWTP on dissemination of ARGs in the aquatic environment.

Microplastic particles adsorb several chemicals like antibiotics, biocides, and heavy metals (Chen et al. 2020; Godoy et al. 2019; Mammo et al. 2020; Wang et al. 2020). The presence of antibiotics and active metabolites from such agents in the environment leads to selection of multidrug resistance among both clinical and environmental bacteria. Similarly, biocides and heavy metals like copper and mercury are known to co-select for antibiotic resistance (Francino 2016; Gullberg et al. 2011, 2014; Imran et al. 2019; Marathe et al. 2013; Seiler and Berendonk 2012). Adsorption of these chemicals on plastic surfaces containing microbial biofilm may lead to selection pressure in the plastisphere, resulting in active selection of antibiotic resistance on microplastic surfaces. In accordance, Imran et al. (2019) has concluded that co-contamination with microplastics and heavy metals results in development and spread of multiple drug-resistant human pathogens through co-selection mechanisms (Imran et al. 2019). Studies have shown that very low levels of antibiotics and biocides not only can select for antibiotic resistance but also can induce horizontal transfer of ARGs (Gullberg et al. 2011; Jutkina et al. 2018; Zhang et al. 2018). Moreover, bacteria in biofilms are more efficient in horizontal gene transfer compared to planktonic bacteria (Abe et al. 2020). Accordingly, studies have shown increased horizontal gene transfer in presence of microplastics via conjugation (Arias-Andres et al. 2018, 2019). Although extensive research on selection of resistance and promotion of horizontal gene transfer by antimicrobial compounds has been carried out, there is limited knowledge on the effect of adsorbed chemicals on plastisphere bacteria, especially, with reference to selection and transfer of antibiotic resistance genes on microplastic particles.

9.5 Conclusions and Directions for Future Research

Microplastics are emerging pollutants that have been detected in a range of environments. With the current trend of plastic consumption and its global production, the environmental pollution and related environmental effects caused by microplastics are expected to increase (Borrelle et al. 2020). Microplastics provide surfaces for the microorganisms to form biofilms (plastisphere) (Zettler et al. 2013). The processes and mechanisms involved in biofilm formation on microplastics are largely unclear. In-depth studies on deciphering the succession of microbes and understanding the effect of different factors that may influence biofilm formation on microplastic particles, such as the environmental conditions and the age of microplastic particles are needed (Su et al. 2020; Yang et al. 2020). Moreover, there are a limited number of studies reporting WGS of bacteria associated with microplastics (Li et al. 2019; Radisic et al. 2021). Bacteria associated with microplastics may play different ecological roles and could also be useful for bioremediation (Debroas et al. 2017). Hence, understanding the metabolic potential of bacteria in plastisphere using WGS is necessary.

Studies have investigated the composition of biofilms on microplastics and shown presence of both fish and human pathogens as well as clinically important antibiotic resistance genes (Dong et al. 2021). However, the risks associated with presence of pathogens in terms of human or fish exposure and the ability of microplastic-associated pathogens for causing infections is not fully understood. In-depth risk assessment studies on the effect of pathogen carrying microplastics on fish and human health are thus warranted. Microplastics originating in different compartments like WWTPs or aquaculture sites may carry different microbiota. WWTPs and aquaculture sites usually have presence of both antibiotic-resistant pathogens and microplastics (Cabello et al. 2016; Rodriguez-Mozaz et al. 2015). There is invariably selection pressure due to presence of antibiotics or biocides along with presence of resistant bacteria in these sites (Cabello et al. 2016; Edo et al. 2020; Yang et al. 2014). These environments could play an important role in enrichment and dispersal of pathogenic bacteria and ARGs to the marine ecosystem via microplastics. Although microplastics have been shown to increase HGT (Arias-Andres et al. 2018, 2019), the impact of microplastics on evolution and dissemination of antibiotic resistance in pathogens and environmental bacteria is largely unknown. In order to understand the indirect effects of microplastics, the relationship and interactions between microplastics and ARGs, as well as the impact of their association on aquatic environment especially on marine environment and sea food safety, needs to be further assessed. Holistic multidisciplinary studies on fate, migration, and potential environmental risks posed by microplastics through dissemination and evolution of antibiotic resistance are needed in the future, for better understanding the indirect effects of microplastic pollution.

References

Abe K, Nomura N, Suzuki S (2020) Biofilms: hot spots of horizontal gene transfer (HGT) in aquatic environments, with a focus on a new HGT mechanism. FEMS Microbiol 96(5):fiaa031

Alcock BP, Raphenya AR, Lau TT, Tsang KK, Bouchard M, Edalatmand A, Huynh W, Nguyen A-LV, Cheng AA, Liu S (2020) CARD 2020: antibiotic resistome surveillance with the comprehensive antibiotic resistance database. Nucleic Acids Res 48(D1):D517–D525

Alexander J, Hembach N, Schwartz T (2020) Evaluation of antibiotic resistance dissemination by wastewater treatment plant effluents with different catchment areas in Germany. Sci Rep 10(1):1–9

Amaral-Zettler LA, Zettler ER, Mincer TJ (2020) Ecology of the plastisphere. Nat Rev Microbiol 18(3):1–13

Amaral-Zettler LA, Zettler ER, Slikas B, Boyd GD, Melvin DW, Morrall CE, Proskurowski G, Mincer TJ (2015) The biogeography of the Plastisphere: implications for policy. Front Ecol Environ 13(10):541–546

Andersson DI, Hughes D (2014) Microbiological effects of sublethal levels of antibiotics. Nat Rev Microbiol 12(7):465–478

Arias-Andres M, Klümper U, Rojas-Jimenez K, Grossart H-P (2018) Microplastic pollution increases gene exchange in aquatic ecosystems. Environ Pollut 237:253–261

Arias-Andres M, Rojas-Jimenez K, Grossart H-P (2019) Collateral effects of microplastic pollution on aquatic microorganisms: an ecological perspective. TrAC Trends Anal Chem 112:234–240

Balcázar JL, Subirats J, Borrego CM (2015) The role of biofilms as environmental reservoirs of antibiotic resistance. Front Microbiol 6:1216

Bank MS, Hansson SV (2019) The plastic cycle: a novel and holistic paradigm for the Anthropocene. Environ Sci Technol 53(13):7177–7179

Bank MS, Ok YS, Swarzenski PW (2020) Microplastic's role in antibiotic resistance. Science 369(6509):1315

Borrelle SB, Ringma J, Law KL, Monnahan CC, Lebreton L, McGivern A, Murphy E, Jambeck J, Leonard GH, Hilleary MA (2020) Predicted growth in plastic waste exceeds efforts to mitigate plastic pollution. Science 369(6510):1515–1518

Boto L (2010) Horizontal gene transfer in evolution: facts and challenges. Proc Biol Sci 277(1683):819–827

Bowley J, Baker-Austin C, Porter A, Hartnell R, Lewis C (2021) Oceanic hitchhikers–assessing pathogen risks from marine microplastic. Trends Microbiol 29(2):107–116

Cabello FC, Godfrey HP, Buschmann AH, Dölz HJ (2016) Aquaculture as yet another environmental gateway to the development and globalisation of antimicrobial resistance. Lancet Infect Dis 16(7):e127–e133

Cassini A, Högberg LD, Plachouras D, Quattrocchi A, Hoxha A, Simonsen GS, Colomb-Cotinat M, Kretzschmar ME, Devleesschauwer B, Cecchini M (2019) Attributable deaths and disability-adjusted life-years caused by infections with antibiotic-resistant bacteria in the EU and the European Economic Area in 2015: a population-level modelling analysis. Lancet Infect Dis 19(1):56–66

Chen X, Gu X, Bao L, Ma S, Mu Y (2020) Comparison of adsorption and desorption of triclosan between microplastics and soil particles. Chemosphere 263:127947

Czekalski N, Díez EG, Bürgmann H (2014) Wastewater as a point source of antibiotic-resistance genes in the sediment of a freshwater lake. ISME J 8(7):1381–1390

Dang H, Lovell CR (2016) Microbial surface colonization and biofilm development in marine environments. Microbiol Mol Biol Rev 80(1):91–138

Debroas D, Mone A, Ter Halle A (2017) Plastics in the North Atlantic garbage patch: a boat-microbe for hitchhikers and plastic degraders. Sci Total Environ 599:1222–1232

Dong H, Chen Y, Wang J, Zhang Y, Zhang P, Li X, Zou J, Zhou A (2021) Interactions of microplastics and antibiotic resistance genes and their effects on the aquaculture environments. J Hazard Mater 403:123961

Donlan RM (2002) Biofilms: microbial life on surfaces. Emerg Infect Dis 8(9):881

Dris R, Gasperi J, Rocher V, Saad M, Renault N, Tassin B (2015) Microplastic contamination in an urban area: a case study in greater Paris. Environ Chem 12(5):592–599

Edo C, González-Pleiter M, Leganés F, Fernández-Piñas F, Rosal R (2020) Fate of microplastics in wastewater treatment plants and their environmental dispersion with effluent and sludge. Environ Pollut 259:113837

Emamalipour M, Seidi K, Vahed SZ, Jahanban-Esfahlan A, Jaymand M, Majdi H, Amoozgar Z, Chitkushev L, Javaheri T, Jahanban-Esfahlan R (2020) Horizontal gene transfer: from evolutionary flexibility to disease progression. Front Cell Dev Biol 8:229

Francino M (2016) Antibiotics and the human gut microbiome: dysbioses and accumulation of resistances. Front Microbiol 6:1543

Godoy V, Blázquez G, Calero M, Quesada L, Martín-Lara M (2019) The potential of microplastics as carriers of metals. Environ Pollut 255:113363

GESAMP (2019) Guidelines or the monitoring and assessment of plastic litter and microplastics in the ocean (Kershaw P.J., Turra A. and Galgani F. eds), IMO/FAO/UNESCO-IOC/UNIDO/WMO/IAEA/UN/UNEP/UNDP/ISA Joint Group of Experts on the Scientific Aspects of Marine Environmental Protection). Rep. Stud. GESAMP No. 99, 130p

Grevskott DH, Salvà-Serra F, Moore ERB, Marathe NP (2020) Nanopore sequencing reveals genomic map of CTX-M-type extended spectrum β-lactamases carried by *Escherichia coli* strains isolated from blue mussels (Mytilus edulis) in Norway. BMC Microbiol 20:134

Gullberg E, Albrecht LM, Karlsson C, Sandegren L, Andersson DI (2014) Selection of a multidrug resistance plasmid by sublethal levels of antibiotics and heavy metals. MBio 5(5):e01918-01914–e01918-01914

Gullberg E, Cao S, Berg OG, Ilback C, Sandegren L, Hughes D, Andersson DI (2011) Selection of resistant bacteria at very low antibiotic concentrations. PLoS Pathog 7(7):e1002158

Guo X-P, Sun X-L, Chen Y-R, Hou L, Liu M, Yang Y (2020) Antibiotic resistance genes in biofilms on plastic wastes in an estuarine environment. Sci Total Environ 745:140916

Guo X-P, Yang Y, Lu D-P, Niu Z-S, Feng J-N, Chen Y-R, Tou F-Y, Garner E, Xu J, Liu M (2018) Biofilms as a sink for antibiotic resistance genes (ARGs) in the Yangtze Estuary. Water Res 129:277–286

Hale RC, Seeley ME, La Guardia MJ, Mai L, Zeng EY (2020) A global perspective on microplastics. J Geophys Res Oceans 125(1):e2018JC014719

Hendriksen RS, Bortolaia V, Tate H, Tyson G, Aarestrup FM, McDermott P (2019) Using genomics to track global antimicrobial resistance. Front Public Health 7:242

Hurley R, Woodward J, Rothwell JJ (2018) Microplastic contamination of river beds significantly reduced by catchment-wide flooding. Nat Geosci 11(4):251–257

Imran M, Das KR, Naik MM (2019) Co-selection of multi-antibiotic resistance in bacterial pathogens in metal and microplastic contaminated environments: an emerging health threat. Chemosphere 215:846–857

Jain R, Rivera MC, Moore JE, Lake JA (2003) Horizontal gene transfer accelerates genome innovation and evolution. Mol Biol Evol 20(10):1598–1602

Jambeck JR, Geyer R, Wilcox C, Siegler TR, Perryman M, Andrady A, Narayan R, Law KL (2015) Plastic waste inputs from land into the ocean. Science 347(6223):768–771

Jutkina J, Marathe N, Flach C-F, Larsson D (2018) Antibiotics and common antibacterial biocides stimulate horizontal transfer of resistance at low concentrations. Sci Total Environ 616:172–178

Karkman A, Pärnänen K, Larsson DJ (2019) Fecal pollution can explain antibiotic resistance gene abundances in anthropogenically impacted environments. Nat Commun 10(1):1–8

Khan ZA, Siddiqui MF, Park S (2019) Current and emerging methods of antibiotic susceptibility testing. Diagnostics 9(2):49

Kirstein IV, Kirmizi S, Wichels A, Garin-Fernandez A, Erler R, Löder M, Gerdts G (2016) Dangerous hitchhikers? Evidence for potentially pathogenic *Vibrio* spp on microplastic particles. Mar Environ Res 120:1–8

Laverty AL, Primpke S, Lorenz C, Gerdts G, Dobbs FC (2020) Bacterial biofilms colonizing plastics in estuarine waters, with an emphasis on Vibrio spp. and their antibacterial resistance. PLoS One 15(8):e0237704

Le T-H, Ng C, Chen H, Yi XZ, Koh TH, Barkham TMS, Zhou Z, Gin KY-H (2016) Occurrences and characterization of antibiotic resistant bacteria and genetic determinants of hospital wastewaters in a tropical country. Antimicrob Agents Chemother 60(12):01556–01516

Li Q, Xu X, He C, Zheng L, Gao W, Sun C, Li J, Gao F (2019) Complete genome sequence of a quorum-sensing bacterium, *Oceanicola* sp. strain D3, isolated from a microplastic surface in coastal water of Qingdao. China Microbiol Resour Announc 8(40):e01022–e01019

Lloyd KG, Steen AD, Ladau J, Yin J, Crosby L (2018) Phylogenetically novel uncultured microbial cells dominate earth microbiomes. MSystems 3(5):e00055–e00018

Lu X-M, Lu P-Z, Liu X-P (2020) Fate and abundance of antibiotic resistance genes on microplastics in facility vegetable soil. Sci Total Environ 709:136276

Mammo F, Amoah I, Gani K, Pillay L, Ratha S, Bux F, Kumari S (2020) Microplastics in the environment: interactions with microbes and chemical contaminants. Sci Total Environ 743:140518

Marathe NP, Berglund F, Razavi M, Pal C, Dröge J, Samant S, Kristiansson E, Larsson DGJ (2019) Sewage effluent from an Indian hospital harbors novel carbapenemases and integron-borne antibiotic resistance genes. Microbiome 7(1):97

Marathe NP, Janzon A, Kotsakis SD, Flach C-F, Razavi M, Berglund F, Kristiansson E, Larsson DJ (2018) Functional metagenomics reveals a novel carbapenem-hydrolyzing mobile beta-lactamase from Indian river sediments contaminated with antibiotic production waste. Environ Int 112:279–286

Marathe NP, Pal C, Gaikwad SS, Jonsson V, Kristiansson E, Larsson DGJ (2017) Untreated urban waste contaminates Indian river sediments with resistance genes to last resort antibiotics. Water Res 1247:388–397

Marathe NP, Regina VR, Walujkar SA, Charan SS, Moore ERB, Larsson DGJ, Shouche YS (2013) A treatment plant receiving waste water from multiple bulk drug manufacturers is a reservoir for highly multi-drug resistant integron-bearing bacteria. PLoS One 8(10):e77310

McCormick A, Hoellein TJ, Mason SA, Schluep J, Kelly JJ (2014) Microplastic is an abundant and distinct microbial habitat in an urban river. Environ Sci Technol 48(20):11863–11871

Murphy F, Ewins C, Carbonnier F, Quinn B (2016) Wastewater treatment works (WwTW) as a source of microplastics in the aquatic environment. Environ Sci Technol 50(11):5800–5808

Oberbeckmann S, Kreikemeyer B, Labrenz M (2018) Environmental factors support the formation of specific bacterial assemblages on microplastics. Front Microbiol 8:2709

Parthasarathy A, Tyler AC, Hoffman MJ, Savka MA, Hudson AO (2019) Is plastic pollution in aquatic and terrestrial environments a driver for the transmission of pathogens and the evolution of antibiotic resistance? Environ Sci Technol 53(4):2

Quainoo S, Coolen JP, van Hijum SA, Huynen MA, Melchers WJ, van Schaik W, Wertheim HF (2017) Whole-genome sequencing of bacterial pathogens: the future of nosocomial outbreak analysis. Clin Microbiol Rev 30(4):1015–1063

Radisic V, Lunestad BT, Sanden M, Bank MS, Marathe NP (2021) Draft genome sequence of multidrug-resistant *Pseudomonas protegens* strain 11HC2, isolated from marine plastic collected from the west coast of Norway. Microbiol Resour Announc 10(2):e01285–e01220

Radisic V, Nimje PS, Bienfait AM, Marathe NP (2020) Marine plastics from Norwegian west coast carry potentially virulent fish pathogens and opportunistic human pathogens harboring new variants of antibiotic resistance genes. Microorganisms 8(8):1200

Richard H, Carpenter EJ, Komada T, Palmer PT, Rochman CM (2019) Biofilm facilitates metal accumulation onto microplastics in estuarine waters. Sci Total Environ 683:600–608

Rizzo L, Manaia C, Merlin C, Schwartz T, Dagot C, Ploy MC, Michael I, Fatta-Kassinos D (2013) Urban wastewater treatment plants as hotspots for antibiotic resistant bacteria and genes spread into the environment: a review. Sci Total Environ 447:345–360

Roberts SC, Zembower TR (2020) Global increases in antibiotic consumption: a concerning trend for WHO targets. Lancet Infect Dis 21(1):2

Rodriguez-Mozaz S, Chamorro S, Marti E, Huerta B, Gros M, Sànchez-Melsió A, Borrego CM, Barceló D, Balcázar JL (2015) Occurrence of antibiotics and antibiotic resistance genes in hospital and urban wastewaters and their impact on the receiving river. Water Res 69:234–242

Rummel CD, Jahnke A, Gorokhova E, Kühnel D, Schmitt-Jansen M (2017) Impacts of biofilm formation on the fate and potential effects of microplastic in the aquatic environment. Environ Sci Technol Lett 4(7):258–267

Seiler C, Berendonk TU (2012) Heavy metal driven co-selection of antibiotic resistance in soil and water bodies impacted by agriculture and aquaculture. Front Microbiol 3:399

Shi J, Wu D, Su Y, Xie B (2020) Selective enrichment of antibiotic resistance genes and pathogens on polystyrene microplastics in landfill leachate. Sci Total Environ 142775

Silva MM, Maldonado GC, Castro RO, de Sá Felizardo J, Cardoso RP, dos Anjos RM, de Araújo FV (2019) Dispersal of potentially pathogenic bacteria by plastic debris in Guanabara Bay, RJ, Brazil. Mar Pollut Bull 141:561–568

Simon C, Daniel R (2011) Metagenomic analyses: past and future trends. Appl Environ Microbiol 77(4):1153–1161

Song J, Jongmans-Hochschulz E, Mauder N, Imirzalioglu C, Wichels A, Gerdts G (2020) The Travelling Particles: investigating microplastics as possible transport vectors for multidrug resistant E. coli in the Weser estuary (Germany). Sci Total Environ 720:137603

Stewart EJ (2012) Growing unculturable bacteria. J Bacteriol 194(16):4151–4160

Su Y, Zhang Z, Zhu J, Shi J, Wei H, Xie B, Shi H (2020) Microplastics act as vectors for antibiotic resistance genes in landfill leachate: the enhanced roles of the long-term aging process. Environ Pollut 270:116278

Sun X, Chen B, Xia B, Li Q, Zhu L, Zhao X, Gao Y, Qu K (2020a) Impact of mariculture-derived microplastics on bacterial biofilm formation and their potential threat to mariculture: a case in situ study on the Sungo Bay, China. Environ Pollut 262:114336

Sun Y, Cao N, Duan C, Wang Q, Ding C, Wang J (2020b) Selection of antibiotic resistance genes on biodegradable and non-biodegradable microplastics. J Hazard Mater 409:124979

Talvitie J, Mikola A, Setälä O, Heinonen M, Koistinen A (2017) How well is microlitter purified from wastewater?–a detailed study on the stepwise removal of microlitter in a tertiary level wastewater treatment plant. Water Res 109:164–172

Thompson FL, Iida T, Swings J (2004) Biodiversity of vibrios. Microbiol Mol Biol Rev 68(3):403–431

van Wijnen J, Ragas AM, Kroeze C (2019) Modelling global river export of microplastics to the marine environment: sources and future trends. Sci Total Environ 673:392–401

Ventola CL (2015) The antibiotic resistance crisis: part 1: causes and threats. Pharm Therapeutics 40(4):277

Viršek MK, Lovšin MN, Koren Š, Kržan A, Peterlin M (2017) Microplastics as a vector for the transport of the bacterial fish pathogen species Aeromonas salmonicida. Mar Pollut Bull 125(1–2):301–309

Wang S, Xue N, Li W, Zhang D, Pan X, Luo Y (2020) Selectively enrichment of antibiotics and ARGs by microplastics in river, estuary and marine waters. Sci Total Environ 708:134594

WHO (2019) No time to wait: securing the future from drug-resistant infections. World Health Organization, Geneva

Wright RJ, Erni-Cassola G, Zadjelovic V, Latva M, Christie-Oleza JA (2020) Marine plastic debris: a new surface for microbial colonization. Environ Sci Technol 54(19):11657–11672

Wu X, Pan J, Li M, Li Y, Bartlam M, Wang Y (2019) Selective enrichment of bacterial pathogens by microplastic biofilm. Water Res 165:114979

Yang K, Chen Q-L, Chen M-L, Li H-Z, Liao H, Pu Q, Zhu Y-G, Cui L (2020) Temporal dynamics of antibiotic resistome in the plastisphere during microbial colonization. Environ Sci Technol 54(18):11322–11332

Yang Y, Li B, Zou S, Fang HH, Zhang T (2014) Fate of antibiotic resistance genes in sewage treatment plant revealed by metagenomic approach. Water Res 62:97–106

Yang Y, Liu G, Song W, Ye C, Lin H, Li Z, Liu W (2019) Plastics in the marine environment are reservoirs for antibiotic and metal resistance genes. Environ Int 123:79–86

Zettler ER, Mincer TJ, Amaral-Zettler LA (2013) Life in the "plastisphere": microbial communities on plastic marine debris. Environ Sci Technol 47(13):7137–7146

Zhang Y, Gu AZ, Cen T, Li X, He M, Li D, Chen J (2018) Sub-inhibitory concentrations of heavy metals facilitate the horizontal transfer of plasmid-mediated antibiotic resistance genes in water environment. Environ Pollut 237:74–82

Zhang Y, Lu J, Wu J, Wang J, Luo Y (2020) Potential risks of microplastics combined with superbugs: enrichment of antibiotic resistant bacteria on the surface of microplastics in mariculture system. Ecotoxicol Environ Saf 187:109852

Zhou W, Han Y, Tang Y, Shi W, Du X, Sun S, Liu G (2020) Microplastics aggravate the bioaccumulation of two waterborne veterinary antibiotics in an edible bivalve species: potential mechanisms and implications for human health. Environ Sci Technol 54(13):8115–8122

10

A Comprehensive Study of UN Basel Convention's Global Plastic Waste Partnership

Susan Wingfield and Melisa Lim

Abstract The pollution of our marine and terrestrial environment by plastic waste is one of the most pressing global environmental challenges faced today. Developing a circular plastic economy and limiting plastic pollution requires multilevel actions from different stakeholders including oil and petrochemical producers, plastic manufacturers, consumer goods companies, retailers, consumers, waste managers, waste management authorities, plastic recyclers and others. As well as cleaning up the enormous quantities of plastic waste already in our oceans and lakes, there is an urgent need to strengthen countries' capacities to prevent, minimize and properly manage this waste. The Basel Convention, the most comprehensive global environmental treaty dealing with hazardous and other wastes, offers an important part of the solution. In addition to its provisions aimed at controlling the exports and imports of hazardous wastes and other wastes generated from households and ensuring their environmentally sound management, the Convention also seeks to tackle the problem at its source through prevention and minimization. With the addition of an amendment to the Convention specifically tackling plastic waste, and

The views expressed in this publication are those of the authors and do not necessarily reflect the views of the Secretariats of the Basel, Rotterdam and Stockholm Conventions (BRS), the United Nations Environment Programme (UNEP), the United Nations (UN) or contributory organizations. BRS, UNEP or the UN do not accept responsibility for the accuracy or completeness of the contents and shall not be liable for any loss or damage that may be occasioned, directly or indirectly, through the use of, or reliance on, the contents of this publication. Mention of the names of companies and commercial products does not imply their endorsement by the UN, UNEP and BRS. The designations employed and the presentation of the materials in this publication do not imply the expression of any opinion whatsoever on the part of BRS, UNEP or the UN, on the geopolitical situations or the legal status of any country, territory, city or area or their authorities, or concerning the delimitation of their frontiers or boundaries.

S. Wingfield (✉) · M. Lim
UNEP Basel, Rotterdam and Stockholm Conventions Secretariat, Geneva, Switzerland
e-mail: susan.wingfield@un.org; melisa.lim@un.org

the establishment of a Plastic Waste Partnership, the Convention is positioned at the forefront in the fight against plastic pollution.

10.1 Introduction

The pollution of our marine and terrestrial environment by plastic waste is one of the most pressing global environmental challenges faced today. Over the last 10 years, we have produced more plastic than during the whole of the last century: global plastic production has increased steadily and reached 320 million tonnes a year in 2015 (Beckman 2018). Of the estimated 6.3 billion tonnes of plastic waste produced since the 1950s, only 9% has been recycled and another 12% incinerated (Parker 2018). An estimated 100 million tonnes of plastic is in our seas, 80–90% of which has come from land-based sources (Miles 2019). The good news is, since an estimated 80% of that land-based waste is due to a lack of efficient collection and management schemes, the problem is solvable (European Commission 2020).

Developing a circular plastic economy and limiting plastic pollution requires multilevel actions from different stakeholders including oil and petrochemical producers, plastic manufacturers, consumer goods companies, retailers, consumers, waste managers, waste management authorities, plastic recyclers and others (Ryberg et al. 2018).

Many governments are already taking considerable steps towards curbing plastic pollution, such as bans on the use of single-use plastic bags in at least 69 countries worldwide (United Nations Environment Programme 2018). Corporations are also taking concerted efforts to reduce their plastic footprints (Ellen Macarthur Foundation 2017). Concerned citizens are also doing their part, saying “no” to plastic straws, relying on reusable water bottles and coming equipped with their own grocery bags to the supermarket. There are also multi-million dollar efforts to clean up rivers, seas and oceans.[1] However, as well as cleaning up the enormous quantities of plastic waste already in our oceans and lakes, there is an urgent need to strengthen countries’ capacities to prevent, minimize and properly manage this waste.

The Basel Convention offers an important part of the solution. In addition to its provisions aimed at controlling the exports and imports of hazardous wastes and wastes generated from households and ensuring their environmentally sound management, the Convention also seeks to tackle the problem at its source through prevention and minimization. With the addition of an amendment to the Convention specifically tackling plastic waste,[2] and the establishment of a Plastic Waste

[1] The Ocean Clean-up. Website. https://theoceancleanup.com/about/

[2] Became effective on 1 January 2021. For further information, see the Basel Convention website: http://www.basel.int/Implementation/Plasticwaste/PlasticWasteAmendments/Overview/tabid/8426/Default.aspx

Partnership, the Convention is positioned at the forefront in the fight against plastic pollution.

10.2 The Evolution of the Basel Convention

The cross-border transport of hazardous wastes seized the public's attention in the 1980s. The misadventures of "toxic ships" such as the Katrin B and the Pelicano, sailing from port to port trying to offload their toxic cargoes made the front-page headlines around the world. These tragic incidents were motivated in good part by tighter environmental regulations in industrialized countries. As the costs of waste disposal skyrocketed, "toxic traders" searching for cheaper solutions started shipping hazardous wastes to Africa, Eastern Europe and other regions. Once on shore, these waste shipments were improperly managed, resulting in profound impacts on human health and the environment. It was against this backdrop that the Basel Convention was negotiated in the late 1980s. Its thrust at the time of its adoption was to combat the "toxic trade", as it was termed.

The Basel Convention, which entered into force in 1992, is the most comprehensive global environmental treaty dealing with hazardous and other wastes requiring special consideration. It has near universal coverage, encompassing 188 Parties as of November 2020. The central objective of the Convention is "to protect, by strict control, human health and the environment against the adverse effects which may result from the generation and management of hazardous wastes and other wastes".[3] The main provisions of the Convention focus on i. the reduction of hazardous and other waste generation and the promotion of environmentally sound management of hazardous and other wastes, wherever the place of disposal; ii. the restriction of transboundary movements of hazardous wastes except where it is perceived to be in accordance with the principles of environmentally sound management; and iii. a regulatory, or prior informed consent (PIC), procedure applying to cases where transboundary movements are permissible (Krueger 1999).

Since its inception, Parties to the Convention have developed a selection of tools to aid in its implementation. Technical guidelines, guidance documents and manuals, developed under the Convention by experts from developed and developing countries, in addition to stakeholders from industry and civil society, have provided countries with practical solutions to prevent and minimize their waste, to inventorize that which is produced and to develop infrastructure and effective techniques to ensure its proper recycling and final disposal. Coupled with this extensive repository of guidance designed to assist Parties and others in implementing the Convention are technical assistance and capacity building efforts undertaken by the Secretariat

[3] Basel Convention on the Control of Transboundary Movements of Hazardous Wastes and their Disposal (Basel Convention), U.N.T.S. vol. 1673, p. 57, Preambular para. 24.

and the Basel and Stockholm Convention's 22 regional centres worldwide.[4] Added to this, a partnership programme under the Convention has a successful track record in tackling problematic waste streams such as end-of-life mobile phones,[5] electrical and electronic waste or e-waste,[6] household waste[7] and, more recently, plastic waste.[8]

10.3 Basel Tackles Plastic Waste

Historically, China has been the largest plastic waste importer. China has imported 72.4 percent of traded plastic waste globally (Brooks et al. 2018). China's "National Sword" policy, enacted in January 2018, banned the import of most plastics and other materials headed for recycling processors, which had handled nearly half of the world's recyclable waste for the past quarter century (Katz 2019). The introduction of these plastic waste import restrictions by China led to an increased amount of plastic waste being sent to countries in Southeast Asia such as Indonesia, Malaysia, Vietnam and the Philippines. These countries, faced with the same challenges that China faced, are subsequently closing their doors (Niranjan 2019).

10.4 The Plastic Waste Amendments

Responding to growing concern and public awareness of the issue of marine plastic litter and microplastics, the Government of Norway submitted in June 2018 a proposal to amend the annexes to the Basel Convention to address plastic waste within its provisions.

The fourteenth meeting of the Conference of the Parties to the Basel Convention (COP-14, 29 April–10 May 2019) adopted amendments to Annexes II,[9] VIII[10] and

[4] For further information, see the Basel Convention website: http://www.basel.int/Partners/RegionalCentres/Overview/tabid/2334/Default.aspx and the Stockholm Convention website: http://chm.pops.int/Partners/RegionalCentres/Overview/tabid/425/Default.aspx

[5] For further information, see the Basel Convention website: http://www.basel.int/Implementation/TechnicalAssistance/Partnerships/MPPI/Overview/tabid/3268/Default.aspx

[6] For further information, see the Basel Convention website: http://www.basel.int/Implementation/TechnicalAssistance/Partnerships/PACE/Overview/tabid/3243/Default.aspx and http://www.basel.int/Implementation/TechnicalAssistance/Partnerships/FollowuptoPACE/tabid/8089/Default.aspx

[7] For further information, see the Basel Convention website: http://www.basel.int/Implementation/HouseholdWastePartnership/Overview/tabid/5082/Default.aspx

[8] For further information, see the Basel Convention website: http://www.basel.int/Implementation/Plasticwaste/PlasticWastePartnership/tabid/8096/Default.aspx

[9] Categories of waste requiring special consideration.

[10] Wastes presumed to be hazardous and therefore subject to the PIC procedure.

IX[11] to the Convention with the objectives of enhancing the control of the transboundary movements of plastic waste and clarifying the scope of the Convention as it applies to such waste. Starting with Annex VIII, the amendment sees insertion of a new entry, A3210, that clarifies the scope of plastic wastes presumed to be hazardous and therefore subject to the PIC procedure. The amendment to Annex IX, with a new entry B3011 replacing existing entry B3010, clarifies the types of plastic wastes that are presumed to not be hazardous and, as such, not subject to the PIC procedure. The wastes listed in entry B3011 include a group of cured resins, non-halogenated and fluorinated polymers, provided the waste is destined for recycling in an environmentally sound manner and almost free from contamination and other types of wastes, and mixtures of plastic wastes consisting of polyethylene (PE), polypropylene (PP) or polyethylene terephthalate (PET), provided they are destined for separate recycling of each material and in an environmentally sound manner and almost free from contamination and other types of wastes. The third amendment is the insertion of a new entry Y48 in Annex II which covers plastic waste, including mixtures of such wastes unless these are hazardous (as they would fall under A3210) or presumed to not be hazardous (as they would fall under B3011). The new entries became effective as of 1 January 2021.

Taken collectively, the ultimate result of the plastic waste amendments is the broadening of the scope of plastic wastes, including mixtures of such wastes, that could be subject to the PIC procedure, meaning that exporting countries will need to formally obtain the consent of importing countries to receive shipments of such plastic waste and ensure that the importing countries have the capacity to manage plastic waste in an environmentally sound manner. It will be up to each Party to take the necessary measures to transpose the new entries into national law, as needed and depending on its legal system. Such measures should be taken in a timely manner to ensure that, on 1 January 2021, each Party is in a position to implement the provisions of the Basel Convention with respect to the plastic wastes listed in entries A3210 and Y48. This includes applying the PIC procedure in case of a transboundary movement of such wastes but also applying the Convention's provisions with respect to minimizing waste generation and ensuring their environmentally sound management.

To assist Parties with these new undertakings, the Conference of the Parties decided on a range of additional steps to ensure that, once the entries became effective, the world would be ready to overcome the plastic waste challenge.[12] To start with, the Plastic Waste Partnership was established to provide a global platform to bring together countries from all over the world, working hand in hand with stakeholders from civil society and the business community to promote the environmentally sound management of plastic waste and prevent and minimize its generation. Additional guidance on how to ensure, more generally, the environmentally sound management of waste as well as its prevention and minimization is available in an

---

[11] Wastes that are presumed to not be hazardous and, as such, not subject to the PIC procedure.

[12] Thanks to generous support from Norway, France, Switzerland, Sweden, Japan, Germany, Canada and the European Union, as well as from the private sector, BRS has been in a position to mobilize around 10 million US dollars over the past 2 years to support these efforts.

ESM toolkit.[13] Equally important is the launch of additional technical and legal work on how to develop an inventory of plastic wastes; on the updating of the technical guidelines on the identification and environmentally sound management of plastic wastes and for their disposal; and to consider whether any additional constituents or characteristics in relation to plastic waste should be added to Annexes I or III, respectively, to the Convention.

Along with supporting these new undertakings, the Basel Convention Secretariat is providing technical assistance to support the preparedness of countries to implement the amendments. The amendments are expected to significantly impact the way in which plastic waste is traded internationally and consequently the extent to which it is generated and how it is managed at the national level. It is also anticipated that the amendments will provide a powerful incentive not only for the private sector – but also for governments and other stakeholders – to strengthen capacities for recycling in an environmentally sound manner. Moreover, by encouraging the expansion of infrastructures for the environmentally sound management of plastic waste, the amendments should also help create jobs and economic opportunities, not least by incentivizing innovation, such as in the design of alternatives to plastic and the phase-out of hazardous additives.[14]

10.5 The Plastic Waste Partnership

The Plastic Waste Partnership, established by the Conference of the Parties to the Basel Convention in May 2019,[15] is a platform that unites stakeholders from governments, international organizations, NGOs and industry towards the common objective of eliminating the leakage of plastic waste into our environment. The Partnership seeks to mobilize its broad stakeholder base to tackle the issue of plastic pollution on multiple fronts: from stimulating the development of strategies to strengthen policy and regulatory frameworks within countries; to developing solutions to improve the collection, separation and sound management of plastic waste; and to stimulating innovations for increasing the durability, reusability, reparability and recyclability of plastics. The Partnership creates a collaborative environment promoting the sharing of experiences, best practices and technologies towards this

[13] For further information, see the Basel Convention website: http://www.basel.int/Implementation/CountryLedInitiative/EnvironmentallySoundManagement/ESMToolkit/Overview/tabid/5839/Default.aspx

[14] Bartley, R. 24 November 2020. Sea of Solutions event (online) on "Tackling Plastic Waste under the Basel Convention". For further information: https://www.unenvironment.org/cobsea/events/virtual-meeting/sea-solutions-2020

[15] At its fourteenth meeting, in its decision BC-14/13, the Conference of the Parties to the Basel Convention on the Control of Transboundary Movements of Hazardous Wastes and Their Disposal, among others, welcomed the proposal to establish the Basel Convention Partnership on Plastic Waste and decided to establish the Partnership and its working group; adopted the terms of reference for the Partnership; and requested the working group to implement its workplan for the biennium 2020–2021.

common objective. Membership is open to Parties and other stakeholders dealing with the different aspects of prevention, minimization and management of plastic waste. It currently stands at 210 representatives from Parties to the Convention, its regional centres, the private sector, civil society and intergovernmental organizations.[16] This robust stakeholder base is representative of all UN geographic regions, from national to local levels of government, from multinational companies to grassroots NGOs.

Adding value to an area of work rich in initiatives attempting to solve a complex problem is a particular challenge for the Partnership. Its uniqueness lies in the broad diversity of its members, the linkage to the normative work undertaken under the Basel Convention and the conduit provided to decision-makers in all 188 Parties to the Convention. Its work is thus expected to inform the work of the Conference of the Parties and could help shape the future decision-making on plastic waste under the Basel Convention. The Plastic Waste Partnership is also uniquely placed to support and prepare Parties, as well as the private sector, in implementing the Plastic Waste Amendments. The work of the Plastic Waste Partnership has been organized around the waste hierarchy, with the underlying idea that tackling sources of plastic waste is the most preferable option, while we also cannot neglect more downstream solutions, such as environmentally sound recycling. A series of pilot projects will be financed through the Plastic Waste Partnership.[17] The first round of applications in 2020 saw 23 projects from 22 countries being selected for implementation. These pilot projects will operationalize the work of the Partnership on the ground and are expected to be replicated in other countries and regions.

The Partnership itself is organized into a working group and four project groups[18] addressing different thematic issues. The working group is mandated by the terms of reference of the Partnership[19] to oversee organizational matters pertaining to the implementation of the activities of the Partnership, including setting priorities and ensuring timely implementation of its workplan; establishing project groups, as necessary, to work on specific tasks; leading awareness raising, outreach, coordination and resource mobilization initiatives; and preparing an annual budget.

Four project groups were established under the Partnership at its first meeting. The first two groups follow the steps of the waste management hierarchy, i.e. focusing on plastic waste prevention and minimization and recycling and other recovery of plastic waste. Another group was established to bring support to the implementation of the Plastic Waste Amendments, and it was deemed important to have a final group focused on outreach, education and awareness raising.

[16] For further information, see the Basel Convention website: http://www.basel.int/Implementation/Plasticwaste/PlasticWastePartnership/Membership/tabid/8098/Default.aspx

[17] For further information, see the Basel Convention website: http://www.basel.int/Implementation/Plasticwaste/PlasticWastePartnership/CallforPWPpilotprojectproposals/tabid/8494/Default.aspx

[18] For further information, see the Basel Convention website: http://www.basel.int/Implementation/Plasticwaste/PlasticWastePartnership/Projectgroupsandactivities/tabid/8410/Default.aspx

[19] COP document UNEP/CHW.14/INF/16/Rev.1 available at:
http://www.basel.int/TheConvention/ConferenceoftheParties/Meetings/COP14/tabid/7520/Default.aspx

The first project group focusing on plastic waste prevention and minimization is looking at issues such as reducing single-use packaging waste, improving the design of plastic products to increase durability, scaling up re-use solutions and biodegradable plastic products. As a first step, the group will compile information on best practices and identify challenges for improving the design of plastic products to increase their durability, reusability, reparability and recyclability, as well as to reduce hazardous constituents in plastic products.

The second group focuses on plastic waste collection, recycling and other recovery including financing and related markets. Topics being addressed by this group include separation, collection and recycling systems, financing schemes, such as Extended Producer Responsibility, innovative technologies, as well as regulatory and voluntary measures. The group has commenced its work by compiling information on best practices and innovative technologies for separating or eliminating hazardous substances from plastic wastes during sorting and recycling.

The third project group focuses on transboundary movements of plastic waste. Its members are exploring means to support customs authorities in their critical work and are gathering information that will help countries to implement the Plastic Waste Amendments. Such information will be presented as, for example, factsheets outlining the roles and responsibilities of key stakeholders in implementing the amendments, including the role of customs in enforcing the related provisions, and a factsheet on the relevant customs codes (WCO Harmonized System codes) on plastics and plastic waste and their relation to the Plastic Waste Amendments.

The final project group, on outreach, education and awareness-raising, has commenced its work to develop a strategy which will shape the messaging, mode and frequency of the Partnership's communications. The group has also commenced development of an electronic and visually based set of materials on the prevention, minimization and ESM of plastic waste, which it intends to make available in multiple languages (subject to available resources).

10.6 Looking Ahead

At its second meeting set to take place in the first half of 2021, the Partnership working group will consider the progress made by its project groups to date in implementing their respective workplans and will provide direction to the work to be embarked on as a next step. In the meantime, the co-chairs of the Partnership, themselves a senior official in the Norwegian government and trade and environment director for an association representing the recycling industry, together with the Secretariat, are promoting the Partnership through their participation in global negotiating forums, conferences and other initiatives hosted by partner organizations. They are striving to strengthen the Partnership's 200-plus stakeholder base in an endeavour to cover all stages of the plastics value chain and to make the fight against plastic waste and pollution a joint success story.

References

Beckman E (13 August 2018). The world's plastic problem in numbers. The Conversation, United Kingdom. Retrieved from https://www.weforum.org/agenda/2018/08/the-world-of-plastics-in-numbers

Brooks A, Wang S, Jambeck J (20 June 2018) The Chinese import ban and its impact on global plastic waste trade. Washington DC: American Association for the Advancement of Science. Retrieved from: https://advances.sciencemag.org/content/4/6/eaat0131

Ellen Macarthur Foundation (2017) Global Commitment: A circular economy for plastic in which it never becomes waste. Ellen Macarthur Foundation, Cowes. Retrieved from: https://www.newplasticseconomy.org/projects/global-commitment

European Commission (2 December 2020) European Commission, Brussels.. Retrieved from: https://ec.europa.eu/environment/marine/good-environmental-status/descriptor-10/index_en.htm

Katz C (7 March 2019) Piling up: How China's ban on importing waste has stalled global recycling. Yale School of the Environment, New Haven.. Retrieved from: https://e360.yale.edu/features/piling-up-how-chinas-ban-on-importing-waste-has-stalled-global-recycling

Krueger J (1999) International trade and the Basel Convention. Earthscan Publications, London

Miles T (10 May 2019) U.N. clinches deal to stop plastic waste ending up in the sea. Reuters, Geneva.. Retrieved from: https://www.reuters.com/article/us-environment-plastic-idUSKCN1SG19S

Niranjan A (4 July 2019) Amid plastic deluge, Southeast Asia refuses Western waste. Bonn, Germany: Deutsche Welle. Retrieved from: https://www.dw.com/en/amid-plastic-deluge-southeast-asia-ref uses-western-waste/a-49467769

Parker L (20 December 2018) A whopping 91% of plastic isn't recycled. National Geographic, London. Retrieved from: https://www.nationalgeographic.com/news/2017/07/plastic-produced-recycling-waste-ocean-trash-debris-environment/

Ryberg M, Laurent A, Hauschild M (2018) Mapping of global plastics value chain and plastics losses to the environment (with a particular focus on marine environment). United Nations Environment Programme, Nairobi

United Nations Environment Programme (2018) Legal limits on single-use plastics and microplastics: a global review of national laws and regulations. United Nations Environment Programme, Nairobi

11

A Conceptual Framework for Dealing with Global Plastic Pollution

Martin Wagner

Abstract There is a broad willingness to act on global plastic pollution as well as a plethora of available technological, governance, and societal solutions. However, this solution space has not been organized in a larger conceptual framework yet. In this essay, I propose such a framework, place the available solutions in it, and use it to explore the value-laden issues that motivate the diverse problem formulations and the preferences for certain solutions by certain actors. To set the scene, I argue that plastic pollution shares the key features of wicked problems, namely, scientific, political, and societal complexity and uncertainty as well as a diversity in the views of actors. To explore the latter, plastic pollution can be framed as a waste, resource, economic, societal, or systemic problem. Doing so results in different and sometimes conflicting sets of preferred solutions, including improving waste management; recycling and reuse; implementing levies, taxes, and bans as well as ethical consumerism; raising awareness; and a transition to a circular economy. Deciding which of these solutions is desirable is, again, not a purely rational choice. Accordingly, the social deliberations on these solution sets can be organized across four scales of change. At the geographic and time scales, we need to clarify where and when we want to solve the plastic problem. On the scale of responsibility, we need to clarify who is accountable, has the means to make change, and carries the costs. At the magnitude scale, we need to discuss which level of change we desire on a spectrum of status quo to revolution. All these issues are inherently linked to value judgments and worldviews that must, therefore, be part of an open and inclusive debate to facilitate solving the wicked problem of plastic pollution.

M. Wagner (✉)
Department of Biology, Norwegian University of Science and Technology (NTNU), Trondheim, Norway
e-mail: martin.wagner@ntnu.no

11.1 Premises and Aims

The scale of plastic pollution and its impacts on nature and societies has been extensively described and discussed in the public and the scientific literature (including this book). While there is much debate on the scale of the problem, the aim of this essay is to explore the solution space for plastic pollution. Therefore, this essay is based on the premise that the case is closed, in such that there is a board consensus that we want to solve it. The relevant question then becomes how to achieve best this. There is abundant literature summarizing potential solutions for plastic pollution (Auta et al. 2017; Eriksen et al. 2018; Löhr et al. 2017; Prata et al. 2019; Sheavly and Register 2007; Tessnow-von Wysocki and Le Billon 2019; Vince and Hardesty 2018). However, many authors focus on specific technological, governance, or economic aspects and some organize solutions in rather arbitrary ways. Such pragmatic collections are certainly useful to get an overview of available options. Nonetheless, they may fall short in addressing the complexity of plastic pollution (e.g., when they present few, specific solutions), the diversity in the perspectives of the multiple actors involved (e.g., when they focus on technological solutions only), and the fundamental aspects driving the preferences for certain solutions. Therefore, the aim of this essay is not to present another collection of technical and policy instruments. Instead, I will first explore the wickedness of the problem because it is important to acknowledge that there is no simple solution to problems that are difficult to define and describe. Secondly, I propose a conceptual framework regarding how specific problem formulations result in diverse and sometimes conflicting sets of solutions. Clarifying distinct problem frames is an important step toward understanding the actors' diverse preferences for solution sets. Thirdly, I lay out a framework for organizing the value judgments inherent in the plastics discourse. Since these are mostly neglected in the public and scientific debate, the aim of this piece is to bring to the surface the value-laden issues underlying the framing of the problem and the preferences for certain solutions.

11.2 Plastic Pollution as Wicked Problem

To contextualize the solutions to plastic pollution, we first need to explore its wickedness. The concept of wicked problems has been used to characterize those problems which defy conventional solutions, including climate change, displacement of people, terrorism, digital warfare, and biodiversity loss (Termeer et al. 2019). Originally introduced to describe "problems which are ill-formulated, where the information is confusing, where there are many clients and decision makers with conflicting values, and where the ramifications in the whole system are thoroughly confusing" (Churchman 1967), Rittel and Webber (1973) provided ten characteristics that define a wicked problem, some of which are shared by plastic pollution (see Table 11.1). Since then, the simple dichotomy of tame vs. wicked problems has

Table 11.1 Characteristics of wicked problems and their applicability to the plastic pollution issue

Characteristics	Applicability to plastic pollution (author's opinion)
(1) Wicked problems are difficult to define. There is no definite formulation.	Yes, there is a diversity in framing the problem.
(2) Wicked problems have no stopping rule.	Yes, we will not know for certain if/when we have solved plastic pollution.
(3) Solutions to wicked problems are not true or false, but good or bad.	Partly, some specific solutions (e.g., stopping pellet loss) can be true. Other solutions require value judgement.
(4) There is no immediate or ultimate test for solutions.	Partly, effectiveness of some local solutions may be testable but is impossible to test for global solutions.
(5) All attempts to solutions have effects that may not be reversible or forgettable.	Unknown but possible, especially when considering the largely unknown environmental impacts of replacements.
(6) These problems have no clear solution, and perhaps not even a set of possible solutions.	Unknown, a broad set of solutions is available in theory or practice, but their potential to actually solve the problem is largely unknown.
(7) Every wicked problem is essentially unique.	Unknown, but strong commonalties of plastic pollution with other global change issues exist.
(8) Every wicked problem may be a symptom of another problem.	Yes, for instance, plastic pollution can be framed as a symptom of a linear economy.
(9) There are multiple explanations for the wicked problem.	Yes, see (1)
(10) The planner (policymaker) has no right to be wrong.	No, there probably is a margin of error, in such that multiple solutions can be tested without (much) regret.

Adapted from Rittel and Webber (1973)

evolved into a view that rather considers degrees of wickedness (Termeer et al. 2019). The question, therefore, is how much wickedness we assign to plastic pollution. The key features of complexity, diversity, and uncertainty (Head and Alford 2013) can be used to do so.

Without question, the issue of plastic pollution is complex, both from a scientific and a societal perspective (SAPEA 2019). The scientific complexity arises from a number of aspects. Firstly, plastic pollution comprises a diverse suite of pollutants with very heterogeneous physicochemical properties (Lambert et al. 2017; Rochman et al. 2019). Secondly, plastics have a multitude of sources, flows, and impacts in nature and societies. Thirdly, plastic pollution is ubiquitous, yet its scale varies in time and space. The combination of these aspects results in complex exposure patterns causing a complex suite of effects on biodiversity and human health, covering all levels of biological organization, as well as on the functioning of ecosystems and societies. To further complicate the matter, these effects will probably not be linear, immediate, obvious, and overt but will be heavily interconnected and aggregate over time scales that are difficult to investigate. Thus, the complexity of plastic pollution – and its underlying causes – cannot be understood with "standard science" based on disciplinary approaches and the assumption of simple cause-effect relationships.

The societal complexity of plastic pollution arises from the fact that plastics are – besides concrete, steel, and fertilizers – one of the main building blocks of modern societies (Kuijpers 2020). They are so closely integrated with many aspects of our lives that modern societies cannot function without plastics. Accordingly, the immense societal benefits of plastics arising from their versatility, light weight, durability, and low costs are very difficult to decouple from their negative impacts caused by just the same properties. The resulting ambiguous relationship of humanity with plastics (Freinkel 2011) in combination with the complex flows of plastics through societies constitutes the societal complexity of plastic pollution.

The public, political, and scientific discourses on plastic pollution are characterized by a high degree of diversity in such that actors take divergent, and sometimes conflicting, views and approaches to the problem and its solutions. Much of that diversity emerges from the fact that the discourse on plastic pollution, just like on many other environmental problems, is a value-laden issue. In such situations, actors will frame the problem and interpret the available evidence differently based on their specific believe systems, values, and agendas.

Finally, plastic pollution is characterized by a high degree of scientific, political, and societal uncertainty. This is not only true for the glaring gaps in our scientific knowledge (SAPEA 2019) but even more so for the nonlinearity and unpredictability of the impacts that plastic pollution (and potential solutions) may have on ecosystems, humans, and societies. As an example of scientific uncertainty, there might be tipping point at which the ecological consequences of increasing pollution might become chaotic and unpredictable. Another, very concrete example of political uncertainty is the need to balance unforeseen benefits of plastics (e.g., massive demand for personal protective gear in case of a pandemic) with the negative impacts of pollution. While continuing research efforts will eventually reduce the scientific uncertainties, "better" evidence will not necessarily reduce the political and societal uncertainty surrounding plastic pollution. This is because the diversity in actors' views and agendas routed in their individual values is unlikely to change when new scientific evidence arrives.

Taken together, plastic pollution comprises a relatively high degree of wickedness because it features scientific and societal complexity, actors with diverse and divergent problem/solution frames and goals, and a high degree of scientific and political uncertainty. Leaving aside the aspects of complexity and uncertainty here, it is worth investigating how divergent problem formulations result in a diversity in solutions and how value judgments inherent in the discourse on solution to plastic pollution can be conceptualized.

11.3 Problem Formulations: Consensus or Dispute?

On the surface, the problem formulation for plastic pollution seems quite straightforward. The accumulation of plastics in nature is a bad thing. Despite many scientific uncertainties, such a statement receives broad support from the scientific community, the public, policymakers, and societal actors (e.g., interest groups) alike. Despite the absence of an overt and coordinated denialism, such as the one for climate change, a closer examination reveals that three aspects of plastic pollution are contested, namely, the risk paradigm, the scale, and the root causes of the problem.

There are two opposing views on what constitutes the risk of plastic pollution. The commonsense perspective is that the sheer presence of plastics in nature represents a risk. Such view is propelled by the attention economy (Backhaus and Wagner 2020) and the scientific uncertainties, in such that scientific ignorance ("we do not know the ecological consequences") becomes a risk itself (Völker et al. 2020). Even though empirical data are absent, this conception of risk is probably very common in the public and is promoted by environmental interest groups. An opposing perspective poses that there are thresholds below which plastic pollution will not be a risk. That more expert view comes from toxicological and regulatory practices which are based on Paracelsus' paradigm of "the dose makes the poison" and risk assessment frameworks to compare the exposure and hazards of synthetic chemicals. The main divergence between the two perspectives is that one claims that there is no "safe" threshold of plastics in nature whereas the other does. This is, in essence, a value-laden question because deciding whether we deem emitting plastics to nature acceptable is a moral, ethical, political, and societal issue rather than a purely scientific one. It may sound provocative, but on a systems level the actors benefiting from environmental action (e.g., environmental interest groups) pursue a "zero pollution" aim whereas the actors benefiting from continued emissions (e.g., plastic industry) push for a "threshold" view.

The scale of the problem of plastic pollution is also a matter of conflicting views, at least among academics and interest groups. This is best exemplified using microplastics as case. Some scientists consider the problem "superficial" (Burton Jr. 2017) and even "distractive" (Stafford and Jones 2019), whereas others consider it "significant" (Rochman et al. 2015) and "urgent" (Xanthos and Walker 2017). Without getting into the details of the different arguments, the main driver of the superficiality perspective is the assumption that environmental problems compete for limited attention and resources (Backhaus and Wagner 2020). Thus, we need to prioritize problems that are deemed more important (e.g., climate change). The opposing view poses that the microplastics problem is part of the larger issue of global change that cannot be viewed in isolation (Kramm et al. 2018) and argues that "we simply do not have the luxury of tackling environmental issues one at a time" (Avery-Gomm et al. 2019). Again, a value-laden question is at the heart of this dispute, namely, whether solving environmental issues is a zero-sum game that requires focusing on the few, most pressing problems or rather represents a win-win situation in which tackling multiple problems at once will yield co-benefits and synergies.

The last area of dispute is the question about the actual causes of plastic pollution. This is essentially a matter of problem framing that will have wide implications for finding solutions. For instance, framing plastic pollution predominantly as a marine litter problem will promote a completely different set of solutions (e.g., ocean cleanup activities) compared to a framing as consumerism problem that would require larger social changes. As with the two areas discussed above, individual values and belief systems will determine how one frames the causes of plastic pollution and which solutions one prefers, accordingly.

11.4 What Are We Trying to Solve?

Investigating the different conceptions of the causes of plastic pollution offers a meaningful way to organize the sets of solutions we have at hand. Importantly, that is not to say that one of the views is true or false but rather to understand why different actors prefer and promote divergent sets of solutions. To start with a commonality, the concerns about the impacts of plastic pollution on nature, human health, and societies are the drivers of all problem-solution frames. However, five different lenses can be used to focus on the problem formulation rendering plastic pollution a waste, resource, economic, societal, and systemic problem (Fig. 11.1).

Importantly, the lack of awareness about these frames can obscure the debate on plastic pollution. For instance, plastics are often used as a proxy to debate other societal issues, such as consumerism. Thus, seemingly scientific controversies become an arena to negotiate political and philosophical issues (Hicks 2017). This is problematic for two reasons. Firstly, scientific debates make a poor proxy for talking about value-laden problems because they are often technical and narrow and, therefore, exclude "nonexpert" opinions and economic and cultural aspects. Secondly, as Hicks puts it "talking exclusively about the science leads us to ignore – and hence fail to address – the deeper disagreement" (Hicks 2017). To make the debate on plastic pollution productive, all involved actors should transparently delineate how they frame the problem, be open to discuss the deeper disagreements

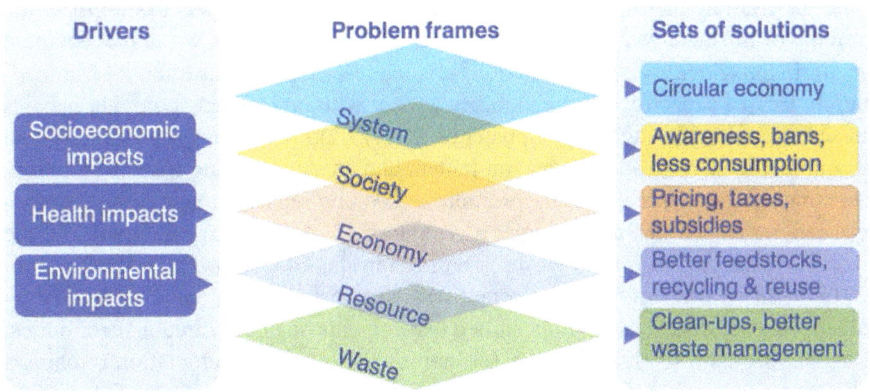

Fig. 11.1 Common drivers result in a diverse framing of the problem of plastic pollution and its causes. This determines the set of preferred solutions

that may be beyond the traditional scope of hard sciences, and be receptive to other arguments and viewpoints (e.g., the cultural value of an unpolluted nature).

11.5 Solving the Waste Problem

The most common approach to plastic pollution is to frame it as a waste problem. From that perspective, the main cause is our inability to effectively manage the plastic waste and prevent its emissions to nature. According to this view, plastic pollution basically becomes an engineering problem that can be fixed with a set of technological solutions.

While not preventive per se, cleanup activities on beaches, rivers, in the open ocean, etc. can be considered part of the set of solutions to the waste problem. Targeted at removing plastic debris from nature, these can range from low-tech solutions involving citizens simply cleaning up polluted places (e.g., organized by Ocean Conservancy, the Nordic Coastal Cleanup, or Fishing for Litter), to medium-tech solutions that collect debris before it enters the oceans (e.g., Mr. Trash Wheel, the Great Bubble Barrier), to high-tech solutions such as the large booms deployed by the Ocean Cleanup or remotely operated underwater vehicles (see Schmaltz et al. 2020 for a comprehensive inventory). Cleanup solutions can be criticized as ineffective and inefficient basically because they represent measures that are the furthest downstream of the sources of plastic pollution. Some technological approaches, such as the Ocean Cleanup booms, might even have negative consequences on marine biota (Clarke 2015). However, these activities may also have benefits that go beyond removing plastics from nature. Engaging volunteers in cleanup activities can increase their awareness of pollution and promote pro-environmental intentions (Wyles et al. 2017, 2019) that may result in a more sustainable change in behaviors.

Improving waste management is at the center of the set of solutions associated with the framing as waste problem. The goal of these activities is to minimize the amount of mismanaged plastic waste "escaping" to nature. The waste management sector in the Global North faces serious challenges, such as infrastructural fragmentation, lack of capacity, and the inability to deal with increasingly complex plastics materials and waste streams (Crippa et al. 2019). Taking the European Union as an example, there is a need to better implement and enforce existing waste legislation, harmonize waste collection, and promote innovation regarding new business models and waste sorting technologies (Crippa et al. 2019). However, most of the worlds' mismanaged plastic waste is emitted in the Global South (Jambeck et al. 2015) with its predominantly informal waste sector where autonomous and organized waste pickers are highly skilled participants in local circular economies. Reconciling their livelihoods with aspirations for industrial automation remains a challenge, and external intervention attempts will likely be unsuccessful without sufficient local capacity building (Velis 2017). The Global North can support such development by sourcing recycled plastics from the informal recycling sector, thereby gradually formalizing this sector (Crippa et al. 2019) and creating socioeconomic benefits for waste pickers (Gall et al. 2020).

Another dimension to look at plastic pollution is the global trade of plastic waste. More than half of the plastic waste intended for recycling has been exported to countries other than the ones producing the waste (Brooks et al. 2018). In the case of the European Union, most exports have been directed toward the Global South (Rosa 2018) with notable shifts since China restricted waste imports in 2017 (European Environment Agency 2019). The concerns over this practice arise from the fact that recipient countries often have low labor and environmental standards resulting in occupational risks and improper waste disposal or recycling (World Economic Forum 2020). In response, the 187 member countries amended the Basel Convention, an international treaty on the transboundary movement of hazardous wastes, to better control the global flows of contaminated, mixed, or unrecyclable plastics (Secretariat of the Basel Convention 2019). While this is promising, the Basel Convention is limited regarding its ability to enforce compliance and monitor progress (Raubenheimer and McIlgorm 2018).

A third approach to tackle the waste problem is to increase the production and use of compostable or biodegradable plastics. The expectation is that such materials will disintegrate on short time scales either in industrial and household settings or in the environment (Crippa et al. 2019; Lambert and Wagner 2017). Compostable and biodegradable plastics would, thus, contribute to decreasing the amount of persistent plastic waste and create biomass to amend soils. While a range of biodegradable plastics from fossil as well as renewable feedstocks is available, their market share remains low, making up less than 0.5% of the global plastic production (Crippa et al. 2019). This is mainly due to their high costs (compared to a limited added value) and technical challenges in scaling up production capacities. Additional challenges arise from misperceptions and misrepresentation regarding what biodegradable plastics can achieve (Crippa et al. 2019, see also the example of oxo-degradable plastics), from a low degradability of available materials in nature, and from the lack of transferability of degradation data from laboratory to field settings (Haider et al. 2019).

Importantly, when choosing to frame plastic pollution as a waste problem, the principles of the waste hierarchy apply that clearly prioritizes the prevention and reuse of waste over its recycling, recovery, or disposal (European Parliament & Council of the European Union 2008). However, contemporary solutions to the plastic waste problem mainly focus on less preferred options, especially on recovery and recycling. As an example, the European Strategy for Plastics in a Circular Economy (European Commission 2018) contains the terms "prevention" and "reuse" only 8 times, each, while it mentions "recycling" 76 times. A reason for that preference might be that the technological approaches to recycling, recovery, and disposal exist within the waste sector, whereas approaches to reduce and reuse plastics would require the inclusion of very different actors, such as social scientists and designers.

11.6 Solving the Resource Problem

Framing plastic pollution as a resource problem is based on the idea that we are losing valuable materials when using plastics in short-lived products, such as packaging and single-use items. Such framing is closely connected to the waste problem as waste management is transforming into resources management. In a broader context, however, this idea can be reformulated as a problem of extractive fossil industries in such that the cause of plastic pollution is indeed the abundance of fossil feedstocks. Both aspects of the resource framing result in divergent sets of solutions.

Approaches to solve the resource problem from a waste perspective basically cover the upper parts of the waste hierarchy, namely, recycling and reuse. The rationale is, of course, to retain the material and functional value of plastics in use and extend the lifetime of materials or products. This would, in turn, reduce waste generation and the need to produce new plastics. The different options fall on a spectrum on which reuse and mechanical recycling preserve best the value of plastics because they avoid the extra costs for breaking up the materials (Fig. 11.2). In contrast, chemical recycling uses chemical or thermal processes (e.g., depolymerization, pyrolysis, gasification) to create purified polymers, oligomers, or monomers which then can be reprocessed into new plastics. This has several advantages over mechanical recycling, such as the higher flexibility and the ability to deal with mixed and contaminated plastics. Nonetheless, chemical recycling currently requires significant improvement regarding their technical and economic feasibility as well as a thorough investigation of its environmental and social impacts (Crippa et al. 2019).

In contrast to set of solutions provided by the recycling plastics, retaining plastic products in use via sharing, repairing, and reusing comes closer to a circular economy ideal. While circular business models for plastics suffer from the lack of economic incentives (see economic problem), the four current types of business models include product as a service ("pay-per-use"), circular supplies (waste of one company becomes the raw material for another), product life extensions (making products durable, repairable, upgradable), and sharing platforms (Accenture 2014).

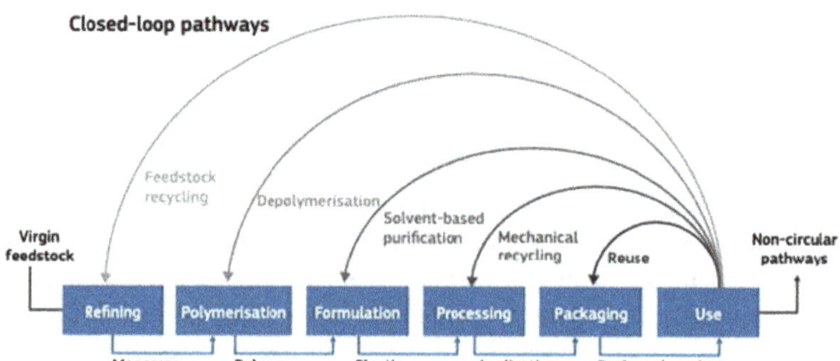

Fig. 11.2 Different loops for the reuse and recycling of plastics. (Source: Crippa et al. 2019)

Such approaches face challenges not only because plastics move so fast through the value chain and are handled by multiple actors but also because they challenge the linear economy paradigm. Here, eco-design guidelines and circularity metrics can help create a more level playing field (Crippa et al. 2019).

A very different solution, namely, the shift to bio-based plastics, emerges when framing plastic pollution as a problem of fossil feedstocks. Here, the idea is to reduce the use of petroleum and natural gas to manufacture plastics and foster the transition to a bio-based economy. Bio-based plastics can be produced from natural polymers (e.g., starch, cellulose), by plants or microbes (e.g., PBS, PHA), and by synthesizing them from biological feedstocks (e.g., ethylene derived from fermented sugarcane) (Lambert and Wagner 2017). As with biodegradable plastics, the market share of bio-based material is rather low for economic reasons, but production capacities and demand are projected to increase in the future (Crippa et al. 2019). The main challenges of shifting to bio-based plastics are their potential environmental and social impacts associated with land and pesticide use. These can be addressed by using feedstocks derived from agricultural, forestry, and food waste as well as from algae (Lambert and Wagner 2017). Eventually, substituting fossil with renewable carbon sources is a laudable aim that can create many co-benefits. However, it is important to realize that this will not solve the problem of plastic pollution.

11.7 Solving the Economic Problem

A very different perspective on the discourses on plastic pollution is the framing as an economic problem. As discussed above, many solutions are not competitive in the marketplace due to their high costs. Accordingly, the low price of virgin plastics which is a result of the low oil and natural gas prices can be considered the major cause of plastic pollution. Taking such view implies that one major benefit of plastics – their low price – is driving consumption which, in turn, results in their emission to nature. It also dictates that solutions should address the economy of plastics.

The goal of economic solutions to plastic pollution is to reduce plastic consumption either directly via financial (dis)incentives or indirectly via creating a level playing field for other solutions, including alternative materials (e.g., bio-based plastics), recycling, and circular business models. The simplest and most widely adopted economic instrument is to place levies on single-use products, especially on plastic bags. For most cases, increasing the price of carrier bag reduced the consumption but the global effect of such policies remains uncertain (Nielsen et al. 2019). In addition, there may be unintended consequences and the ecological impacts of replacements in particular often remain neglected.

Plastic taxes follow the same logic as levies and fees but target a wider range of products. While there is no literature on the implementation of plastic taxes across countries, the European Union, for instance, plans to implement a plastic tax on non-recycled plastic packaging waste (European Council 2020). Similar initiatives exist in the US State of California (Simon 2020). In principle, such taxes can be raised at the counter to change consumer behavior and/or directed toward plastic producers (see Powell 2018 for in-depth discussion). The latter aims at internalizing the external costs of plastics in such that their negative environmental impacts are reflected in their pricing, in line with the idea of extended producer responsibility. Although the actual external costs of plastics are far from clear and depend on the specific context, ecosystems services approaches, valorizing the supporting, provisioning, regulating, and cultural services nature provides, can be used to estimate those. According to a recent assessment, plastic pollution results in an annual loss of $500–2500 billion in marine natural capital, or $3300–33,000 per ton plastic in the ocean (Beaumont et al. 2019).

The benefit of taxing plastic producers would be twofold. If targeting the sale or purchase of non-recycled plastic monomers or resins, a tax would incentivize recycling. If the tax revenue would be collected in a dedicated fund, this could be used to subsidize other solutions, such as innovation in materials, products and business models, or awareness campaigns. General plastic taxes could be modeled after carbon taxation following the polluter pays principle. However, the latter requires a value judgment regarding who the polluter indeed is, and different actors would certainly disagree where to place responsibility along the life cycle of plastics. An additional challenge can be that the taxes are absorbed by the supply chain and, thus, not achieve the desired aim (Powell 2018).

Apart from levies and taxes on specific products, broader plastic taxation has not been implemented so far. However, the price of virgin plastics is expected to decrease further due to the oil industry shifting their production away from fuels and massively increase their capacity to produce new plastics (Pooler 2020). Such technology lock-in will further decrease the pricing of virgin plastics, propel plastic consumption, and render solving the plastics problem uneconomic. At the same time, the surge in production may increase the public pressure and political willingness to implement taxation that mitigates the negative impacts on recycling (Lim 2019) and of increasing waste exports (Tabuchi et al. 2020) and aggregated greenhouse gas emissions (Gardiner 2019).

11.8 Solving the Societal Problem

In contrast to the techno-economic problem-solution frames discussed above, a very different perspective attributes plastic pollution to a deeper-rooted cause, namely, consumerism and capitalism. Accordingly, plastic pollution is a result of humanity's overconsumption of plastics that is, in turn, driven by our capitalist system. In this way, it becomes a societal problem. It remains unclear how pervasive such views are, but the idea that we are consuming too much is one center piece of environmentalism, arguably one of the few remaining major ideologies. The problem with this framing is that often it remains implicit in the discourse on plastic pollution. Thus, plastic becomes a proxy to debate larger, value-laden topics, such as industrialization, economic materialism and growth, globalization, and, eventually, capitalism. The set of solutions promoted by framing plastic pollution as a societal problem are manifold. Interestingly, there is a dichotomy regarding who is responsible: When viewed as a consumption problem, solutions should motivate individuals to change their behaviors. When framed as a capitalist issue, more collective and systemic change is desired.

Plastic consumption behavior is affected by a range of factors, among others, sociodemographic variables, convenience, habits, social factors, and environmental attitudes (Heidbreder et al. 2019). The ban of plastic products, especially of single-use items, such as carrier bags, straws, cutlery, and tableware, targets the convenience and habits of consumers simply by limiting their choice. Plastic bag bans are now implemented in more than 30 countries, and bans on other single-use products are in effect in 12 countries (Schnurr et al. 2018). While generally considered effective and publicly acceptable, plastic bag bans have been criticized to disproportionally affect low-income and homeless persons. The major criticism concerns the environmental impacts of replacements made of natural materials (paper, cotton, linen) due to their higher resource demand and greenhouse gas emissions (Schnurr et al. 2018).

Social factors, including norms and identities, are the drivers for plastic avoidance, another way to reduce plastic consumption. On the one hand, social pressure and guilt can motivate individuals to not use plastics (Heidbreder et al. 2019). On the other hand, a person can practice plastic avoidance, a plastic-free lifestyle being its most intense form, to affirm their identity as environmentally conscious (Cherrier 2006). Notably, it is exactly those social norms and identities that environmental interest groups and similarly motivated actors tap into. On the business side, the marketing of "ethical" plastic products (e.g., made from ocean plastics) applies similar mechanisms, sometimes criticized as greenwashing. Interestingly, all those solutions are based on the idea of ethical consumerism, emphasizing individual responsibility, all the while staying firmly within the realm of capitalism.

As a more collective solution, activities that raise awareness regarding plastic pollution and consumption (e.g., communication campaigns) target at changing environmental attitudes and encourage pro-environmental behaviors on a wider scale. Behavior change interventions range from policies (bans, levies, see above), information campaigns, educational programs, point-of-sale interventions (e.g., asking if customers want plastic bags rather than handing them out), and the participation in cleanup activities (Heidbreder et al. 2019; Pahl et al. 2020). Importantly, Pahl et al. (2020) note that it "is advisable [to] build on personal and social norms and values, as this could lead to spillover into other pro-environmental domains and behaviours." This goes in line with the idea that awareness of plastic pollution is a gateway to wider pro-environmental attitudes (Ives 2017).

11.9 Solving the Systemic Problem

In contrast to framing plastic pollution as a waste, resource, economic, or societal problem, it can be viewed as a composite of some or all of these facets; it becomes a systemic problem. The latter view acknowledges that plastic pollution is multi-causal and that the individual causes are strongly interconnected. In other words, such systems perspective takes the wickedness of plastic pollution into account. Intuitively, this seems like the most holistic approach to the problem since it is quite apparent that plastic pollution is the result of multiple failures at multiple levels of the "plastic ecosystem."

However, the main challenge with framing this as a systemic problem is that the problem formulation becomes much less tangible compared to other perspectives. For instance, the framings as waste, resource, or economic problem are much clearer with regard to their intervention points. They also provide sets of solutions that require an engineering approach in such that technologies, processes, and functions need to be redesigned and optimized. Thus, solutions appear relatively straightforward and easy to implement. Such promises of easy wins might be one reason why the idea to engineer our way out of plastic pollution is so popular. In contrast, solutions to the systemic problem are diverse, interconnected, and at times conflicting. This makes them appear as much harder to implement. At the same time, this renders the systems view somewhat immune to criticism as individual solutions (and their limitations) will always be just a small piece of the larger approach.

Arguably, the concept of a circular economy has recently gained most momentum to tackle plastic pollution systemically. Promoted by powerful actors, including the World Economic Forum, Ellen MacArthur Foundation, McKinsey & Company, and the European Union, the vision of a circular economy is to "increase prosperity, while reducing demands on finite raw materials and minimizing negative externalities" (World Economic Forum et al. 2016). While there are multiple definitions of the meaning of circular economy (Kirchherr et al. 2017), it is basically a reincarnation of the "3Rs principle" of reduce, reuse, recycle and of the idea of sustainable design. Accordingly, a circular economy "requires innovations in the way industries produce, consumers use and policy makers legislate" (Prieto-Sandova et al. 2018). Applied to plastic pollution, the circular economy concept identifies the linear economic model as root cause of the problem.

Accordingly, it promotes designing closed loop systems that prevent plastic from becoming waste as the key solution. Whereas this seems to reiterate the solution set to the waste problem, the circular economy concept integrates the solutions supported by all other problem frames. A report by the Pew Trust and SYSTEMIQ predicts that the future plastic emissions to the ocean can only be significantly reduced with systemic change (Lau et al. 2020; The Pew Charitable Trusts and SYSTEMIQ 2020). Highlighting that there is no single solution to plastic pollution, such scenario requires the concurrent and global implementation of measures to reduce production and consumption and increase the substitution with compostable materials, recycling rates, and waste collection (The Pew Charitable Trusts and

SYSTEMIQ 2020). As such, the circular approach is, thus, a composite of the waste, resource, and societal framing combined with the prospect of economic co-benefits through innovation. The latter is indeed why repacking the other solution sets in a circular economy context has become so successful that it, as an example, has been rapidly adopted by the European Union (European Commission 2018). In addition to the economic angle, the focus on technological and societal innovation provides a powerful narrative of a better future that makes the circular economy ideology even more appealing. However, two important aspects need to be considered: Firstly, it is unclear whether a circular economy is able to deliver the promised environmental benefits (Manninen et al. 2018). Secondly, we need to realize that the ideology is not as radical as it claims, given that it further promotes the current model of business-led economic growth (Clube and Tennant 2020; Hobson and Lynch 2016). Thus, more radical and utopian solutions to plastic pollution remain out of sight.

11.10 The Four Scales of Solutions

Discussing and evaluating the solutions derived from the different problem frames outlined above requires value-based judgments regarding their relative importance, desirability, costs, and social consequences. These values should be made transparent and open in the discourse on plastic pollution to mitigate the proxy politics problem. This is important because making the debate about larger value-laden issues that remain implied can result in polarization and entrenchment and, in turn, would make solving the problem much harder.

While there is a multitude of dimensions to consider when evaluating solutions to plastic pollution, there are four basic scales of change that require value judgment and social deliberation. These cover the geography, time, responsibility, and magnitude of/for change desired by different actors (Fig. 11.3).

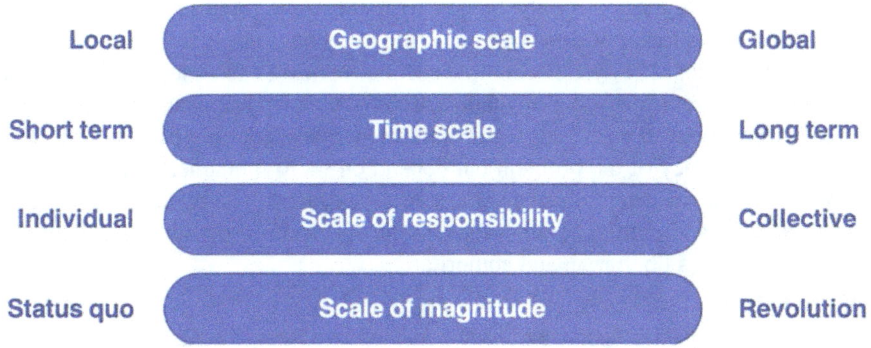

Fig. 11.3 Conceptual framework to facilitate deliberation on the scales of changes needed to solve plastic pollution

The scales of geography and time do not appear very contentious. However, the preference for local, national, regional, or global solutions to plastic pollution very much depends on which geographic unit actors most trust for developing and implementing effective measures. Some actors might be localists valuing small- over the large-scale approaches a globalist might prefer. Whereas there seems to be consensus that plastic problem is a global problem (implying a preference for global action), very focused solutions (e.g., at emission hotspots) might be very effective in a local context and much faster to implement.

The time scales desired for implementing measures and achieving their ends depends on perception of the immediacy of the problem. While a general notion of urgency to solve plastic pollution is prevailing and requires instant action, a very different standpoint may be that there is sufficient time to better understand the problem because the negative impacts are not immanent. Such view would be supported by calls for more and better research. While part of that question can be addressed scientifically, for instance, by prospective risk assessment or modeling approaches, decisions on the urgency of action remain value laden and context dependent.

At the scale of responsibility, we need to address the question who has the agency and means to implement solutions and who has to carry the burden of costs and consequences. This is as well a matter of individual vs. collective action as of which actors across the plastic life cycle have most responsibility. Some actors, especially the plastics industry, emphasize the individual consumer's responsibility. However, the systems view places much more focus on collective action. Others, especially environmental interest groups, want to hold the plastic industry accountable. However, one could also prefer to assign the burden of action to the retail or waste sectors, making it a matter of up- or downstream solutions. While it is very obvious that all actors in the plastic system share responsibility, the question of where to allocate how much accountability is open to debate.

The magnitude of desired changes is probably the most difficult aspect to agree upon because it touches not only on powerful economic interests but also on the fundamental question of whether one prefers to keep the status quo or wants to revolutionize individual lifestyles, economic sectors, or whole societies. It also covers preferences for very focused, pragmatic actions (e.g., easy wins that are sometimes tokenistic) or for systemic change. Such preferences are not only linked to perceptions of the urgency of the problem but depend on more fundamental worldviews. As with all other scales of changes, preferences will be driven by cultural context, social identity, and political orientations on the spectrum of conservative and progressive as well as libertarian and authoritarian.

11.11 How to Solve the Wicked Problem of Plastic Pollution?

Per definition, it is difficult or even impossible to solve wicked problems with conventional instruments and approaches. As argued above, plastic pollution is characterized by a relatively high degree of wickedness. At the same time, contemporary, mainstream solutions come from the standard toolbox, and it is rather the combination of all those instruments that is considered "transformative." Implementing such combinatorial approach is appealing but can be complicated by the different underlying problem formulations and sometimes conflicting value judgments regarding the relative effectiveness of individual tools.

Thus, we need to organize an inclusive, open, and probably uncomfortable conversation about the scales of change we desire and the individual values that motivate those preferences. Such debate should not be reserved for the usual actors (i.e., experts, activists, and lobbyists) but must include (marginalized) groups that are most affected by plastic pollution and carry the burden of solutions (e.g., waste pickers). The debate must be open in the sense that, for instance, instead of fighting over bans of plastic straws, we should be clear on which issues these are proxies for (e.g., consumerism). Importantly, this is not to say that we need to create an all-encompassing consensus. Instead, the current plurality in problem-solution formulations is beneficial as it acknowledges that plastic pollution is multicausal, prevents a polarization and entrenchment, and enables tackling the problem from a systems perspective.

While we will have to face a multitude of technological, governance, and societal challenges on our road to solve plastic pollution, there are some conditions that will facilitate that journey. This includes robust evidence from the natural and social sciences regarding the effectiveness of different solutions, a broad willingness to solve the problem, and an acceptance of shared responsibility.

Acknowledgments M.W. acknowledges the support by the German Federal Ministry for Education and Research (02WRS1378I, 01UU1603), the Norwegian Research Council (301157), and the European Union's Horizon 2020 programme (860720).

References

Accenture (2014) Circular advantage – innovative business models and technologies to create value in a world without limits to growth

Auta HS, Emenike CU, Fauziah SH (2017) Distribution and importance of microplastics in the marine environment: a review of the sources, fate, effects, and potential solutions. Environ Int 102:165–176. https://doi.org/10.1016/j.envint.2017.02.013

Avery-Gomm S, Walker TR, Mallory ML, Provencher JF (2019) There is nothing convenient about plastic pollution. Rejoinder to Stafford and Jones "Viewpoint – ocean plastic pollution: a convenient but distracting truth?". Marine Policy, 106. https://doi.org/10.1016/j.marpol.2019.103552

Backhaus T, Wagner M (2020) Microplastics in the environment: much ado about nothing? A debate. Global Chall 4(6):1900022. https://doi.org/10.1002/gch2.201900022

Beaumont NJ, Aanesen M, Austen MC, Börger T, Clark JR, Cole M, Hooper T, Lindeque PK, Pascoe C, Wyles KJ (2019) Global ecological, social and economic impacts of marine plastic. Mar Pollut Bull 142:189–195. https://doi.org/10.1016/j.marpolbul.2019.03.022

Brooks AL, Wang S, Jambeck JR (2018) The Chinese import ban and its impact on global plastic waste trade. Sci Adv *4*(6):eaat0131. https://doi.org/10.1126/sciadv.aat0131

Burton GA Jr (2017) Stressor exposures determine risk: so, why do fellow scientists continue to focus on superficial microplastics risk? Environ Sci Technol 51(23):13515–13516. https://doi.org/10.1021/acs.est.7b05463

Cherrier H (2006) Consumer identity and moral obligations in non-plastic bag consumption: a dialectical perspective. Int J Consum Stud 30(5):514–523

Churchman CW (1967) Wicked problems. Manag Sci 14(4):B141–B142

Clarke C (2015) 6 Reasons that floating ocean plastic cleanup Gizmo is a horrible idea. Retrieved 26 January 2021 from https://www.kcet.org/redefine/6-reasons-that-floating-ocean-plastic-cleanup-gizmo-is-a-horrible-idea

Clube RKM, Tennant M (2020) The circular economy and human needs satisfaction: promising the radical, delivering the familiar. Ecol Econ 177. https://doi.org/10.1016/j.ecolecon.2020.106772

Crippa M, De Wilde B, Koopmans R, Leyssens J, Muncke J, Ritschkoff A-C, Van Doorsselaer K, Velis C, Wagner M (2019) A circular economy for plastics – insights from research and innovation to inform policy and funding decisions

Directive 2008/98/EC of the European Parliament and of the Council of 19 November 2008 on waste and repealing certain Directives, (2008)

Eriksen M, Thiel M, Prindiville M, Kiessling T (2018) Microplastic: what are the solutions? In: Wagner M, Lambert S (eds) Freshwater microplastics, pp 273–298. https://doi.org/10.1007/978-3-319-61615-5_13

European Commission (2018) A European strategy for plastics in a circular economy. Retrieved 26 February 2018 from http://ec.europa.eu/environment/circular-economy/pdf/plastics-strategy.pdf

European Council (2020) Special meeting of the European Council (17, 18, 19, 20 and 21 July 2020) – conclusions

European Environment Agency (2019) The plastic waste trade in the circular economy. Retrieved 16 November 2020, from https://www.eea.europa.eu/themes/waste/resource-efficiency/the-plastic-waste-trade-in

Freinkel S (2011) Plastic: a toxic love story. Houghton Mifflin Harcourt

Gall M, Wiener M, Chagas de Oliveira C, Lang RW, Hansen EG (2020) Building a circular plastics economy with informal waste pickers: Recyclate quality, business model, and societal impacts. Resour Conserv Recycl 156. https://doi.org/10.1016/j.resconrec.2020.104685

Gardiner B (2019) The plastics pipeline: a surge of new production is on the way. YaleEnvironment360. Retrieved 16 November 2020, from https://e360.yale.edu/features/the-plastics-pipeline-a-surge-of-new-production-is-on-the-way

Haider TP, Volker C, Kramm J, Landfester K, Wurm FR (2019) Plastics of the future? The impact of biodegradable polymers on the environment and on society. Angew Chem Int Ed 58(1):50–62. https://doi.org/10.1002/anie.201805766

Head BW, Alford J (2013) Wicked problems: implications for public policy and management. Adm Soc 47(6):711–739. https://doi.org/10.1177/0095399713481601

Heidbreder LM, Bablok I, Drews S, Menzel C (2019) Tackling the plastic problem: a review on perceptions, behaviors, and interventions. Sci Total Environ 668:1077–1093. https://doi.org/10.1016/j.scitotenv.2019.02.437

Hicks DJ (2017) Scientific controversies as proxy politics. Issues Sci Technol 33(2) (Winter 2017)

Hobson K, Lynch N (2016) Diversifying and de-growing the circular economy: radical social transformation in a resource-scarce world. Futures 82:15–25. https://doi.org/10.1016/j.futures.2016.05.012

Ives D (2017) The gateway plastic. Retrieved 16 November 2020, from https://www.globalwildlife.org/blog/the-gateway-plastic/

Jambeck JR, Geyer R, Wilcox C, Siegler TR, Perryman M, Andrady A, Narayan R, Law KL (2015) Marine pollution. Plastic waste inputs from land into the ocean. Science 347(6223):768–771. https://doi.org/10.1126/science.1260352

Kirchherr J, Reike D, Hekkert M (2017) Conceptualizing the circular economy: an analysis of 114 definitions. Resour Conserv Recycl 127:221–232. https://doi.org/10.1016/j.resconrec.2017.09.005

Kramm J, Volker C, Wagner M (2018) Superficial or substantial: why care about microplastics in the anthropocene? Environ Sci Technol 52(6):3336–3337. https://doi.org/10.1021/acs.est.8b00790

Kuijpers M (2020) The materials that build our world are also destroying it. What are the alternatives? The Correspondent. https://thecorrespondent.com/665/the-materials-that-build-our-world-are-also-destroying-it-what-are-the-alternatives/828687075370-d68b2d25

Lambert S, Wagner M (2017) Environmental performance of bio-based and biodegradable plastics: the road ahead. Chem Soc Rev 46(22):6855–6871. https://doi.org/10.1039/c7cs00149e

Lambert S, Scherer C, Wagner M (2017) Ecotoxicity testing of microplastics: considering the heterogeneity of physicochemical properties. Integr Environ Assess Manag 13(3):470–475. https://doi.org/10.1002/ieam.1901

Lau WWY, Shiran Y, Bailey RM, Cook E, Stuchtey MR, Koskella J, Velis CA, Godfrey L, Boucher J, Murphy MB, Thompson RC, Jankowska E, Castillo AC, Pilditch TD, Dixon B, Koerselman L, Kosior E, Favoino E, Gutberlet J, Baulch S, Atreya ME, Fischer D, He KK, Petit MM, Sumaila UR, Neil E, Bernhofen MV, Lawrence K, Palardy JE (2020) Evaluating scenarios toward zero plastic pollution. Science 369(6510):1455–1461. https://doi.org/10.1126/science.aba9475

Lim X (2019) How Fossil Fuel Companies Are Killing Plastic Recycling. HuffPost. Retrieved 16 November 2020, from https://www.huffpost.com/entry/plastic-recycling-oil-companies-landfill_n_5d8e4916e4b0e9e7604c832e

Löhr A, Savelli H, Beunen R, Kalz M, Ragas A, Van Belleghem F (2017) Solutions for global marine litter pollution. Curr Opin Environ Sustain 28:90–99. https://doi.org/10.1016/j.cosust.2017.08.009

Manninen K, Koskela S, Antikainen R, Bocken N, Dahlbo H, Aminoff A (2018) Do circular economy business models capture intended environmental value propositions? J Clean Prod 171:413–422. https://doi.org/10.1016/j.jclepro.2017.10.003

Nielsen TD, Holmberg K, Stripple J (2019) Need a bag? A review of public policies on plastic carrier bags – where, how and to what effect? Waste Manag 87:428–440. https://doi.org/10.1016/j.wasman.2019.02.025

Pahl S, Richter I, Wyles K (2020) Human perceptions and behaviour determine aquatic plastic pollution. In: Stock F, Reifferscheid G, Brennholt N, Kostianaia E (eds) Plastics in the aquatic environment – Part II: Stakeholders role against pollution. https://doi.org/10.1007/698_2020_672

Pooler M (2020) Surge in plastics production defies environmental backlash. Financial Times. https://www.ft.com/content/4980ec74-4463-11ea-abea-0c7a29cd66fe

Powell D (2018) The price is right… or is it? The case for taxing plastics. Rethink Plastic Alliance

Prata JC, Silva ALP, da Costa JP, Mouneyrac C, Walker TR, Duarte AC, Rocha-Santos T (2019) Solutions and integrated strategies for the control and mitigation of plastic and microplastic pollution. Int J Environ Res Public Health 16(13). https://doi.org/10.3390/ijerph16132411

Prieto-Sandova V, Jaca C, Ormazabal M (2018) Towards a consensus on the circular economy. J Clean Prod 179:605–615. https://doi.org/10.1016/j.jclepro.2017.12.224

Raubenheimer K, McIlgorm A (2018) Can the Basel and Stockholm conventions provide a global framework to reduce the impact of marine plastic litter? Mar Policy 96:285–290. https://doi.org/10.1016/j.marpol.2018.01.013

Rittel HWJ, Webber MM (1973) Dilemmas in a general theory of planning. Policy Sci 4(2):155–169. https://doi.org/10.1007/Bf01405730

Rochman CM, Kross SM, Armstrong JB, Bogan MT, Darling ES, Green SJ, Smyth AR, Verissimo D (2015) Scientific evidence supports a ban on microbeads. Environ Sci Technol 49(18):10759–10761. https://doi.org/10.1021/acs.est.5b03909

Rochman CM, Brookson C, Bikker J, Djuric N, Earn A, Bucci K, Athey S, Huntington A, McIlwraith H, Munno K, De Frond H, Kolomijeca A, Erdle L, Grbic J, Bayoumi M, Borrelle SB, Wu TN, Santoro S, Werbowski LM, Zhu X, Giles RK, Hamilton BM, Thaysen C, Kaura A, Klasios N, Ead L, Kim J, Sherlock C, Ho A, Hung C (2019, Apr) Rethinking microplastics as a diverse contaminant suite. Environ Toxicol Chem 38(4):703–711. https://doi.org/10.1002/etc.4371

Rosa F (2018) Europe at crossroads: after the Chinese ban on plastic waste imports, what now? Retrieved 16 November 2020, from https://zerowasteeurope.eu/2018/02/europe-after-chinese-plastic-ban/

SAPEA (2019) A scientific perspective on microplastics in nature and society

Schmaltz E, Melvin EC, Diana Z, Gunady EF, Rittschof D, Somarelli JA, Virdin J, Dunphy-Daly MM (2020) Plastic pollution solutions: emerging technologies to prevent and collect marine plastic pollution. Environ Int 144:106067. https://doi.org/10.1016/j.envint.2020.106067

Schnurr REJ, Alboiu V, Chaudhary M, Corbett RA, Quanz ME, Sankar K, Srain HS, Thavarajah V, Xanthos D, Walker TR (2018) Reducing marine pollution from single-use plastics (SUPs): a review. Mar Pollut Bull 137:157–171. https://doi.org/10.1016/j.marpolbul.2018.10.001

Secretariat of the Basel Convention (2019) Basel convention plastic waste amendments. Retrieved 16 November 2020, from http://www.basel.int/Implementation/Plasticwaste/PlasticWasteAmendments/Overview/tabid/8426/Default.aspx

Sheavly SB, Register KM (2007) Marine debris & plastics: environmental concerns, sources, impacts and solutions. J Polym Environ 15(4):301–305. https://doi.org/10.1007/s10924-007-0074-3

Simon M (2020) Should governments slap a tax on plastic? Retrieved 16 November 2020, from https://www.wired.com/story/should-governments-slap-a-tax-on-plastic/

Stafford R, Jones PJS (2019) Viewpoint – ocean plastic pollution: a convenient but distracting truth? Mar Policy 103:187–191. https://doi.org/10.1016/j.marpol.2019.02.003

Tabuchi H, Corkery M, Mureithi C (2020) Big oil is in trouble. Its plan: flood Africa with plastic. The New York Times. Retrieved 16 November 2020, from https://www.nytimes.com/2020/08/30/climate/oil-kenya-africa-plastics-trade.html

Termeer CJAM, Dewulf A, Biesbroek R (2019) A critical assessment of the wicked problem concept: relevance and usefulness for policy science and practice. Polic Soc 38(2):167–179. https://doi.org/10.1080/14494035.2019.1617971

Tessnow-von Wysocki I, Le Billon P (2019) Plastics at sea: treaty design for a global solution to marine plastic pollution. Environ Sci Pol 100:94–104. https://doi.org/10.1016/j.envsci.2019.06.005

The Pew Charitable Trusts, & SYSTEMIQ (2020) Breaking the plastic wave: a comprehensive assessment of pathways towards stopping ocean plastic pollution

Velis C (2017) Waste pickers in Global South: informal recycling sector in a circular economy era. Waste Manag Res 35(4):329–331. https://doi.org/10.1177/0734242X17702024

Vince J, Hardesty BD (2018) Governance solutions to the tragedy of the commons that marine plastics have become. Front Mar Sci 5. https://doi.org/10.3389/fmars.2018.00214

Völker C, Kramm J, Wagner M (2020) On the creation of risk: framing of microplastics risks in science and media. Global Chall 4(6). https://doi.org/10.1002/gch2.201900010

World Economic Forum (2020) Plastics, the circular economy and global trade

World Economic Forum, Ellen MacArthur Foundation, & McKinsey & Company. (2016). The new plastics economy: Rethinking the future of plastics

Wyles KJ, Pahl S, Holland M, Thompson RC (2017) Can beach cleans do more than clean-up litter? Comparing beach cleans to other coastal activities. Environ Behav 49(5):509–535. https://doi.org/10.1177/0013916516649412

Permissions

All chapters in this book were first published by Springer; hereby published with permission under the Creative Commons Attribution License or equivalent. Every chapter published in this book has been scrutinized by our experts. Their significance has been extensively debated. The topics covered herein carry significant information for a comprehensive understanding. They may even be implemented as practical applications or may be referred to as a beginning point for further studies.

The contributors of this book come from diverse backgrounds, making this book a truly international effort. We would like to thank all the contributing authors for lending their expertise to make the book truly unique. They have played a crucial role in the development of this book. Without their invaluable contributions this book wouldn't have been possible. They have made vital efforts to compile up to date information on the varied aspects of this subject to make this book a valuable addition to the collection of many professionals and students.

This book was conceptualized with the vision of imparting up-to-date and integrated information in this field. To ensure the same, a matchless editorial board was set up. Every individual on the board went through rigorous rounds of assessment to prove their worth. After which they invested a large part of their time researching and compiling the most relevant data for our readers.

The editorial board has been involved in producing this book since its inception. They have spent rigorous hours researching and exploring the diverse topics which have resulted in the successful publishing of this book. They have passed on their knowledge of decades through this book. To expedite this challenging task, the publisher supported the team at every step. A small team of assistant editors was also appointed to further simplify the editing procedure and attain best results for the readers.

Apart from the editorial board, the designing team has also invested a significant amount of their time in understanding the subject and creating the most relevant covers. They scrutinized every image to scout for the most suitable representation of the subject and create an appropriate cover for the book.

The publishing team has been an ardent support to the editorial, designing and production team. Their endless efforts to recruit the best for this project, has resulted in the accomplishment of this book. They are a veteran in the field of academics and their pool of knowledge is as vast as their experience in printing. Their expertise and guidance has proved useful at every step. Their uncompromising quality standards have made this book an exceptional effort. Their encouragement from time to time has been an inspiration for everyone.

The publisher and the editorial board hope that this book will prove to be a valuable piece of knowledge for students, practitioners and scholars across the globe.

Index

www.ingramcontent.com/pod-product-compliance
Lightning Source LLC
LaVergne TN
LVHW051049220926
848485LV00001BA/54

* 9 7 8 1 6 4 7 4 0 3 8 1 2 *